Elements of Mathematics

Third printing, and first paperback printing, 2018
Paper ISBN 978-0-691-17854-7

The Library of Congress has cataloged the cloth edition of this book as follows:
Names: Stillwell, John.
Title: Elements of mathematics : from Euclid to Gödel / John Stillwell.
Description: Princeton : Princeton University Press, [2016] |
Includes bibliographical references and index.
Identifiers: LCCN 2015045022 | ISBN 9780691171685 (hardcover : alk. paper)
Subjects: LCSH: Mathematics—Study and teaching (Higher)
Classification: LCC QA11.2 .S8485 2016 | DDC 510.71/1—dc23 LC record available at
http://lccn.loc.gov/2015045022

British Library Cataloging-in-Publication Data is available

This book has been composed in Minion Pro

Printed on acid-free paper. ∞

Typeset by Nova Techset Pvt Ltd, Bangalore, India
Printed in the United States of America

7 9 10 8

To Hartley Rogers Jr.
In Memoriam

ᔐ Contents ᔐ

ᔰ Preface ᔰ

This book grew from an article I wrote in 2008 for the centenary of Felix Klein's *Elementary Mathematics from an Advanced Standpoint*. The article reflected on Klein's view of elementary mathematics, which I found to be surprisingly modern, and made some comments on how his view might change in the light of today's mathematics. With further reflection I realized that a discussion of elementary mathematics today should include not only some topics that are elementary from the twenty-first-century viewpoint, but also a more precise explanation of the term "elementary" than was possible in Klein's day.

So, the first goal of the book is to give a bird's eye view of elementary mathematics and its treasures. This view will sometimes be "from an advanced standpoint," but nevertheless as elementary as possible. Readers with a good high school training in mathematics should be able to understand most of the book, though no doubt everyone will experience some difficulties, due to the wide range of topics. Bear in mind what G. H. Hardy (1942) said in a review of the excellent book *What is Mathematics?* by Courant and Robbins (1941): "a book on mathematics without difficulties would be worthless."

The second goal of the book is to explain what "elementary" means, or at least to explain why certain pieces of mathematics seem to be "more elementary" than others. It might be thought that the concept of "elementary" changes continually as mathematics advances. Indeed, some topics now considered part of elementary mathematics are there because some great advance *made* them elementary. One such advance was the use of algebra in geometry, due to Fermat and Descartes. On the other hand, some concepts have remained persistently difficult. One is the concept of real number, which has caused headaches since the time of Euclid. Advances in logic in the twentieth century help to explain why the real numbers remain an "advanced" concept, and this idea will be gradually elaborated in the second half of the book. We will see how elementary mathematics collides with the real number concept from various directions, and how logic identifies the advanced nature of the real numbers—and, more generally, the nature of *infinity*—in various ways.

Those are the goals of the book. Here is how they are implemented. Chapter 1 briefly introduces eight topics that are important at the elementary level—arithmetic, computation, algebra, geometry, calculus, combinatorics, probability, and logic—with some illustrative examples. The next eight chapters develop these topics in more detail, laying down their basic principles, solving some interesting problems, and making connections between them. Algebra is used in geometry, geometry in arithmetic, combinatorics in probability, logic in computation, and so on. Ideas are densely interlocked, even at the elementary level! The mathematical details are supplemented by historical and philosophical remarks at the end of each chapter, intended to give an overview of where the ideas came from and how they shape the concept of elementary mathematics.

Since we are exploring the scope and limits of elementary mathematics we cannot help crossing the boundary into advanced mathematics on occasion. We warn the reader of these incursions with a star (*) in the titles of sections and subsections that touch upon advanced concepts. In chapter 10 we finally cross the line in earnest, with examples of *non*-elementary mathematics in each of the eight topics above. The purpose of these examples is to answer some questions that arose in the elementary chapters, showing that, with just small steps into the infinite, it is possible to solve interesting problems beyond the reach of elementary methods.

What is new in this book—apart from a hopefully fresh look at elementary mathematics—is a serious study of what it means for one theorem to be "more advanced" or "deeper" than others. In the last 40 years the subject of *reverse mathematics* has sought to classify theorems by the strength of axioms needed to prove them, measuring "strength" by how much the axioms assume about infinity. With this methodology, reverse mathematics has classified many theorems in basic analysis, such as the completeness of the real numbers, Bolzano-Weierstrass theorem, and Brouwer fixed point theorem. We can now say that these theorems are definitely "more advanced" than, say, elementary number theory, because they depend on stronger axioms.

So, if we wish to see what lies just beyond elementary mathematics, the first place to look is analysis. Analysis clarifies not only the scope of elementary calculus, but also of other fields where infinite processes occur. These include algebra (in its fundamental theorem) and combinatorics (in the Kőnig infinity lemma, which is also important in topology and logic). Infinity may not be the only characteristic that defines advanced mathematics, but it is probably the most important, and the one we understand the best.

Lest it seem that logic and infinity are formidable topics for a book about elementary mathematics, I hasten to add that we approach them very gently and gradually. Deeper ideas will appear only when they are needed, and the logical foundations of mathematics will be presented only in chapter 9—at which stage I hope that the reader will understand their value. In this respect (and many others) I agree with Klein, who said:

> In fact, mathematics has grown like a tree, which does not start at its tiniest rootlets and grow merely upward, but rather sends its roots deeper and deeper and at the same time and rate that its branches and leaves are spreading upward.
>
> Klein (1932), p.15

In chapter 9 we pursue the roots of mathematics deep enough to see, I hope, those that nourish elementary mathematics, and some that nourish the higher branches.

I expect that this book will be of interest to prospective mathematics students, their teachers, and to professional mathematicians interested in the foundations of our discipline. To students about to enter university, this book gives an overview of things that are useful to know before proceeding further, together with a glimpse of what lies ahead. To those mathematicians who teach at university level, the book can be a refresher course in the topics we want our students to know, but about which we may be (ahem) a little vague ourselves.

Acknowledgments. For the germ of the idea that led to this book, credit should go to Vagn Lundsgaard Hansen and Jeremy Gray, who commissioned my article on Klein, and later suggested that I write a book of a similar kind. I thank my wife, Elaine, as ever, for her tireless proofreading and general encouragement. Thanks also go to Derek Holton, Rossella Lupacchini, Marc Ryser, and two anonymous referees for corrections and helpful comments. I am indebted to the University of San Francisco for their continuing support, and to Cambridge University DPMMS for the use of their facilities while several chapters of the book were being written. Finally, special thanks go to Vickie Kearn and her team at Princeton University Press for masterly coordination of all aspects of the production of this book.

John Stillwell
Cambridge, July 2, 2015

Elements of Mathematics

1

⟿

Elementary Topics

PREVIEW

The present chapter introduces the fields of mathematics that will be considered "elementary" in this book. They have all been considered "elementary" at some stage in the history of mathematics education, and they are all still taught at school level in some places today. But even "elementary" topics have their mysteries and difficulties, which call for explanation from a "higher standpoint." As we will show, this applies to the topics considered by Klein (1908)—arithmetic, algebra, analysis, and geometry—plus a few other topics that existed only in embryonic form in 1908 but are quite mature today.

Thus we have sections on arithmetic, algebra, and geometry, as Klein did, plus his "analysis" interpreted as "calculus," and the new topics of computation, combinatorics, probability, and logic, which matured only in the last century.

It is clear that computation looms over mathematics today, at all levels, and that this should include the elementary level. Combinatorics is a close relative of computation, and it has some very elementary aspects, so it should be included for that reason alone. A second, more classical reason, is that combinatorics is a gateway to probability theory—another topic with elementary roots.

Finally, there is the topic of logic. Logic is the heart of mathematics, yet logic is not viewed as a mathematical topic by many mathematicians. This was excusable in 1908—when few if any theorems *about* logic were known—but not today. Logic contains some of the most interesting theorems of mathematics, and it is inextricably connected with computation and combinatorics. The new trio computation-combinatorics-logic now deserves to be taken as seriously in elementary mathematics as the old trio arithmetic-algebra-geometry.

1.1 Arithmetic

Elementary mathematics begins with counting, probably first with the help of our fingers, then by words "one," "two," "three," ..., and in elementary school by symbols 1, 2, 3, 4, 5, 6, 7, 8, 9, 10, ... This symbolism, of *base* 10 *numerals*, is already a deep idea, which leads to many fascinating and difficult problems about numbers. Really? Yes, really. Just consider the meaning of a typical numeral, say 3671. This symbol stands for three thousands, plus six hundreds, plus seven tens, plus one unit; in other words:

$$3671 = 3 \cdot 1000 + 6 \cdot 100 + 7 \cdot 10 + 1$$
$$= 3 \cdot 10^3 + 6 \cdot 10^2 + 7 \cdot 10 + 1.$$

Thus to know the meaning of decimal numerals, one already has to understand addition, multiplication, and exponentiation!

Indeed, the relationship between numerals and the numbers they represent is our first encounter with a phenomenon that is common in mathematics and life: *exponential growth*. Nine positive numbers (namely, 1, 2, 3, 4, 5, 6, 7, 8, 9) are given by numerals of one digit, 90 (namely 10, 11, 12, ..., 99) by numerals of two digits, 900 by numerals of three digits, and so on. Adding one digit to the numeral multiplies by 10 the number of positive numbers we can represent, so a small number of digits can represent any number of physical objects that we are likely to encounter. Five or six digits can represent the capacity of any football stadium, eight digits the population of any city, ten digits the population of the world, and perhaps 100 digits suffices to represent the number of elementary particles in the known universe. Indeed, it is surely because the world teems with large numbers that humans developed a system of notation that can express them.

It is a minor miracle that large numbers can be encoded by small numerals, but one that comes at a price. Large numbers can be added and multiplied only by operating on their numerals, and this is not trivial, though you learned how to do it in elementary school. Indeed, it is not uncommon for young students to feel such a sense of mastery after learning how to add and multiply decimal numerals, that they feel there is not much else to learn in math. Maybe just bigger numbers. It is lucky that we gloss over exponentiation, because exponentiation of large numbers is practically impossible! Thus it takes only a few seconds to work out $231 + 392 + 537$ by hand, and a few minutes to work out $231 \times 392 \times 537$. But the numeral for

$$231^{392^{537}}$$

is too long to be written down in the known universe, with digits the size of atoms.

Even with numerals of more modest length—say, those that can be written on a single page—there are problems about multiplication that we do not know how to solve. One such is the problem of *factorization*: finding numbers whose product is a given number. If the given number has, say, 1000 digits, then it may be the product of two 500-digit numbers. There are about 10^{500} such numbers, and we do not know how to find the right ones substantially faster than trying them all.

Here is another problem in the same vein: the problem of recognizing *prime* numbers. A number is prime if it is greater than 1 and not the product of smaller numbers. Thus the first few prime numbers are

$$2, \quad 3, \quad 5, \quad 7, \quad 11, \quad 13, \quad 17, \quad 19, \quad 23, \quad 29, \quad 31, \quad \ldots.$$

There are infinitely many prime numbers (as we will see in chapter 2) and it seems relatively easy to find large ones. For example, by consulting the Wolfram Alpha website one finds that

$$\text{next prime after } 10^{10} = 10^{10} + 19,$$

$$\text{next prime after } 10^{20} = 10^{20} + 39,$$

$$\text{next prime after } 10^{40} = 10^{40} + 121,$$

$$\text{next prime after } 10^{50} = 10^{50} + 151,$$

$$\text{next prime after } 10^{100} = 10^{100} + 267,$$

$$\text{next prime after } 10^{500} = 10^{500} + 961,$$

$$\text{next prime after } 10^{1000} = 10^{1000} + 453.$$

Thus we can readily find primes with at least 1000 digits. Even more surprising, we can test any number with 1000 digits and decide *whether* it is prime. The surprise is not only that it is feasible to recognize large primes (a problem not solved until recent decades) but that it is feasible to recognize *non*-prime numbers without finding their factors. Apparently, it is harder to find factors—as we said above, we do not know how to do this for 1000-digit numbers—than to prove that they exist.

These recent discoveries about primes and factorization underline the mysterious nature of elementary arithmetic. If multiplication can be this difficult, what other surprises may be in store? Evidently, a complete understanding of elementary arithmetic is not as easy as it seemed in elementary school. Some "higher standpoint" is needed to make arithmetic clearer, and we will search for one in the next chapter.

1.2 Computation

As we saw in the previous section, working with decimal numerals requires some nontrivial computational skills, even to add and multiply whole numbers. The rules, or *algorithms*, for adding, subtracting, and multiplying decimal numerals are (I hope) sufficiently well known that I need not describe them here. But it is well to recall that they involve scores of facts: the sums and products of possible pairs of digits, plus rules for properly aligning digits and "carrying." Learning and understanding these algorithms is a significant accomplishment!

Nevertheless, we will usually assume that algorithms for addition, subtraction, and multiplication are given. One reason is that the decimal algorithms are fast, or "efficient," in a sense we will explain later, so any algorithm that is "efficient" in its use of addition, subtraction, and multiplication is "efficient" in some absolute sense. Such algorithms have been known since ancient times, before decimal numerals were invented. The original and greatest example is the *Euclidean algorithm* for finding the greatest common divisor of two numbers.

The Euclidean algorithm takes two positive whole numbers and, as Euclid put it, "repeatedly subtracts the smaller from the larger." For example, if one begins with the pair 13, 8 then repeated subtraction gives the following series of pairs

$$13, 8 \to 8, 13 - 8 = 8, 5$$
$$\to 5, 8 - 5 = 5, 3$$
$$\to 3, 5 - 3 = 3, 2$$
$$\to 2, 3 - 2 = 2, 1$$
$$\to 1, 2 - 1 = 1, 1$$

—at which point the two numbers are equal and the algorithm halts. The terminal number, 1, is indeed the greatest common divisor (gcd) of 13 and 8, but why should the gcd be produced in this way? The first point is: *if a number d divides two numbers a and b, then d also divides a − b*. In particular, the *greatest* common divisor of a and b is also a divisor of $a − b$, and hence of all numbers produced by the sequence of subtractions. The second point is: *subtraction continually decreases the maximum member of the pair, and hence the algorithm eventually halts, necessarily with a pair of equal numbers*. From this it follows that the terminal number equals the gcd of the initial pair.

The Euclidean algorithm is an admirable algorithm because we can easily prove that it does what it is supposed to, and with a little more work we can prove that it is fast. To be more precise, if the initial numbers are given as decimal numerals, and if we replace repeated subtractions of b from a by division of a by b with remainder, then the number of divisions needed to obtain gcd(a, b) is roughly proportional to the total number of digits in the initial pair.

Our second example of an algorithm is more modern—apparently dating from the 1930s—and again involving elementary arithmetic operations. The so-called *Collatz* algorithm takes an arbitrary positive whole number n, replacing it by $n/2$ if n is even and by $3n + 1$ if n is odd, then repeats the process until the number 1 is obtained. Amazingly, we do not know whether the algorithm always halts, despite the fact that it has halted for every number n ever tried. The question whether the Collatz algorithm always halts is known as the *Collatz* or $3n + 1$ *problem*.

Here is what the Collatz algorithm produces for the inputs 6 and 11:

$$6 \rightarrow 3 \rightarrow 10 \rightarrow 5 \rightarrow 16 \rightarrow 8 \rightarrow 4 \rightarrow 2 \rightarrow 1.$$
$$11 \rightarrow 34 \rightarrow 17 \rightarrow 52 \rightarrow 26 \rightarrow 13 \rightarrow 40 \rightarrow 20 \rightarrow 10 \rightarrow 5 \rightarrow 16 \rightarrow$$
$$8 \rightarrow 4 \rightarrow 2 \rightarrow 1.$$

A century ago there was no theory of algorithms, because it was not known that the concept of "algorithm" could be made mathematically precise. Quite coincidentally, the Collatz problem arrived at about the same time as a formal concept of algorithm, or *computing machine*, and the discovery that the general halting problem for algorithms is *unsolvable*. That is, there is no algorithm which, given an algorithm A and input i, will decide whether A halts for input i. This result has no known implications for the Collatz problem, but it has huge implications for both computation and logic, as we will see in later chapters.

In the 1970s the theory of computation underwent a second upheaval, with the realization that *computational complexity* is important. As pointed out in the previous section, some computations (such as exponentiation of large numbers) cannot be carried out in practice, even though they exist in principle. This realization led to a reassessment of the whole field of computation, and indeed to a reassessment of all fields of mathematics that *involve* computation, starting with arithmetic. In the process, many puzzling new phenomena were discovered, which as yet lack a clear explanation. We have already mentioned one in the previous section: it is feasible to decide whether 1000-digit numbers have factors, but apparently *not* feasible to find the factors. This is a troubling development for those who believe that existence of a mathematical object should imply the ability to *find* the object.

It remains to be seen exactly how computational complexity will affect our view of elementary mathematics, because the main problems of computational complexity are not yet solved. In chapter 3 we will explain what these problems are, and what they mean for the rest of mathematics.

1.3 Algebra

Elementary algebra has changed considerably since the time of Klein. In his day, the term meant mainly the manipulation of polynomials—solving equations up to the fourth degree, solving systems of linear equations in several unknowns and related calculations with determinants, simplifying complicated rational expressions, and studying the curves defined by polynomials in two variables—skills which were developed to a high level. Formidable examples can be found in the "pre-calculus" books of 100 years ago, such as the *Algebra* of Chrystal (1904) and the *Pure Mathematics* of Hardy (1908).

For example, Chrystal's very first exercise set asks the student to simplify

$$\left(x+\frac{1}{x}\right)\left(y+\frac{1}{y}\right)\left(z+\frac{1}{z}\right)-\left(x-\frac{1}{x}\right)\left(y-\frac{1}{y}\right)\left(z-\frac{1}{z}\right),$$

and by the third exercise set (immediately after addition and multiplication of fractions have been defined) the student is expected to show that the following expression is independent of x:

$$\frac{x^4}{a^2 b^2} + \frac{(x^2-a^2)^2}{a^2(a^2-b^2)} - \frac{(x^2-b^2)^2}{b^2(a^2-b^2)}.$$

Today, just entering these expressions into a computer algebra system would probably be considered a challenging exercise. But if hand computation has suffered, abstraction has gained, and there is now a "higher standpoint" from which elementary algebra looks entirely different.

This is the standpoint of *structure* and *axiomatization*, which identifies certain algebraic laws and classifies algebraic systems by the laws they satisfy. From this standpoint, the above exercises in Chrystal are simply consequences of the following algebraic laws, now known as the *field axioms*:

$$a + b = b + a, \qquad ab = ba$$
$$a + (b + c) = (a + b) + c, \qquad a(bc) = (ab)c$$
$$a + 0 = a, \qquad a \cdot 1 = a$$
$$a + (-a) = 0, \qquad a \cdot a^{-1} = 1 \quad \text{for } a \neq 0$$
$$a(b + c) = ab + ac.$$

The object of algebra now is not to do a million exercises, but to understand the axiom system that encapsulates them all. The nine field axioms encapsulate the arithmetic of numbers, high school algebra, and many other algebraic systems. Because these systems occur so commonly in mathematics, they have a name—*fields*—and an extensive theory. As soon as we recognize that a system satisfies the nine field axioms, we know that it satisfies all the known theory of fields (including, if necessary, the results in Chrystal's exercises). We also say that a system satisfying the field axioms has the *structure* of a field. The first field that we all meet is the system $\mathbb{Q}$ of *rational numbers*, or fractions, but there are many more.

With the explosion of mathematical knowledge over the last century, identifying structure, or "encapsulation by axiomatization," has become one of the best ways of keeping the explosion under control. In this book we will see that there are not only axiom systems for parts of algebra, but also for geometry, number theory, and for *mathematics as a whole*. It is true that the latter two axiom systems are not complete—there are some mathematical facts that do not follow from them—but it is remarkable that an axiom system can even come close to encapsulating all of mathematics. Who would have thought that *almost everything, in the vast world of mathematics, follows from a few basic facts?*

To return to algebraic structures, if we drop the axiom about a^{-1} from the field axioms (which effectively allows the existence of fractions) we get axioms for a more general structure called a *ring*. The first ring that we all meet is the system $\mathbb{Z}$ of *integers*. (The letter $\mathbb{Z}$ comes from the German word "Zahlen" for "numbers.") Notice that the number system we started with, the *positive integers*

$$\mathbb{N} = \{1, 2, 3, 4, 5, \ldots\},$$

is neither a ring nor a field. We get the ring $\mathbb{Z}$ by throwing in the *difference* $m - n$ for any m and n in $\mathbb{N}$, and then we get the field $\mathbb{Q}$ by throwing in the *quotient* m/n of any m and $n \neq 0$ in $\mathbb{Z}$. (This is presumably where the letter $\mathbb{Q}$ comes from.)

Thus $\mathbb{N}$, $\mathbb{Z}$, and $\mathbb{Q}$ can be distinguished from each other not only by their axiomatic properties, but also by *closure properties*:

- $\mathbb{N}$ is closed under $+$ and $\times$; that is, if m and n are in $\mathbb{N}$ then so are $m + n$ and $m \times n$.

- $\mathbb{Z}$ is closed under $+$, $-$, and $\times$. In particular, $0 = a - a$ exists and $0 - a$, or $-a$, is meaningful for each a in $\mathbb{Z}$.

- $\mathbb{Q}$ is closed under $+$, $-$, $\times$, and $\div$ (by a nonzero number). In particular, $a^{-1} = 1 \div a$ is meaningful for each nonzero a in $\mathbb{Q}$.

It is not immediately clear why $\mathbb{Z}$ and $\mathbb{Q}$ are more useful than $\mathbb{N}$, since all properties of integers or rational numbers are inherited from properties of positive integers. The reason must be that they have "better algebraic structure" in some sense. Ring structure seems to be a good setting for discussing topics such as divisibility and primes, while field structure is good for many things—not only in algebra, but also in geometry, as we will see in the next section.

1.4 Geometry

Over the last century there has been much debate about the place of geometry in elementary mathematics, and indeed about the meaning of "geometry" itself. But let's start with something that has been an indisputable part of geometry for over 2000 years: the *Pythagorean theorem*. As everyone knows, the theorem states that the square on the hypotenuse of a right-angled triangle is equal (in area) to the sum of the squares on the other two sides. Figure 1.1 shows the squares in question, with the square on the hypotenuse in gray and the squares on the other two sides in black.

Figure 1.1: The Pythagorean theorem.

Figure 1.2: Proof of the Pythagorean theorem.

The theorem is hardly obvious, yet there is a surprisingly simple proof, shown in figure 1.2. The left half of the figure shows that the square on the hypotenuse equals a certain big square minus four copies of the triangle.

The right half shows that the sum of the squares on the other two sides is the same: the big square minus four copies of the triangle. QED!

Given that the Pythagorean theorem belongs in any treatment of geometry, the question remains: how best to "encapsulate" geometry so that the centrality of the Pythagorean theorem is clear? The traditional answer was by the axioms in Euclid's *Elements*, which yield the Pythagorean theorem as the climax of Book I. This approach was universal until the nineteenth century, and still has advocates today, but 100 years ago it was known to be lacking in rigor and universality. It was known that Euclid's axiom system has gaps, that filling the gaps

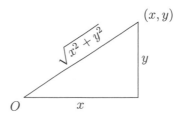

Figure 1.3: Distance from the origin.

requires quite a large number of extra axioms, and that there are other geometries which require further modifications of the axiom system.

It seemed to Klein, for example, that the axiomatic approach should be abandoned and that geometry should be based on the algebraic approach pioneered by Descartes in the seventeenth century. In algebraic geometry, points in the plane are given by ordered pairs (x, y) of numbers, and lines and curves are given by polynomial equations in x and y. Since the point (x, y) lies at horizontal distance x and vertical distance y from the origin O, we define its distance from O to be $\sqrt{x^2 + y^2}$, motivated by the Pythagorean theorem (see figure 1.3).

It follows that the unit circle, consisting of the points at distance 1 from O, has equation $x^2 + y^2 = 1$. More generally, the circle with center (a, b) and radius r has equation $(x - a)^2 + (y - b)^2 = r^2$.

The problem with this algebraic approach is that it goes too far: there is no natural restriction on the equations that yields precisely the geometric concepts in Euclid. If we stop at linear equations we get only lines; if we stop at quadratic equations we get all the conic sections—ellipses, parabolas, and hyperbolas—whereas Euclid has only circles. However, there is a different algebraic concept that stops at precisely the right place: the concept of a *vector space with an inner product*. We will not give the general definition of a vector space here (see chapter 4), but instead describe the particular vector space $\mathbb{R}^2$ that is suitable for Euclidean plane geometry.

This space consists of all the ordered pairs (x, y), where x and y belong to $\mathbb{R}$, the set of real numbers (we say more about $\mathbb{R}$ in the next section; geometrically it is the set of points on the line). We are allowed

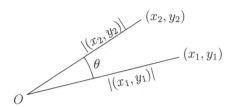

Figure 1.4: The angle between two vectors.

to add pairs by the rule

$$(x, y) + (a, b) = (x + a, y + b)$$

and to multiply a pair by any real number c using the rule

$$c(x, y) = (cx, cy).$$

These operations have natural geometric interpretations: Adding (a, b) to each (x, y) means *translating* the plane; namely, shifting all its points through distance a horizontally and distance b vertically. Multiplying each (x, y) by c means *magnifying* the whole plane by the factor c. As we will see in chapter 5, even in this simple setting we can prove some geometrically interesting theorems. But to capture all of Euclid's geometry we need an extra ingredient: the *inner product* (also called the dot product) defined by

$$(x_1, y_1) \cdot (x_2, y_2) = x_1 x_2 + y_1 y_2.$$

Notice that

$$(x, y) \cdot (x, y) = x^2 + y^2 = |(x, y)|^2,$$

where $|(x, y)|$ denotes the distance of (x, y) from the origin O. Thus the inner product gives a definition of distance agreeing with the Pythagorean theorem. Once we have the concept of distance, we can also obtain the concept of angle, because it turns out that

$$(x_1, y_1) \cdot (x_2, y_2) = |(x_1, y_1)||(x_2, y_2)| \cos \theta,$$

where θ is the angle "between" (x_1, y_1) and (x_2, y_2) as shown in figure 1.4.

The main advantages of using the concept of a vector space with an inner product, rather than Euclid-style axioms, are familiarity and universality. The rules for calculating with vectors are similar to traditional algebra; also, vector spaces and inner products occur in many parts of mathematics, so they are worth learning as general-purpose tools.

1.5 Calculus

Calculus differs from elementary arithmetic, algebra, and geometry in a crucial way: the presence of *infinite processes*. Maybe the gulf between finite and infinite is so deep that we should use it to separate "elementary" from "non-elementary," and to exclude calculus from elementary mathematics. However, this is not what happens in high schools today. A century ago, calculus *was* excluded, but infinite processes certainly were not: students became familiar with infinite series in high school before proceeding to calculus at university. And way back in 1748, Euler wrote a whole book on infinite processes, *Introductio in analysin infinitorum* (Introduction to the analysis of the infinite), without mentioning differentiation and integration. This is what "pre-calculus" used to mean!

So, it is probably not wise to exclude infinity from elementary mathematics. The question is whether infinity should be explored *before* calculus, in a study of infinite series (and perhaps other infinite processes), or after.

In my opinion there is much to be said for looking at infinity first. Infinite series arise naturally in elementary arithmetic and geometry, and indeed they were used by Euclid and Archimedes long before calculus was invented. Also coming before calculus, albeit by a narrower historical margin, was the concept of *infinite decimals*, introduced by Stevin (1585a). Infinite decimals are a particular kind of infinite series, extending the concept of decimal fraction, so they are probably the infinite process most accessible to students today.

Indeed, an infinite decimal arises from almost any ordinary fraction when we attempt to convert it to a decimal fraction. For example

$$1/3 = 0.333333\ldots.$$

So, in some ways, infinite decimals are familiar. In other ways they are puzzling. Many students dislike the idea that

$$1 = 0.999999\ldots,$$

because $0.999999\ldots$ seems somehow (infinitesimally?) less than 1. Examples like this show that the limit concept can, and probably should, be discussed long before it comes up in calculus. But before getting to the precise meaning of infinite decimals, there is plenty of fun to be had with them. In particular, it is easy to show that any periodic infinite decimal represents a rational number. For example, given

$$x = 0.137137137137\ldots$$

we can shift the decimal point three places to the right by multiplying by 1000, so

$$1000x = 137.137137137\ldots = 137 + x.$$

We can then solve for x, obtaining $x = 137/999$. A similar argument works with any decimal that is *ultimately* periodic, such as

$$y = 0.31555555\ldots.$$

In this case $1000y = 315.555555\ldots$ and $100y = 31.555555\ldots$, so that

$$1000y - 100y = 315 - 31,$$

which means $900y = 284$ and hence $y = 284/900$.

Conversely, any rational number has an ultimately periodic decimal (perhaps ultimately all zeros). This is because only finitely many remainders are possible in the division process that produces the successive decimal digits, so eventually a repetition will occur.

The infinite decimals above are examples of the *geometric series*

$$a + ar + ar^2 + ar^3 + \cdots \quad \text{with} \quad |r| < 1.$$

For example,

$$\frac{1}{3} = \frac{3}{10} + \frac{3}{10^2} + \frac{3}{10^3} + \cdots,$$

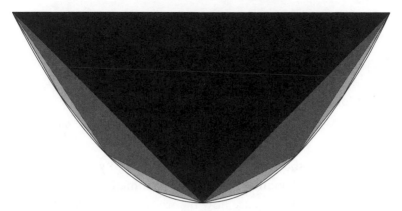

Figure 1.5: Filling the parabolic segment with triangles.

which has $a = 3/10$ and $r = 1/10$. There is no compelling reason to call these series "geometric," but they do arise in geometry. One of the first examples was given by Archimedes: finding the *area of a parabolic segment*. This problem, which today would be solved by calculus, can be reduced to summation of a geometric series as follows.

The idea is to fill the parabolic segment by infinitely many triangles, and to sum their areas. It turns out, with the very simple choice of triangles shown in figure 1.5, that the areas form a geometric series. The first triangle has two vertices at the ends of the parabolic segment, and its third vertex at the bottom of the parabola. The next two triangles lie under the lower sides of the first triangle, with their third vertices on the parabola at horizontal distance half-way between their first two, and so on.

Figure 1.5 shows the first three stages of the filling process for the segment of the parabola $y = x^2$ between $x = -1$ and $x = 1$. The first triangle (black) obviously has area 1. It can be checked that the next two (dark gray) each have area 1/8, so together they have area 1/4. The next four (light gray) have total area $1/4^2$, and so on. Hence the area of the parabolic segment is

$$A = 1 + \left(\frac{1}{4}\right) + \left(\frac{1}{4}\right)^2 + \cdots .$$

We can find A by multiplying both sides of this equation by 4, obtaining

$$4A = 4 + 1 + \left(\frac{1}{4}\right) + \left(\frac{1}{4}\right)^2 + \cdots,$$

whence it follows by subtraction that

$$3A = 4 \quad \text{and therefore} \quad A = 4/3.$$

This example shows that, with a little ingenuity, a problem normally solved by integration reduces to summation of a geometric series. In chapter 6 we will see how far we can go with an elementary minimum of calculus (integration and differentiation of powers of x) when infinite series are given a greater role. In particular, we will see that the geometric series is the main ingredient in such celebrated results as

$$\ln 2 = 1 - \frac{1}{2} + \frac{1}{3} - \frac{1}{4} + \cdots$$

and

$$\frac{\pi}{4} = 1 - \frac{1}{3} + \frac{1}{5} - \frac{1}{7} + \cdots.$$

1.6 Combinatorics

A fine example of a combinatorial concept is the so-called *Pascal's triangle*, which has historical roots in several mathematical cultures. Figure 1.6 shows an example from China in 1303.

Figure 1.7 shows the same numbers as ordinary Arabic numerals.

The Chinese knew that the numbers in the $(n + 1)$st row are the coefficients in the expansion of $(a + b)^n$. Thus

$$(a+b)^1 = a+b$$
$$(a+b)^2 = a^2 + 2ab + b^2$$
$$(a+b)^3 = a^3 + 3a^2b + 3ab^2 + b^3$$
$$(a+b)^4 = a^4 + 4a^3b + 6a^2b^2 + 4ab^3 + b^4$$
$$(a+b)^5 = a^5 + 5a^4b + 10a^3b^2 + 10a^2b^3 + 5ab^4 + b^5$$

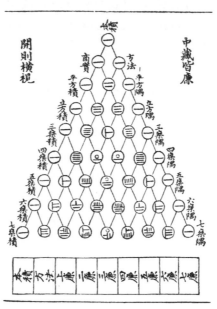

Figure 1.6: "Pascal triangle" of Zhu Shijie (1303).

$$
\begin{array}{ccccccccccccc}
 & & & & & & 1 & & & & & & \\
 & & & & & 1 & & 1 & & & & & \\
 & & & & 1 & & 2 & & 1 & & & & \\
 & & & 1 & & 3 & & 3 & & 1 & & & \\
 & & 1 & & 4 & & 6 & & 4 & & 1 & & \\
 & 1 & & 5 & & 10 & & 10 & & 5 & & 1 & \\
1 & & 6 & & 15 & & 20 & & 15 & & 6 & & 1 \\
\end{array}
$$
$$1 \quad 7 \quad 21 \quad 35 \quad 35 \quad 21 \quad 7 \quad 1$$
$$\cdots$$

Figure 1.7: Arabic numeral Pascal triangle.

$$(a+b)^6 = \quad a^6 + 6a^5b + 15a^4b^2 + 20a^3b^3 + 15a^2b^4 + 6ab^5 + b^6$$

$$(a+b)^7 = \quad a^7 + 7a^6b + 21a^5b^2 + 35a^4b^3 + 35a^3b^4 + 21a^2b^5 + 7ab^6 + b^7$$

Because they arise from the "binomial" $a + b$, the numbers in the $(n+1)$st row of the triangle are called *binomial coefficients*. They are denoted by $\binom{n}{0}, \binom{n}{1}, \ldots, \binom{n}{n}$. Looking back at figure 1.7, we notice that each binomial coefficient $\binom{n}{k}$ in row $n+1$ is the sum of the two above it, $\binom{n-1}{k-1}$ and $\binom{n-1}{k}$, in row n. This famous property of the binomial coefficients is easily explained by algebra. Take $\binom{6}{3}$ for example. On the one hand, by definition

$$\binom{6}{3} = \text{coefficient of } a^3 b^3 \text{ in } (a+b)^6.$$

On the other hand, $(a+b)^6 = a(a+b)^5 + b(a+b)^5$, so there are two ways that $a^3 b^3$ arises in $(a+b)^6$: from the first term, as $a \cdot a^2 b^3$, and from the second term, as $b \cdot a^3 b^2$. Because of this

$$\binom{6}{3} = \text{coefficient of } a^2 b^3 \text{ in } (a+b)^5 + \text{coefficient of } a^3 b^2 \text{ in } (a+b)^5$$

$$= \binom{5}{2} + \binom{5}{3}.$$

This argument is already a little bit "combinatorial," because we consider how $a^3 b^3$ terms arise as *combinations* of terms from $a(a+b)^5$ and $b(a+b)^5$. Now let's get really combinatorial, and consider how $a^k b^{n-k}$ terms can arise from the n factors $a+b$ in $(a+b)^n$.

To get $a^k b^{n-k}$ we must choose a from k of the factors and b from the remaining $n-k$ factors. Thus the number of such terms,

$$\binom{n}{k} = \text{number of ways of choosing } k \text{ items from a set of } n \text{ items.}$$

As a reminder of this fact, we pronounce the symbol $\binom{n}{k}$ as "n choose k." The combinatorial interpretation gives us an explicit formula for $\binom{n}{k}$, namely

$$\binom{n}{k} = \frac{n(n-1)(n-2) \cdots (n-k+1)}{k!}.$$

To see why, imagine making a sequence of k choices from a set of n items.

The first item can be chosen in n ways, then $n - 1$ items remain,
Next, the second item can be chosen in $n - 1$ ways, and $n - 2$ items remain.
Next, the third item can be chosen in $n - 2$ ways, and $n - 3$ items remain.
$$\vdots$$
Finally, the kth item can be chosen in $n - k + 1$ ways.

Thus there are $n(n - 1)(n - 2) \cdots (n - k + 1)$ *sequences* of choices. However, we do not care about the order in which items are chosen— only the *set* of k items finally obtained—so we need to divide by the number of ways of arranging k items in a sequence. This number, by the argument just used, is

$$k! = k(k - 1)(k - 2) \cdots 3 \cdot 2 \cdot 1.$$

This is how we arrive at the formula for the binomial coefficient $\binom{n}{k}$ above.

Combining this evaluation of the binomial coefficients with their definition as the coefficients in the expansion of $(a + b)^n$, we obtain the so-called *binomial theorem*:

$$(a + b)^n = a^n + na^{n-1}b + \frac{n(n - 1)}{2}a^{n-2}b^2$$
$$+ \frac{n(n - 1)(n - 2)}{3 \cdot 2}a^{n-3}b^3 + \cdots + nab^{n-1} + b^n.$$

This name is also used for the special case with $a = 1$ and $b = x$, namely

$$(1 + x)^n = 1 + nx + \frac{n(n - 1)}{2}x^2 + \frac{n(n - 1)(n - 2)}{3 \cdot 2}x^3 + \cdots + nx^{n-1} + x^n.$$

We now have two ways to compute the binomial coefficients $\binom{n}{k}$: by explicit formulas and by the process of forming successive rows in Pascal's triangle. We also have a very concise *encapsulation* of the sequence $\binom{n}{0}$, $\binom{n}{1}$, ..., $\binom{n}{n}$: as the coefficients in the expansion of $(1 + x)^n$. A function such as $(1 + x)^n$, which encapsulates a sequence of numbers as the coefficients of powers of x, is called a *generating function* for the sequence. Thus $(1 + x)^n$ is a generating function for the sequence of binomial coefficients $\binom{n}{0}$, $\binom{n}{1}$, ..., $\binom{n}{n}$.

In chapter 7 we will find generating functions for other sequences of numbers that arise in combinatorics. In many cases these are infinite sequences. So combinatorics, like calculus, draws on the theory of infinite series.

Combinatorics is sometimes called "finite mathematics" because, at least at the elementary level, it deals with finite objects. However, there are infinitely many finite objects, so to prove anything about *all* finite objects is to prove something about infinity. This is the ultimate reason why elementary mathematics cannot exclude infinity, and we say more about it in section 1.8.

1.7 Probability

Given two players each of whom lacks a certain
number of games to complete the set, to find by the
arithmetic triangle what the division should be (if
they wish to separate without playing) in the light of
the games each lacks.

Pascal (1654), p. 464

The concept of probability has been in the air for as long as human beings have gambled, yet until a few hundred years ago it was thought too lawless for mathematics to handle. This belief began to change in the sixteenth century, when Cardano wrote an elementary book on games of chance, the *Liber de ludo aleae*. However, Cardano's book was not published until 1663, by which time mathematical probability theory had begun in earnest, with the Pascal (1654) solution of the problem of division of stakes, and the first published book on probability theory by Huygens (1657).

We can illustrate Pascal's solution with a simple example. Suppose players I and II agree to flip a fair coin a certain number of times, with the winner agreed to be the first to call the outcome correctly a certain number of times. For some reason (police knocking at the door?) the game has to be called off with n plays remaining, at which stage player I needs k more correct calls to win. How should the players divide the money they have staked on the game?

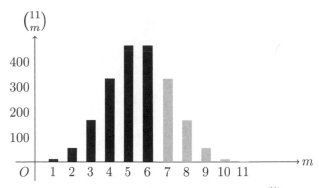

Figure 1.8: Graph of the binomial coefficients $\binom{11}{m}$.

Pascal argued that the stakes should be divided in the ratio

probability of a win for I : probability of a win for II.

Further, since each play of the game is equally likely to be a win for I or II, these probabilities are in the ratio

how often I has $\geq k$ wins in n plays : how often I has $< k$ wins in n plays.

The problem is now reduced to a problem in combinatorics: in how many ways can $\geq k$ things be chosen from a set of n things? And the binomial coefficients give the answer:

$$\binom{n}{n} + \binom{n}{n-1} + \cdots + \binom{n}{k}.$$

Thus the ratio of probabilities, which is the ratio in which the stakes should be divided, is:

$$\binom{n}{n} + \binom{n}{n-1} + \cdots + \binom{n}{k} : \binom{n}{k-1} + \binom{n}{k-2} + \cdots + \binom{n}{0}.$$

For even moderate values of n and k, this ratio would be difficult to compute, or even express, without the binomial coefficients. Suppose, for example, that $n = 11$ and $k = 7$. Figure 1.8 shows a bar graph of the values of $\binom{11}{m}$ for $m = 0$ to 11. They range in value from 1 to 462, with those for $m \geq 7$ shown in gray. Thus the ratio in this case is the ratio of the gray area to the black area.

And in fact

$$\binom{11}{7} + \binom{11}{8} + \binom{11}{9} + \binom{11}{10} + \binom{11}{11} = 330 + 165 + 55 + 11 + 1$$

$$= 562.$$

The sum of all the binomial coefficients $\binom{11}{k}$ is $(1 + 1)^{11} = 2^{11} = 2048$, so the other side of the ratio is $2048 - 562 = 1486$. Thus, in this case, $562/2048$ of the stake should go to player I and $1486/2048$ to player II.

With larger values of n and k the binomial coefficients rapidly become larger; indeed their total 2^n grows exponentially. However, an interesting thing happens as n increases. The *shape* of the graph of binomial coefficients, when suitably scaled in the vertical direction, approaches that of the continuous curve

$$y = e^{-x^2}.$$

This is advanced probability theory, which involves calculus, but we will say a little more about it in chapter 8 and give a proof in section 10.7.

1.8 Logic

The most distinctive feature of mathematics is that it *proves* things, by logic; however, we postpone the details until chapter 9. Here we discuss only the most *mathematical* part of logic: mathematical induction, which is the simplest principle for reasoning about infinity. Mathematical induction is also known as *complete* induction to distinguish it from the "incomplete induction" in daily use, which guesses a general conclusion (often incorrectly) from a few special cases. Proof by induction owes its existence to the *inductive property* of the natural numbers 0, 1, 2, 3, 4, 5, ...; namely, that any natural number can be reached by starting at 0 and repeatedly adding 1.

Figure 1.9: The towers of Hanoi.

It follows from the inductive property that any property $\mathcal{P}$ true of all natural numbers can be proved in two steps:

1. Prove that $\mathcal{P}$ holds for 0 (the *base step*).
2. Prove that $\mathcal{P}$ "propagates" from each number to the next; that is, if $\mathcal{P}$ holds for n then $\mathcal{P}$ holds for $n+1$ (the *induction step*).

Obviously, it is not essential to start at 0. If we wish to prove that some property $\mathcal{P}$ holds for all natural numbers from, say, 17 onwards then the base step will be to prove that $\mathcal{P}$ holds for 17.

Induction is not only a natural (and indeed inevitable) method of proof, it is often remarkably efficient, because it "hides" the details of why $\mathcal{P}$ holds for each n. We only have to understand why $\mathcal{P}$ holds for the starting value, and why it propagates from each number to the next. Here is an example: the classic combinatorial problem known as the *towers of Hanoi* (figure 1.9).

We are given a board with three pegs, on one of which is a stack of n disks whose radii decrease with height. (The disks are pierced in the center so that they can slip onto a peg.) The problem is to move all the disks onto another peg, one at a time, in such a way that a larger disk never rests on top of a smaller one.

First suppose that $n = 1$. With only one disk we can obviously solve the problem by moving the disk to any other peg. Thus the problem is solved for $n = 1$. Now suppose that it is solved for $n = k$ disks and consider what to do with $k+1$ disks. First, use the solution for k disks to shift the top k disks of the stack onto another peg; say, the middle peg. This leaves just the bottom disk of the stack on the left peg, and we can move it onto the empty right peg. Then use the solution for k disks again to shift the stack of k disks on the middle peg onto the right peg. Done!

It is a great virtue of this proof that we do not have to know *how* to shift a stack of n disks—only that it can be done—because it is quite complicated to shift stacks of only three or four. In fact, it takes $2^n - 1$ moves to shift a stack of n disks, and the proof is by a similar induction:

Base step. It clearly takes $1 = 2^1 - 1$ move to shift a stack of 1 disk.

Induction step. If it takes $2^k - 1$ moves to shift a stack of k disks, consider what it takes to shift a stack of $k + 1$. However this is done, we must first shift the top k disks, which takes $2^k - 1$ moves. Then we must move the bottom disk to a different peg (one move), because it cannot rest on top of any other disk. Finally we must shift the stack of k disks back on top of the bottom disk, which takes $2^k - 1$ moves. Therefore, the minimum number of moves to shift a stack of $k + 1$ disks is

$$(2^k - 1) + 1 + (2^k - 1) = 2^{k+1} - 1,$$

as required.

To bolster my claim that induction is "inevitable," let me point out its role in arithmetic. As we have already seen, the natural numbers 0, 1, 2, 3, 4, 5, ... arise from 0 by repeated applications of the *successor function* $S(n) = n + 1$. What is more remarkable is that all computable functions can be built from $S(n)$ by *inductive definitions* (also called recursive definitions). Here is how to obtain addition, multiplication, and exponentiation.

The base step in the definition of addition is

$$m + 0 = m,$$

which defines $m + n$ for all m and for $n = 0$. The induction step is

$$m + S(k) = S(m + k),$$

which defines $m + n$ for all m and for $n = S(k)$, given that $m + k$ is already defined. So it follows by induction that $m + n$ is defined for all natural numbers m and n. Essentially, induction formalizes the idea that addition is repeated application of the successor function.

Now that addition is defined, we can use it to define multiplication by the following equations (base step and induction step, respectively):

$$m \cdot 0 = 0, \qquad m \cdot S(k) = m \cdot k + m.$$

This definition formalizes the idea that multiplication is repeated addition. And then, with multiplication defined, we can define exponentiation by

$$m^0 = 1, \qquad m^{S(k)} = m^k \cdot m,$$

which formalizes the idea that exponentiation is repeated multiplication.

Induction has been present in mathematics, in some form, since the time of Euclid (see the Historical Remarks below). However, the idea of using induction as the foundation of arithmetic is comparatively recent. The inductive definitions of addition and multiplication were introduced by Grassmann in 1861, and were used by him to inductively prove all the ring properties of the integers given in section 1.3. These imply the field structure of the rational numbers, and with it the field structure of the real (see chapter 6) and complex numbers. Thus induction is not only the basis for counting but also for algebraic structure.

1.9 Historical Remarks

Once upon a time in America, Euclid was a revered figure who gave his name to many a Euclid Avenue across the country. (This was part of the nineteenth-century classical renaissance, during which many place names were chosen from the Greek and Roman classics.) For example, there is Euclid Avenue in Cleveland which became "millionaire's row," and Euclid Avenue in Brooklyn which became a stop on the route of the A train. Figure 1.10 gives a glimpse of Euclid Avenue in San Francisco, with some appropriate geometric figures.

In nineteenth-century America, as in most of the Western world, Euclid's *Elements* was regarded as a model presentation of mathematics and logic: essential knowledge for any educated person. One such

Figure 1.10: Euclid Avenue, San Francisco.

person was Abraham Lincoln. Here is what he, and one of his biographers, said about Lincoln's study of Euclid.

> He studied and nearly mastered the six books of Euclid since he was a member of Congress. He regrets his want of education, and does what he can to supply the want.
>
> Abraham Lincoln (writing of himself), *Short Autobiography*

> He studied Euclid until he could demonstrate with ease all the propositions in the six books.
>
> Herndon's *Life of Lincoln*

So what is the *Elements*, this book that cast such a long shadow over mathematics and education? The *Elements* is a compilation of the mathematics known in the Greek civilization of Euclid's time, around 300 BCE. It contains elementary geometry and number theory, much as they are understood today, except that numbers are not applied to geometry, and there is very little algebra. There are actually thirteen books in the *Elements*, not six, but the first six contain the elementary geometry for which the *Elements* is best known. They also contain the very subtle Book V which tackles (what we would now call) the problem of describing real numbers in terms of rational numbers. If Lincoln really mastered Book V he was a mathematician!

The Greeks did not have a written notation for numbers such as decimals, so the *Elements* contains nothing about algorithms for addition and multiplication. Instead, there is quite a sophisticated introduction to the abstract theory of numbers in Books VII to IX, with numbers denoted by letters as if they were line segments. These books contain the basic theory of divisibility, the Euclidean algorithm, and prime numbers that remains the starting point of most number theory courses today. In particular, Book IX contains a famous proof that there are infinitely many primes.

We say more about the *Elements* in later chapters, because it has influenced elementary mathematics more than any other book in history. Indeed, as the name suggests, the *Elements* have a lot to do with the very meaning of the word "elementary." Since we will often be referring to particular propositions in the *Elements*, it will be useful to have a copy handy. For English-speaking readers, the best edition (because of its extensive commentary) is still Heath (1925). Another useful version is *The Bones* by Densmore (2010), which lists all the definitions and propositions of the *Elements* in durable and compact form.

Decimal numerals developed in India and the Muslim world. They were introduced to Europe in medieval times, most famously (though not first) by the Italian mathematician Leonardo Pisano in his book *Liber abaci* of 1202. Leonardo is better known today by his nickname Fibonacci, and the title of his book refers to the abacus, which until then was synonymous with calculation in Europe. His highly influential book had the paradoxical effect of associating the word "abaci" with calculation *not* by the abacus. (Though in fact the abacus remained competitive with pencil and paper calculation until both were superseded by electronic calculators in the 1970s.) The famous *Fibonacci numbers*

1, 2, 3, 5, 8, 13, 21, 34, 55, 89, 144, 233, 377, 610, 987, 1597, 2584, . . . ,

each of which is the sum of the previous two, were introduced in the *Liber abaci* as an exercise in addition. Fibonacci could not have

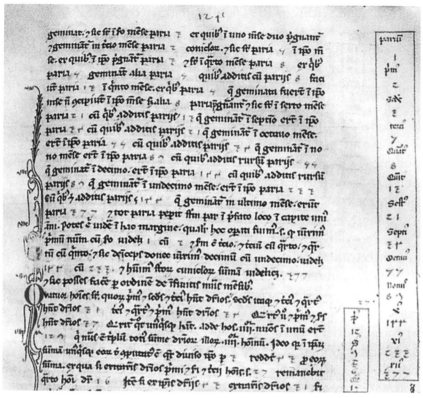

Figure 1.11: Fibonacci numbers in the *Liber abaci*.

known that his numbers would have a long career in number theory and combinatorics, though perhaps he wondered whether there was an explicit formula for the *n*th number. This question was not answered for more than 500 years. Finally, in the 1720s, Daniel Bernoulli and Abraham de Moivre showed that

$$F_n = \frac{1}{\sqrt{5}} \left[\left(\frac{1 + \sqrt{5}}{2} \right)^n - \left(\frac{1 - \sqrt{5}}{2} \right)^n \right],$$

where F_n is the *n*th Fibonacci number (starting, for convenience, with $F_0 = 0$ and $F_1 = 1$). For more about this, see chapter 7.

Figure 1.11 shows an early manuscript of the *Liber abaci*, from the Biblioteca Nazionale in Florence. It is open at the page where the Fibonacci numbers from 1 to 377 are displayed, in a column on the right. (Notice that the digits for 3, 4, and 5 are rather different from ours.)

Algebra also developed in the Muslim world, but it received a huge boost with the solution of the cubic equation in Italy in the early 1500s. Over the next century, European mathematicians developed great fluency in handling polynomial equations. This led in turn to the application of algebra to geometry by Fermat and Descartes in the 1620s, then to the development of calculus by Newton in the 1660s and Leibniz in the 1670s.

But despite all this ferment in the mathematical community, which created most of the mathematics taught in schools today, educated people in the 1700s could be abysmally ignorant of mathematics. (Some things never change ...) The English diarist Samuel Pepys, a Cambridge-educated man who later held positions such as chief secretary to the Admiralty and president of the Royal Society, at one point hired a tutor to teach him the multiplication table! Here is what Pepys wrote in his diary entry for July 4, 1662 (when he was aged 29):

> By and by comes Mr. Cooper, mate of the Royall Charles, of whom I intend to learn mathematiques, and do begin with him to-day, he being a very able man, and no great matter, I suppose, will content him. After an hour's being with him at arithmetique (my first attempt being to learn the multiplication-table); then we parted till to-morrow.

A week later, he was able to report some progress:

> Up by 4 a-clock, and hard at my multiplicacion[*sic*] table, which I am now almost maister of.

At about the same time, in France, Pascal's triangle made its European debut in a small book by Pascal called *The arithmetic triangle*. Working independently of the Asian mathematicians who discovered the triangle some centuries earlier, Pascal took quite a different tack. He proved about 20 arithmetic properties of the triangle, by induction, then took up "Use of the arithmetical triangle for combinations." Here he proved that

$$\binom{n}{k} = \frac{n(n-1)(n-2)\cdots(n-k+1)}{k!}$$

is the number of combinations of n things, taken k at a time. And he made the first application of this result to *probability theory*; namely, to the problem of dividing the stakes in a game of chance which is interrrupted before completion. As we saw in section 1.7, the problem can be solved by forming a ratio of sums of the terms $\binom{n}{k}$.

By the 1700s, the mathematics had changed so much under the influence of calculus that the concept of "elementary mathematics" had to be revised. Some of the most eminent mathematicians took part in this revision.

In 1707, Newton published *Universal Arithmetick*, originally in Latin and later in English. He described it as a book on the "science of computing," but by "computing" he meant quite a general theory that encompassed both arithmetic and algebra. His opening paragraph made clear this new and general viewpoint (here "computation by Species" means computation with letters for unknowns, as in high school algebra):

> Computation is either performed by *Numbers*, as in Vulgar Arithmetick, or by *Species*, as is usual among Algebraists. They are both built on the same Foundations, and aim at the same End, *viz. Arithmetic* Definitely and Particularly, *Algebra* Indefinitely and Universally; so that almost all Expressions that are found out by this Computation, and particularly Conclusions, may be called *Theorems*. But Algebra is particularly excellent in this, that whereas in Arithmetick Questions are only resolv'd by proceeding from given Quantities to the Quantities sought; Algebra proceeds in a retrograde Order, from the Quantities sought, as if they were given, to the Quantities given, as if they were sought, to the end that we may, some way or other, come to a Conclusion or Equation, from which we may bring out the Quantity sought. And after this Way the most difficult Problems are resolved, the Resolutions whereof would be sought in vain from only common Arithmetick. Yet Arithmetick in all its Operations is so subservient to Algebra, as that they seem both but to make one perfect *Science of Computing*; and therefore I will explain them both together.

In 1770 Euler wrote *Elements of Algebra* from a similar viewpoint as Newton's, though with somewhat different content. At the low end, Euler omits the algorithms for basic operations on decimal numerals, and at the high end, he has much more number theory. Indeed, Euler proves some difficult results for the first time, including a claim of Fermat that $y = 3$ and $x = 5$ is the only positive integer solution of $y^3 = x^2 + 2$. Euler's solution is one of the first steps in *algebraic number theory* (see chapter 2).

In the late 1700s, the French Revolution brought big changes to mathematics education in France. A new institution for advanced studies, the *École Normale*, was founded and the top mathematicians were enlisted to modernize the mathematics curriculum. Among them was Lagrange, who in 1795 wrote a book on elementary mathematics based on lectures at the *École Normale*. A century later it was still popular, and was translated into English as *Lectures on Elementary Mathematics*. Like Newton and Euler, Lagrange views algebra as a "universal arithmetic," to be studied alongside traditional arithmetic. And like Euler, he takes arithmetic to include what is now called "number theory"—primes, divisibility, and the solution of equations in integers.

In the 1800s, Germany took the initiative in number theory, both at the advanced level (Gauss, Dirichlet, Kummer, Dedekind) and at the level of foundations (Grassmann and Dedekind). Perhaps the most surprising event was the discovery of Grassmann, mentioned in the previous section, that the basic functions and theorems of arithmetic are based on induction. Grassmann, who was a high school teacher and a leading Sanskrit scholar, attempted to spread his ideas through a high school textbook, the *Lehrbuch der Arithmetik*. Not surprisingly, this attempt failed, but the ideas took hold the second time round when rediscovered by Dedekind (1888).

In his *Elementary Mathematics from an Advanced Standpoint* Klein (1908) credits Grassmann with establishing the foundation of arithmetic by induction. He also mentions some "fundamental rules of reckoning" that follow from it, such as $a + b = b + a$ and $ab = ba$. However, he stops short of discussing algebraic *structures*, such as rings and fields. Klein viewed algebra as mainly the study of polynomial equations, enriched by the algebraic geometry of curves and surfaces. Linear algebra did not yet exist as a separate discipline, because its fundamental concepts were buried under the concept of determinant—a concept that is now considered relatively advanced.

Geometry was Klein's favorite subject, and it is present throughout his first volume (*Arithmetic, Algebra, Analysis*) before becoming the sole subject of his second. He takes the late nineteenth-century view that all mathematics should be "arithmetized" (based on numbers), so he bases geometry entirely on coordinates. He generally disparages Euclid's *Elements* as elementary and pedantic, and ignores the striking new results on the axiomatic geometry found by Hilbert in 1899 (for more on this, see chapter 5). However, it must be said that the twentieth century was generally in favor of the arithmetization of geometry, and that linear algebra gives a particularly efficient approach to Euclid. Klein's own approach to elementary geometry is not quite linear algebra as we know it (because the very concept of linearity is still obscured by the concept of determinant), but it is evolving in that direction.

Thus Klein's book seems to foreshadow the modern era of mathematics, in which induction, abstract algebra, and linear algebra play important roles. We continue our discussion of this era in later chapters.

1.10 Philosophical Remarks

The various samples of mathematics above are considered elementary by most mathematicians and teachers of mathematics. They are taught at high school level or below in most parts of the world, though usually not all in the same school. Nevertheless, though all of them have been considered elementary in some schools, at some times, we must admit that some are less elementary than others. This raises the question: how far can we go before mathematics ceases to be elementary? Is there a clear borderline between elementary and advanced mathematics?

Unfortunately, no. There is no sharp separation between elementary and advanced mathematics, but certain characteristics become more prominent as mathematics becomes more advanced. The most obvious ones, which we will highlight in this book, are

- infinity,

- abstraction,

- proof.

Some mathematics programs have been tempted to use one or more of these characteristics to separate elementary from advanced mathematics. In particular, in the United States it is thought possible to postpone "introduction to proof" until the later stages of undergraduate education. This, I believe, is a delusion.

It is no doubt a good idea to withhold the *theory* of proof from junior undergraduates, but *examples* of proof should be part of mathematics from high school level[1]—as soon as students encounter statements that are not obvious, such as the Pythagorean theorem. In the beginning, of course, proofs will not be very formal. Indeed, most mathematicians dislike completely formal proofs and resist the basic idea of mathematical logic, which is to view proofs themselves as mathematical objects. This is why I postpone logic until the last chapter of this book, after giving enough examples to suggest that a theory of proof might be useful, though perhaps beyond the scope of elementary mathematics.

Similar thoughts arise with infinity and abstraction.

Mathematics without infinity is a lot more worthwhile than mathematics without proof, so it could be a candidate for "elementary mathematics." But it seems unfair to exclude harmless and easily understood infinite objects, such as the infinite decimal

$$1/3 = 0.33333\cdots,$$

so it becomes a question of deciding how much infinity is "elementary." There is an ancient way to answer this question: by distinguishing between "potential" and "actual" infinity.

The infinity of natural numbers 0, 1, 2, 3, . . . is a potential infinity in the sense that we can view it as an unending *process*: start with 0 and repeatedly add 1. There is no need to believe that this process can be completed—it suffices that it produce each natural number at some finite step. The totality of real numbers, on the other hand, cannot be viewed as a potential infinity. As we will see in chapter 9, there is no step-by-step process for generating numbers that will produce each real

---

[1] Some of my mathematical colleagues think that proof should be introduced earlier than this. In San Francisco a school called *Proof School* has recently opened, for students at middle school level and above.

number at some finite step. We have to view the real numbers as a *completed* or *actual* infinity (as indeed we do when we view them as a line, or as the path of a continuously moving point).

Thus, by allowing potential infinities and excluding actual infinities we create a borderline between elementary and advanced mathematics, with the natural numbers on the elementary side and the (totality of) real numbers on the advanced side. This borderline is still somewhat fuzzy—what about individual real numbers, such as $\sqrt{2}$?—but useful. We will find, particularly when we discuss calculus, that mathematics often becomes advanced to the degree that it involves the real numbers.

Finally, abstraction. Here it is even harder to draw a borderline.

If there is such a thing as mathematics without abstraction $(1 + 1 = 2$ maybe?), it is *too* elementary to cover all the mathematics usually called "elementary." We must at least include identities such as

$$a^2 - b^2 = (a + b)(a - b), \tag{*}$$

where a and b can be interpreted as any numbers. In fact, as I have argued in section 1.3, I would go so far as to include certain axiom sets, such as those for rings and fields, since they efficiently encode all the identities that one can prove for numbers. Indeed, justifying the steps in the calculation

$$(a + b)(a - b) = a(a - b) + b(a - b)$$
$$= a^2 - ab + ba - b^2$$
$$= a^2 - ab + ab - b^2$$
$$= a^2 - b^2$$

by the ring axioms is a nice example of a proof that can be carried out at high school level (and one that helps develop awareness of the role of axioms).

It will be seen in chapters 4 and 5 that the ring and field axioms (and the related vector space axioms) unify a lot of elementary algebra and geometry, so I favor their inclusion in elementary mathematics. But I leave it open whether there is any clear boundary between "elementary" and "advanced" abstraction.

2

⌒

Arithmetic

PREVIEW

To most people, "arithmetic" is the humble domain of addition, subtraction, multiplication, and division of decimal numbers and fractions. It is learned in elementary school and quickly becomes just a painful memory, once we can outsource such calculations to various electronic devices.

But there is a "higher arithmetic," or *number theory*, concerned with discovering general properties of numbers, and proving them. Number theory is an endlessly fascinating and difficult subject, which has been cultivated by mathematicians since the time of Euclid, and which today draws on the resources of almost all of mathematics. In later chapters we will see how number theory permeates elementary mathematics.

The purpose of this chapter is to introduce some perennial themes of elementary number theory—prime numbers and solving equations in integers—and some elementary methods of proof. These include the principle of *induction*, introduced by Euclid in a form called "descent," and some simple algebra and geometry. Descent allows us to find the greatest common divisor of any two positive integers by the *Euclidean algorithm*, and to prove that any positive integer factorizes uniquely into primes.

Algebra and *algebraic numbers* come to light when we search for positive integer solutions of the equations $y^3 = x^2 + 2$ and $x^2 - 2y^2 = 1$. Surprisingly, it helps to introduce numbers of the form $a + b\sqrt{-2}$ and $a + b\sqrt{2}$, where a, b are ordinary integers, and to pretend that these new numbers behave like ordinary integers. The pretense is actually justifiable, and it allows us to develop and exploit a theory of "primes" in the new numbers very like the ordinary theory of prime numbers.

2.1 The Euclidean Algorithm

Given a fraction, say $\frac{1728941}{4356207}$, how do we know that it is in *reduced form*; that is, with no common divisor in the numerator and denominator? To answer this question, we need to find the greatest common divisor of 1728941 and 4356207, which looks difficult. Even finding the divisors of 1728941 seems difficult, and indeed no good method for finding the divisors for large numbers is yet known.

Remarkably, it can be much easier to find the *common* divisors of two numbers than to find the divisors of either one of them. For example, I immediately know that the greatest common divisor of 10000011 and 10000012 is 1, without knowing the divisors of either number. Why? Well, if d is a common divisor of 10000011 and 10000012 we have

$$10000011 = dp \quad \text{and} \quad 10000012 = dq,$$

for some positive integers p and q. And therefore

$$10000012 - 10000011 = d(q - p),$$

so d also divides the difference of 10000011 and 10000012, which is 1. But the only positive integer that divides 1 is 1, so $d = 1$. More generally, if d is a common divisor of two numbers a and b, then d also divides $a - b$. In particular, the *greatest* common divisor of a and b is also a divisor of $a - b$.

This simple fact is the basis of an efficient algorithm for finding the greatest common divisor. It is called the *Euclidean algorithm* after Euclid, who described it over 2000 years ago in the *Elements*, Book VII, Proposition 2. In Euclid's words, one "continually subtracts the lesser number from the greater." More formally, one computes a sequence of pairs of numbers.

Euclidean algorithm. Starting with a given pair a, b, where $a > b$, each new pair consists of the smaller member of the previous pair, and the difference of the previous pair. The algorithm terminates with a pair of equal numbers, each of which is the greatest common divisor of a and b.

For example, if we start with the pair $a = 13, b = 8$, the algorithm leads to the pair $(1, 1)$, as we found in section 1.2, so 1 is the greatest common divisor of 13 and 8.

The main reason that the Euclidean algorithm works is the fact noted above: that greatest common divisor of a and b is also a divisor of $a - b$. If we denote the greatest common divisor by gcd, then $\gcd(a, b) = \gcd(b, a - b)$ and in the above example we have

$$\gcd(13, 8) = \gcd(8, 5) = \gcd(5, 3) = \gcd(3, 2)$$

$$= \gcd(2, 1) = \gcd(1, 1) = 1.$$

(Notice that, when we start with the consecutive Fibonacci numbers 13 and 8, subtraction gives all the preceding Fibonacci numbers, ending inevitably at the number 1. It is the same with any pair of consecutive Fibonacci numbers, so the gcd of any such pair is 1.)

A secondary reason, also important, is that the algorithm continually produces smaller numbers, and hence it *terminates* (necessarily with equal numbers) because *positive integers cannot decrease forever.* This "no infinite descent" principle is obvious, and Euclid often used it, but it is nevertheless profound. It is the first expression of *proof by induction*, which underlies all of number theory, as we will see in section 9.4.

Finally, we return to our implicit claim that the Euclidean algorithm is a fast way of finding common divisors—faster than any known way of finding the divisors of a single number. When we use only subtraction, as Euclid did, this is not quite right. For example, if we attempt to find $\gcd(101, 10^{100} + 1)$ by repeated subtraction, we will have to subtract 101 from $10^{100} + 1$ nearly 10^{98} times. This is *not* fast.

However, repeatedly subtracting b from a until the difference r becomes less than b is the same as *dividing a by b* and obtaining the *remainder r*. This gives the following fact, on which we will base the Euclidean algorithm from now on.

Figure 2.1: Visualizing quotient and remainder.

Division property. For any natural numbers a and $b \neq 0$ there are natural numbers q and r ("quotient" and "remainder") such that

$$a = qb + r \quad \text{where} \quad |r| < |b|.$$

The division property is also visually obvious, from figure 2.1, because any natural number a must lie between successive multiples of b. In particular, its distance r from the lower multiple qb is less than the distance b between them.

The advantage of division with remainder is that it is at least as fast, and generally much faster, than repeated subtraction. Each division by a k-digit number knocks about k digits off the number being divided and leaves a remainder with at most k digits. So the number of divisions is at most the total number of digits in the numbers we begin with. That's fast enough to find the gcd of numbers with thousands of digits.

2.2 Continued Fractions

The Euclidean algorithm, like any algorithm, produces a sequence of events. Each event depends in a simple way on the previous event, but one does not expect to capture the whole sequence in a single formula. Yet in fact there is such a formula, the so-called *continued fraction*.

For example, when we apply the Euclidean algorithm to the pair 117, 25 it produces the sequence of quotients 4, 1, 2, 8. This sequence is captured by the equation

$$\frac{117}{25} = 4 + \cfrac{1}{1 + \cfrac{1}{2 + \cfrac{1}{8}}}.$$

The fraction on the right arises by the following process, which reflects the divisions with remainder in the Euclidean algorithm by calculations with fractions.

$$\frac{117}{25} = 4 + \frac{17}{25} \qquad \text{(quotient 4, remainder 17)}$$

$$= 4 + \frac{1}{25/17} \qquad \text{(use remainder as new divisor)}$$

$$= 4 + \cfrac{1}{1 + \cfrac{8}{17}} \qquad \text{(quotient 1, remainder 8)}$$

$$= 4 + \cfrac{1}{1 + \cfrac{1}{17/8}} \qquad \text{(use remainder as new divisor)}$$

$$= 4 + \cfrac{1}{1 + \cfrac{1}{2 + \cfrac{1}{8}}} \qquad \text{(quotient 2, remainder 1)}$$

—at which stage the process halts because the remainder 1 divides the previous divisor 8 exactly.

Since the continued fraction algorithm perfectly simulates the Euclidean algorithm, it produces numbers that decrease in size and hence always halts. It follows that *any positive rational number has a finite continued fraction.* And conversely, *if a ratio of numbers produces an infinite continued fraction, then the ratio is irrational.* Until now, we had not contemplated applying the Euclidean algorithm to numbers not obviously in a rational ratio, but this observation prompts us to try. The outcome is remarkably simple and satisfying when we apply the continued fraction algorithm to $\sqrt{2} + 1$ and 1.

To allow the algorithm to run with a minimum of commentary, we point out in advance that

$$(\sqrt{2} + 1)(\sqrt{2} - 1) = 1, \quad \text{so that} \quad \sqrt{2} - 1 = \frac{1}{\sqrt{2} + 1}.$$

Here now is what happens:

$$\sqrt{2}+1 = 2+(\sqrt{2}-1) \quad \text{(separating into integer and fractional part)}$$

$$= 2+\frac{1}{\sqrt{2}+1} \qquad \left(\text{because } \sqrt{2}-1 = \tfrac{1}{\sqrt{2}+1}\right)$$

No need to go any further! The denominator $\sqrt{2}+1$ on the right-hand side can now be replaced by $2+\frac{1}{\sqrt{2}+1}$, in which $\sqrt{2}+1$ occurs again, and so on. Hence *the continued fraction algorithm will never halt.*

(This may remind you of one of those products in a box that contains an image of itself, like the one shown in figure 2.2. The situation is similar.)

It follows that $\sqrt{2}+1$ is *irrational*, and hence so is $\sqrt{2}$. The ancient Greeks knew that $\sqrt{2}$ is irrational, so it is tempting to wonder whether they knew this proof. Certainly, Euclid was aware that non-halting of the Euclidean algorithm implies irrationality. He says so in the *Elements*, Book X, Proposition 2, and his Proposition 5 of Book XIII implies non-halting of the Euclidean algorithm on the pair $\frac{1+\sqrt{5}}{2}$, 1. Thus it is possible that irrationality was first discovered this way, as Fowler (1999) suggests. Another way arises more directly from the study of divisibility, as we will see in the next section.

2.3 Prime Numbers

The prime numbers are perhaps the most wonderful objects of mathematics: easy to define, yet difficult to understand. They are the positive integers greater than 1 that are not products of smaller positive integers. Thus the sequence of prime numbers begins with

2, 3, 5, 7, 11, 13, 17, 19, 23, 29, 31, 37, 41, 43, 47, 53, 59, 61, 67, 71, 73, 79, 83, 89, 97, ...

Every positive integer n can be factorized into primes, because if n is not prime itself it is a product of smaller positive integers a and b, and we can repeat the argument with a and b: if either of them is not a prime then it is the product of smaller positive integers, and so on.

Figure 2.2: Containing an image of itself.

Since infinite descent through the positive integers is impossible, this process must eventually halt—necessarily with a factorization of n into primes. Thus all positive integers greater than 1 can be built by multiplying primes together.

This is hardly the simplest way to understand the positive integers—building them by *adding* 1 repeatedly seems a much better idea—but it does help us to understand the primes.

In particular, it enables us to see why there are *infinitely many* primes—something that is not obvious just by looking at the sequence above. The first proof of this is one of the great results of Euclid's *Elements*, and a modern version of Euclid's proof goes as follows.

Given the primes 2, 3, 5, . . . , p up to p, it will suffice to find a new prime, as this will show that there is no end to the sequence of primes.

Well, given the primes 2, 3, 5, . . . , p, consider the number

$$n = (2 \cdot 3 \cdot 5 \cdot \; \cdots \; \cdot p) + 1.$$

Then n is not divisible by any of the primes 2, 3, 5, . . . , p, because they all leave remainder 1. But n is divisible by *some* prime q, since it has a factorization into primes, so q is a new prime.

Thus the infinitude of the primes is something we can prove from the (easy) fact that every natural number has a prime factorization. A harder, and more powerful, fact is that the *prime factorization is unique.* More precisely: *each prime factorization of n involves the same primes, with each appearing the same number of times.* To illustrate, consider what happens when we break the number 60 down into smaller factors. There are several ways to do this, but they all end with the same primes. For example,

$$60 = 6 \cdot 10 = (2 \cdot 3) \cdot (2 \cdot 5) = 2^2 \cdot 3 \cdot 5,$$
$$60 = 2 \cdot 30 = 2 \cdot (2 \cdot 3 \cdot 5) = 2^2 \cdot 3 \cdot 5.$$

There are various ways to prove unique prime factorization, but none of them is obvious, so we will opt for a proof that at least uses familiar machinery: the Euclidean algorithm.

Recall from section 2.1 that we can find $\gcd(a, b)$ by starting with a and b and making a series of subtractions. Each of these subtractions produces an *integer combination of a and b, $ma + nb$* for some integers m and n. This is because we start with the integer combinations— namely $a = 1 \cdot a + 0 \cdot b$ and $b = 0 \cdot a + 1 \cdot b$—and the difference of any two integer combinations is again an integer combination. It follows, in particular, that

$$\gcd(a, b) = ma + nb \quad \text{for some integers } m \text{ and } n.$$

This fact now enables us to prove something about primes: *if a prime p divides a product ab of positive integers, then p divides a or p divides b.* To prove this *prime divisor property,* we suppose that p does *not* divide a, and try to see why p must divide b.

Well, if p does not divide a we have, since p has no divisors but itself and 1,

$$1 = \gcd(a, p) = ma + np.$$

Multiplying both sides of this equation by b, we get

$$b = mab + npb.$$

Now p divides both terms on the right, because it divides ab by assumption, and pb obviously. It follows that p divides the sum of these two terms; that is, b.

The prime divisor property was proved by Euclid, and unique prime factorization easily follows. Suppose, for the sake of argument, that some number has two different prime factorizations. By removing all common prime factors from these two factorizations, we get a certain product of primes (starting with p say) equal to a product of entirely different primes. But if p divides a product it divides one of the factors, by Euclid's proposition, and *this is impossible*, since all the factors are primes different from p. Thus there cannot be a positive integer with two different prime factorizations.

It is hard to believe, at first, that the unique prime factorization of positive integers should not be obvious. To appreciate why it is not obvious, it may help to look at a similar system where unique prime factorization *fails*. This is the case for the system of even numbers 2, 4, 6, 8, 10, ... This system is quite similar to the system of positive integers: the sum and product of any two members is a member, and properties like $a + b = b + a$ and $ab = ba$ are inherited from the positive integers.

In this system, we would call an even number an "even prime" if it is not the product of smaller even numbers. For example, 2, 6, and 10 are "even primes." It follows, by the descent argument we used for positive integers, that every even number has a factorization into "even primes." But notice that the number 60 has two *different* factorizations into "even primes":

$$60 = 6 \cdot 10,$$

$$60 = 2 \cdot 30.$$

(Of course, we can explain these two factorizations by the "hidden" odd primes 3 and 5 in the "even prime" factors, but the world of even numbers does not know about these.) Thus unique prime factorization must depend on something that the positive integers do not share with the even numbers. We will get a better idea of what this is when we study another system of numbers in section 2.6.

Irrationality of $\sqrt{2}$, Again

Unique prime factorization gives a different, and very simple, explanation of the irrationality of $\sqrt{2}$. Suppose, on the contrary, that $\sqrt{2}$ is rational, and imagine $\sqrt{2}$ written as a fraction in lowest terms. This means that the numerator and denominator of the fraction have no prime factor in common. So, in the square of this fraction (which equals 2), the numerator and denominator again have no common prime factor. But then the denominator cannot divide the numerator, and we have a contradiction.

Thus it is wrong to assume that $\sqrt{2}$ is rational.

2.4 Finite Arithmetic

There is an old rule, probably almost as old as decimal notation, called "casting out nines." The rule is that a number is divisible by 9 if and only if the sum of its digits is divisible by 9. For example, 711 is divisible by 9 because $7 + 1 + 1 = 9$. Indeed, more than this is true, the *remainder* when a number is divided by 9 equals the remainder when the sum of the digits is divided by 9.

For example, 823 leaves remainder 4 on division by 9, because $8 + 2 + 3$ leaves remainder 4. This happens because $823 = 8 \cdot 10^2 + 2 \cdot 10 + 3$ and 10^2 and 10 both leave remainder 1 on division by 9, since $10^2 = 99 + 1$ and $10 = 9 + 1$. We might say that 1, 10, 10^2 (and, similarly, higher powers of 10) are the "same, ignoring multiples of 9," and therefore 823 is the "same" as $8 + 2 + 3$. This notion of "sameness" is called *congruence modulo 9*.

In general, we say that integers a and b are *congruent modulo n*, written

$$a \equiv b \,(\mathrm{mod}\ n),$$

if $a - b$ is a multiple of n (so a and b are the "same, up to multiples of n"). The numbers that are *distinct* mod n are $0, 1, 2, \ldots, n-1$, and every other integer is congruent, mod n, to one of these. Also, we can add and multiply these n numbers mod n, by adding them in the ordinary way and then taking the remainder on division by n. Addition and multiplication mod n inherit the usual algebraic properties from the integers, in the sense that $a + b = b + a$ becomes $a + b \equiv b + a$ (mod n) and so on, so we can speak of *arithmetic mod n*. (This is not quite obvious, so you may wish to consult section 4.2 for the details.)

Arithmetic Mod 2

The simplest and smallest example is arithmetic mod 2, in which the only two distinct numbers are 0 and 1. This is the same as the arithmetic of "even" and "odd," because all the even numbers are congruent to 0 mod 2 and all the odd numbers are congruent to 1. Replacing $\equiv$ by $=$ and omitting the mod 2 symbols for simplicity, the rules for adding and multiplying 0 and 1 are

$$0+0=0, \quad 0+1=1, \quad 1+0=1, \quad 1+1=0,$$
$$0 \cdot 0 = 0, \quad 0 \cdot 1 = 0, \quad 1 \cdot 0 = 0, \quad 1 \cdot 1 = 1.$$

In particular, the rule $1 + 1 = 0$ reflects the fact that "odd" + "odd" = "even."

Since addition and multiplication mod 2 satisfy the usual rules of algebra, we can manipulate equations as we normally do, as long as we bear in mind that $1 + 1 = 0$. For example, we can solve the equation

$$x^2 + xy + y^2 = 0 \quad (\mathrm{mod}\ 2)$$

by substituting all possible values of x and y. It is easy to see that the only pair of values that satisfies this equation is $x = 0$, $y = 0$.

Solving polynomial equations mod 2 is easy in principle, because we need only substitute the finitely many sets of values of the variables and write down those, if any, that satisfy the equation. However, it may be hard in practice, because there are 2^m sequences of values for m variables (two for the first variable; for each of these, two for the second variable; and so on). So the number of possibilities quickly becomes astronomical as m increases. And no known method for solving m-variable equations, mod 2, is substantially faster than trying all possibilities.

In fact, no fast method is known to decide even *existence* of solutions for m-variable equations, mod 2. This is a fundamental unsolved problem in computation and logic, as we will explain further in chapters 3 and 9.

2.5 Quadratic Integers

Let us begin with a curious fact about the numbers 25 and 27. 25 is a square and 27 is a cube, so these numbers give a solution of the equation $y^3 = x^2 + 2$; namely, $x = 5$ and $y = 3$. Nearly 2000 years ago, Diophantus specifically mentioned this equation and this solution in his *Arithmetica*, Book VI, Problem 17. After reading this passage in Diophantus, Fermat (1657) claimed that it is the *only* solution in the positive integers. Why the equation caught their attention we do not really know. But it was a turning point in the development of number theory when Euler (1770), p. 401, proved Fermat's claim by a new and audacious method.

Observing that $x^2 + 2 = (x + \sqrt{-2})(x - \sqrt{-2})$, Euler's attention was drawn to numbers of the form $a + b\sqrt{-2}$, where a and b are integers. Leaving aside the meaning of the "imaginary" $\sqrt{-2}$, these new numbers are also "integers" in some sense. This is because the sum, difference, and product of any two of them is another number of the same kind and they satisfy the algebraic rules of a ring (mentioned in section 1.3). The audacious part of Euler's solution was to assume that they also have more doubtful properties, such as unique prime factorization. To see why this assumption comes up, we follow Euler's train of thought.

Assume that x and y are ordinary integers such that

$$y^3 = x^2 + 2 = (x + \sqrt{-2})(x - \sqrt{-2}).$$

Then one might think, if the factors on the right behave like ordinary integers, that $\gcd(x + \sqrt{-2}, x - \sqrt{-2}) = 1$ and (imagining $x + \sqrt{-2}$ and $x - \sqrt{-2}$ factorized into primes, with no prime common to both) that each of them is a cube, since their product is the cube y^3. This is not quite right, but it is true when x is a solution of $y^3 = x^2 + 2$, as we will see in the next subsection.

Then, writing $x + \sqrt{-2}$ as a cube, we get

$$x + \sqrt{-2} = (a + b\sqrt{-2})^3$$

$$= a^3 + 3a^2 \cdot b\sqrt{-2} + 3a \cdot (b\sqrt{-2})^2 + (b\sqrt{-2})^3$$

$$= a^3 - 6ab^2 + (3a^2 b - 2b^3)\sqrt{-2}.$$

It follows, by equating "real and imaginary parts," that

$$x = a^3 - 6ab^2 \quad \text{and} \quad 1 = 3a^2 b - 2b^3 = b(3a^2 - 2b^2).$$

Now if $1 = b(3a^2 - 2b^2)$ then b divides 1 and therefore $b = \pm 1$. If $b = -1$ then $1 = -(3a^2 - 2)$, which is impossible because $3a^2 - 2$ is either -2 (for $a = 0$) or positive. We must therefore have $b = 1$, in which case $1 = 3a^2 - 2$. So $a^2 = 1$ and hence $a = \pm 1$.

Substituting $a = 1$, $b = 1$ in $x = a^3 - 6ab^2$ gives the negative value $x = -5$, so the only remaining option is $a = -1$, $b = 1$. This gives the value $x = 5$ claimed by Fermat, and with it the value $y = 3$.

What Is Going on in Euler's Proof?

Euler came up with his proof before the "imaginary" number $\sqrt{-2}$ was well understood, not to mention the concept of the "integers" $a + b\sqrt{-2}$ and their greatest common divisors and primes. Today we know that it is valid to picture $a + b\sqrt{-2}$ as a point in the plane with real coordinate a and imaginary coordinate $b\sqrt{2}$, so that the distance of this point from the origin is $\sqrt{a^2 + 2b^2}$, by the Pythagorean theorem (figure 2.3). We also call this distance the *absolute value* of $a + b\sqrt{-2}$, and denote it by $|a + b\sqrt{-2}|$.

The square $|a + b\sqrt{-2}|^2$ of this distance is the ordinary integer $a^2 + 2b^2$, which is called the *norm* of $a + b\sqrt{-2}$. With its help we can

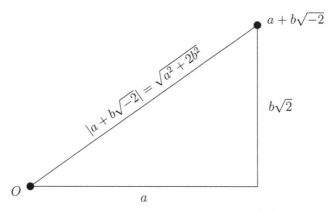

Figure 2.3: Distance to the point $a + b\sqrt{-2}$.

reduce questions about divisibility of these *quadratic integers* to questions about divisibility in the ordinary integers. The magical property of the norm that makes this possible is

$$|uv|^2 = |u|^2 |v|^2, \tag{*}$$

or "norm of a product equals the product of the norms." We say more about this *multiplicative property* in the next section, but we first explain how it enables us to prove that $\gcd(x + \sqrt{-2}, x - \sqrt{-2}) = 1$ when x and y are integers such that $y^3 = x^2 + 2$.

The magical property (*) means that if v divides w (which means in turn that $w = uv$ for some u) then $|v|^2$ divides $|w|^2$. Also, the only numbers $a + b\sqrt{-2}$ with norm 1 are ± 1, because $a^2 + 2b^2 = 1$ only if $a = \pm 1$, $b = 0$. Thus it suffices to prove that *any* common divisor of $x + \sqrt{-2}, x - \sqrt{-2}$ has norm 1 when $y^3 = x^2 + 2$.

First notice that

$$y^3 \equiv 0, 1, \text{ or } 3 \pmod 4.$$

This can be seen by cubing the four possible values of y mod 4, namely 0, 1, 2, 3. On the other hand, for the even values of x mod 4 (0 and 2) we have

$$x^2 + 2 \equiv 2 \pmod 4.$$

So $y^3 = x^2 + 2$ only for x odd, and then the norm $x^2 + 2$ of $x \pm \sqrt{-2}$ is odd.

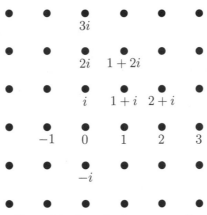

Figure 2.4: Gaussian integers near 0.

Now a divisor of $x + \sqrt{-2}$ and $x - \sqrt{-2}$ also divides their difference $2\sqrt{-2}$, which has norm 8. The gcd of the norms 8 and the odd number $x^2 + 2$ is 1, so any divisor of $x + \sqrt{-2}$, $x - \sqrt{-2}$ is a number whose norm divides 1; that is, 1 itself.

2.6 The Gaussian Integers

In the previous section we saw how a problem about ordinary integers draws our attention to weird "integers" involving the quantity $\sqrt{-2}$. Moreover, if we play along with these weird integers—granting them attributes such as gcd and primes—they enable us to solve the original problem very simply. It seems as though the factorization $x^2 + 2 = (x + \sqrt{-2})(x - \sqrt{-2})$ explains the behavior of $x^2 + 2$ better than we can in the language of ordinary integers.

However, we have yet to make sense of the concept of "prime" and "unique prime factorization" in the world of the quadratic integers $a + b\sqrt{-2}$. To pave the way for this, we look first at the simplest quadratic integers: those of the form $a + b\sqrt{-1}$, or $a + bi$, where a and b are ordinary integers. These are called the *Gaussian integers* because they were first studied by Gauss (1832). The Gaussian integers form a square grid in the plane of complex numbers, part of which is shown in figure 2.4.

Like the integers $a + b\sqrt{-2}$, the Gaussian integers form a ring because the sum, difference, and product of Gaussian integers are again Gaussian integers, and the various algebraic laws are easily checked. The distance from the origin to $a + bi$ (also known as the *absolute value*, $|a + bi|$) is $\sqrt{a^2 + b^2}$ by the Pythagorean theorem, and the square of this distance, $a^2 + b^2$, is an ordinary integer called the *norm* of $a + bi$. Again it is true that "the norm of a product equals the product of the norms," but this time we will check.

Consider Gaussian integers $a + bi$ and $c + di$, whose norms $|a + bi|^2$ and $|c + di|^2$ are $a^2 + b^2$ and $c^2 + d^2$, respectively. Their product is

$$(a + bi)(c + di) = ac + adi + bci + bdi^2$$
$$= (ac - bd) + (ad + bc)i \quad \text{because} \quad i^2 = -1.$$

It follows that the norm of the product is $(ac - bd)^2 + (ad + bc)^2$, and miraculously

$$(ac - bd)^2 + (ad + bc)^2 = (a^2 + b^2)(c^2 + d^2)$$

—the product of the norms. This identity can be checked by multiplying out both sides, and finding that they both equal $a^2c^2 + a^2d^2 + b^2c^2 + b^2d^2$.

Notice that this calculation does not assume that a, b, c, d are integers, so in fact it is true for any complex numbers u, v that

$$|uv|^2 = |u|^2|v|^2.$$

This is why "norm of a product equals product of the norms" is also true for the integers $a + b\sqrt{-2}$.[1] As for the latter integers, it follows for the Gaussian integers that if v divides w then $|v|^2$ divides $|w|^2$, so divisibility of Gaussian integers depends on divisibility of ordinary integers.

It now makes sense for us to define a *Gaussian prime* to be a Gaussian integer, of norm greater than 1, which is not the product of

---

[1] It also follows that $|uv| = |u||v|$, a fact which has some geometric implications for the complex numbers, as we will see in section 10.3.

Gaussian integers of smaller norm. Since norms are ordinary integers, the process of factorizing a Gaussian integer into Gaussian integers of smaller norm must eventually halt, necessarily with Gaussian primes. So *every Gaussian integer has a Gaussian prime factorization.*

Examples of Gaussian Prime Factorizations

The smallest example is $2 = (1+i)(1-i)$. This is a factorization into numbers of smaller norm, because $|1+i|^2 = |1-i|^2 = 2$ and $|2|^2 = 4$. Also, neither $1+i$ nor $1-i$ can be split into factors of smaller norm, because their norm 2 is not the product of smaller integers.

It is the same story for any ordinary prime that is the sum of two squares, such as $37 = 6^2 + 1^2$. The ordinary prime splits into Gaussian factors of smaller norm, in this case

$$6^2 + 1^2 = (6+i)(6-i), \quad \text{where} \quad |6+i|^2 = |6-i|^2 = 37,$$

$$\text{whereas} \quad |37|^2 = 37^2.$$

But neither factor splits into factors of smaller norm, because 37 is not a product of smaller ordinary integers. Thus any ordinary prime that is a sum of two squares is a product of two Gaussian primes.

If we are given a random Gaussian integer, such as $3+i$, we can find its Gaussian prime factors from the ordinary prime factors of its norm. In this case

$$|3+i|^2 = 3^2 + 1^2 = 10 = 2 \cdot 5.$$

Thus we are looking for Gaussian factors of norms 2 and 5, and they will necessarily be Gaussian primes because their norms are ordinary primes. We already know some Gaussian primes of norm 2; namely, $1+i$ and $1-i$. Apart from changing the sign of 1, these are the only ones. And the only Gaussian primes of norm 5 are $2+i$ and similar numbers obtained by changing signs (since the only squares with sum 5 are 4 and 1). Among this small number of possibilities, we soon find that

$$3+i = (1-i)(2+i).$$

Division with Remainder

We now consider a rather different problem: *division with remainder*. For ordinary integers we have the visually evident *division property* that $a = qb + r$ with $|r| < |b|$, as we saw in section 2.1. Given a Gaussian integer, say $5 + 3i$, and a Gaussian integer of smaller norm, say $3 + i$, we would like to be able to divide $5 + 3i$ by $3 + i$ and obtain a remainder r of smaller norm than $3 + i$. This means finding Gaussian integers q and r ("quotient" and "remainder") with

$$5 + 3i = (3 + i)q + r, \quad \text{such that} \quad |r| < |3 + i|.$$

This can be done by looking at the multiples $3 + i$ in the neighborhood of $5 + 3i$, and picking the one, $(3 + i)q$, nearest to $5 + 3i$. Then the difference

$$r = 5 + 3i - (3 + i)q$$

is as small as possible, and hopefully it is smaller than $3 + i$.

Indeed, this always works, and for a rather striking reason: *the multiples of $3 + i$ form a grid of squares of side $|3 + i|$, and the distance from any point in a square to the nearest corner is less than the length of the side.*

Why squares? Well, any Gaussian integer multiple of $3 + i$, by $a + bi$ say, is the sum of a copies of $3 + i$ and b copies of $i(3 + i) = -1 + 3i$. The points $3 + i$ and $-1 + 3i$ lie at distance $|3 + 1| = \sqrt{10} = |-1 + 3i|$ from 0, and *in perpendicular directions*. Thus the points 0, $3 + i$, and $i(3 + i) = -1 + 3i$ form two sides of a square. Adding further copies of $3 + i$ and $-1 + 3i$ creates further squares, all of side length $|3 + i|$, as shown in figure 2.5.

We see from figure 2.5 that the multiple of $3 + i$ nearest to $5 + 3i$ is $2(3 + i)$, and that their difference is

$$r = 2(3 + i) - (5 + 3i) = 1 - i, \quad \text{with absolute value } \sqrt{2} < \sqrt{10}.$$

More generally, it is easy to prove (say, using the Pythagorean theorem) that the distance from any point in a square to the nearest corner is less than the side of the square. Also, the multiples uq of any Gaussian integer u form a grid of squares of side length $|u|$ by

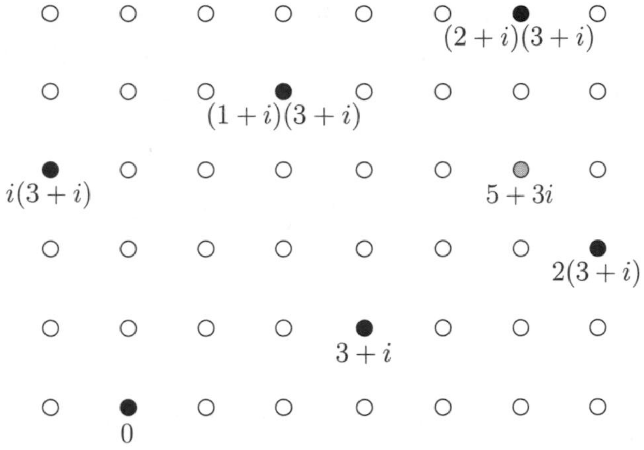

Figure 2.5: Multiples of $3 + i$ near $5 + 3i$.

the argument just used for $u = 3 + i$, hence any Gaussian integer v has difference

$$r = v - uq \quad \text{of absolute value } < |u|$$

from some uq. In other words, we have the:

Division property. For any Gaussian integers u, v there are Gaussian integers q, r such that

$$v = uq + r \quad \text{with} \quad |r| < |u|.$$

Since the remainder is smaller than the divisor u, the Euclidean algorithm by repeated division with remainder will terminate for any pair s, t of Gaussian integers, giving us their gcd in the form

$$\gcd(s, t) = ms + nt \quad \text{for some Gaussian integers } m \text{ and } n.$$

Then, by the same trick used for ordinary integers in section 2.3 we can prove:

Prime divisor property. If a Gaussian prime p divides a product uv of Gaussian integers, then p divides u or p divides v.

As we know from section 2.3, *unique prime factorization* now follows. The only difference is that it is slightly "less unique." This is because, when one Gaussian prime divides another, the quotient is not necessarily 1: it could also be -1 or $\pm i$. For this reason we say that *Gaussian prime factorization is unique up to factors of ± 1 and $\pm i$.*

2.7 Euler's Proof Revisited

Now we know how to talk about "primes" among the integers $a + b\sqrt{-2}$ brought to light in Euler's solution of $y^3 = x^2 + 2$.

We measure the size of $a + b\sqrt{-2}$ by its norm $|a + b\sqrt{-2}|^2 = a^2 + 2b^2$, and say that $a + b\sqrt{-2}$ is *prime* if its norm is greater than 1 and $a + b\sqrt{-2}$ is not the product of integers of smaller norm. (Incidentally, the only integers $a + b\sqrt{-2}$ of norm 1 are ± 1, because $a^2 + 2b^2 = 1$ only for $a = \pm 1$ and $b = 0$.)

We also know that we can prove unique prime factorization for the integers $a + b\sqrt{-2}$ if we can prove the *division property*: for any integers u, v of the form $a + b\sqrt{-2}$ there are integers q, r (also of this form) such that

$$v = uq + r \quad \text{with} \quad |r| < |u|.$$

And we know how to find q and r: look at the multiples uq of u, and let

$$r = v - (\text{multiple } uq \text{ nearest to } v).$$

Thus it remains only to get a view of the multiples of u—clear enough that we can see that r is smaller than u.

As with the Gaussian integers, we will illustrate the idea with a particular example: in this case $u = 1 + \sqrt{-2}$, $v = 5 + \sqrt{-2}$. Figure 2.6 shows some of the relevant points in the plane of complex numbers.

The integers $a + b\sqrt{-2}$ form a grid of rectangles of width 1 and height $\sqrt{2}$. The multiples of $1 + \sqrt{-2}$ form a grid of rectangles of the *same shape*, but magnified by $|1 + \sqrt{-2}| = \sqrt{5}$ and rotated so that their short sides lie in the direction of $1 + \sqrt{-2}$.

The multiple of $1 + \sqrt{-2}$ nearest to $5 + \sqrt{-2}$ is

$$(2 - \sqrt{-2})(1 + \sqrt{-2}) = 4 + \sqrt{-2},$$

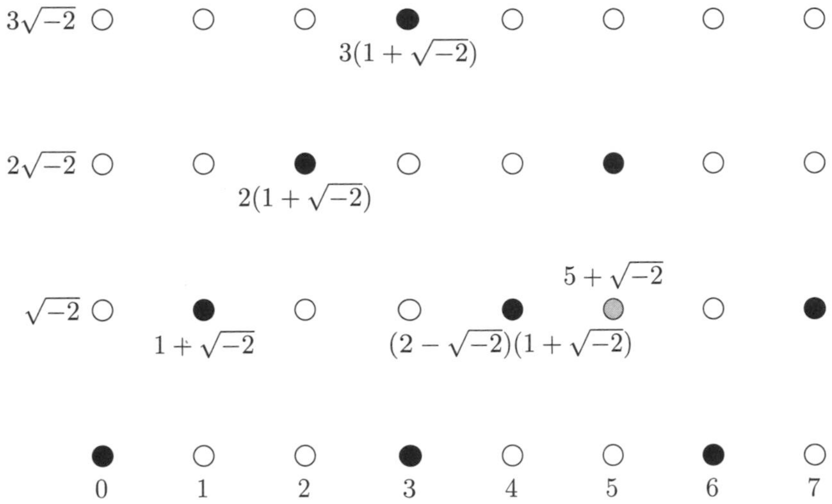

$3\sqrt{-2}$ $3(1+\sqrt{-2})$

$2\sqrt{-2}$ $2(1+\sqrt{-2})$

$\sqrt{-2}$ $1+\sqrt{-2}$ $(2-\sqrt{-2})(1+\sqrt{-2})$ $5+\sqrt{-2}$

0   1   2   3   4   5   6   7

Figure 2.6:   Multiples of $1+\sqrt{-2}$ near $5+\sqrt{-2}$.

so we get $q = 2 - \sqrt{-2}$ and $r = 1$. This r is certainly of smaller absolute value than the divisor $1 + \sqrt{-2}$.

In general, the multiples of an integer $u = a + b\sqrt{-2}$ form a grid of rectangles of the same shape as the original grid, but magnified by $|u|$ and rotated so that their short sides are in the direction of u. Since the size $|r|$ of the remainder is the distance from v to the nearest multiple of u, it remains to show that *in a rectangle with short side $|u|$ and long side $\sqrt{2}|u|$, the distance from any point to the nearest corner is less than $|u|$.*

To see why this is true, consider the worst-case scenario shown in figure 2.7, where v is at the center of the rectangle.

By the Pythagorean theorem,

$$|r| = \sqrt{\frac{|u|^2}{2^2} + \frac{|u|^2}{2}} = \sqrt{\frac{3}{4}|u|^2} = \frac{\sqrt{3}}{2}|u| < |u|.$$

This finally proves the division property for the integers $a + b\sqrt{-2}$. As with ordinary and Gaussian integers, this gives the Euclidean algorithm, prime divisor property, and unique prime factorization. In fact, prime factorization is unique up to factors ± 1 here, since the only integers $a + b\sqrt{-2}$ of norm 1 are ± 1.

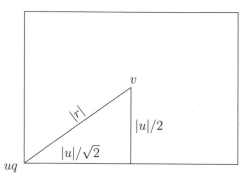

Figure 2.7: The point most distant from the corners.

Now we should recall *why* we want unique prime factorization for the integers $a + b\sqrt{-2}$. In section 2.5 we had

$$y^3 = x^2 + 2 = (x + \sqrt{-2})(x - \sqrt{-2}),$$

and we established that $x + \sqrt{-2}$ and $x - \sqrt{-2}$ have no common prime factor. But each prime factor p in y occurs as p^3 in y^3, so each prime factor p in $x + \sqrt{-2}$ occurs as $\pm p^3 = (\pm p)^3$, and each prime factor q in $x - \sqrt{-2}$ occurs as $\pm q^3 = (\pm q)^3$. In other words, $x + \sqrt{-2}$ *is a product of cubes and hence is a cube itself; $x - \sqrt{-2}$ is a product of cubes and hence is a cube itself.*

Euler's calculation and proof, starting with $x + \sqrt{-2} = (a + b\sqrt{-2})^3$, is now justified. The justification is probably far from any Euler had in mind, but it is based on ideas that now belong to elementary mathematics:

- The theory of divisibility and primes, based on the division property and the Euclidean algorithm.

- The multiplicative property of absolute value, $|uv| = |u||v|$, for complex numbers.

- Geometric representation of complex numbers and the Pythagorean theorem.

2.8 $\sqrt{2}$ and the Pell Equation

Another famous equation for which integer solutions are sought is

$$x^2 - 2y^2 = 1. \tag{*}$$

This equation was of interest to the ancient Greeks because, if x and y are large integers satisfying the equation, then x^2/y^2 is very close to 2. This implies that the fraction x/y is very close to $\sqrt{2}$.

In this section we will show that indeed there are arbitrarily large solutions of (*) and we will find a simple algorithm for producing them (and hence for finding arbitrarily close rational approximations to $\sqrt{2}$). We begin with the obvious solution $x = 3$, $y = 2$, which is the smallest positive integer solution.

Now since

$$1 = 3^2 - 2 \cdot 2^2$$

then

$$1 = (3^2 - 2 \cdot 2^2)^n \quad \text{for any positive integer } n.$$

Also,

$$3^2 - 2 \cdot 2^2 = (3 + 2\sqrt{2})(3 - 2\sqrt{2}),$$

so we have

$$1 = (3 + 2\sqrt{2})^n (3 - 2\sqrt{2})^n.$$

This prompts us to look at the positive integers x_n and y_n such that

$$(3 + 2\sqrt{2})^n = x_n + y_n\sqrt{2}. \tag{**}$$

For example,

$$(3 + 2\sqrt{2})^2 = 3^2 + 2 \cdot 3 \cdot 2\sqrt{2} + (2\sqrt{2})^2 = 17 + 12\sqrt{2},$$

so $x_2 = 17$ and $y_2 = 12$. We notice that $x = 17$, $y = 12$ is also a solution of (*).

In fact $x = x_n$, $y = y_n$ is a solution of (*) for each positive integer n. We prove this by induction.

Base step. The statement is true for $n = 1$ because $x_1^2 - 2y_1^2 = 3^2 - 2 \cdot 2^2 = 1$.

Induction step. We prove that if the statement is true for $n = k$ (that is, $x_k^2 - 2y_k^2 = 1$), then it is true for $n = k + 1$. By the definition (**) of x_{k+1} and y_{k+1},

$$x_{k+1} + y_{k+1}\sqrt{2} = (3 + 2\sqrt{2})^{k+1}$$
$$= (3 + 2\sqrt{2})^k (3 + 2\sqrt{2})$$
$$= (x_k + y_k\sqrt{2})(3 + 2\sqrt{2})$$
$$\text{(by definition of } x_k, \, y_k)$$
$$= 3x_k + 4y_k + (2x_k + 3y_k)\sqrt{2}.$$

Equating "rational and irrational parts" we get

$$x_{k+1} = 3x_k + 4y_k, \qquad y_{k+1} = 2x_k + 3y_k. \qquad (***)$$

It follows that

$$x_{k+1}^2 - 2y_{k+1}^2 = (3x_k + 4y_k)^2 - 2(2x_k + 3y_k)^2$$
$$= 9x_k^2 + 24x_k y_k + 16y_k^2 - 2(4x_k^2 + 12x_k y_k + 9y_k^2)$$
$$= x_k^2 - 2y_k^2$$
$$= 1, \qquad\qquad \text{(by induction hypothesis)}$$

which completes the induction.

Thus $x_n^2 - 2y_n^2 = 1$ for all positive integers n.

It follows from the equations (***) that x_n and y_n increase with n (in fact, very rapidly). So x_n/y_n rapidly approaches $\sqrt{2}$ as n increases.

The first few approximations are

$$x_1/y_1 = 3/2 = 1.5$$

$$x_2/y_2 = 17/12 = 1.416\cdots$$

$$x_3/y_3 = 99/70 = 1.41428\cdots$$

$$x_4/y_4 = 577/408 = 1.4142156\cdots,$$

which are correct in the first 1, 3, 5, 7 decimal places, respectively.

The Pell Equation

The equation $x^2 - 2y^2 = 1$ is a special case of the equation $x^2 - my^2 = 1$, for a nonsquare positive integer m, called the *Pell equation*. A proof similar to the one above shows that, if $x = x_1$, $y = y_1 \neq 0$ is one solution to the Pell equation, then we can find arbitrarily large solutions $x = x_n$, $y = y_n$ by the formula

$$x_n + y_n\sqrt{m} = (x_1 + y_1\sqrt{m})^n.$$

For example, $x = 2$, $y = 1$ is a solution of $x^2 - 3y^2 = 1$, and we get further solutions by the formula

$$x_n + y_n\sqrt{3} = (2 + \sqrt{3})^n.$$

In particular, using $(2 + \sqrt{3})^2 = 7 + 4\sqrt{3}$, we find the second solution $x = 7$, $y = 4$.

This is easy and fun to do, but two questions present themselves.

- How do we know that a solution of $x^2 - my^2 = 1$ *exists*?

- Given the smallest positive solution $x = x_1$ and $y = y_1$, does

$$x_n + y_n\sqrt{m} = (x_1 + y_1\sqrt{m})^n$$

give *all* positive integer solutions?

These questions are most easily answered by importing some ideas from outside number theory (the "pigeonhole principle" and some deeper algebra), so we will defer the answers until section 10.1.

2.9 Historical Remarks

As with many branches of mathematics, the story of number theory begins with Euclid. The *Elements* contains the first known proofs by descent (existence of prime factorization, termination of the Euclidean algorithm), the first proof that there are infinitely many primes, and the first extensive study of irrational numbers. Euclid also made a breakthrough in a topic that has seen very little progress since: primes of the form $2^n - 1$ and perfect numbers.

Prime Numbers and Perfect Numbers

A positive integer is called *perfect* if it is the sum of its divisors less than itself (its "proper divisors" or "aliquot parts"). For example, the proper divisors of 6 are 1, 2, and 3 and $6 = 1 + 2 + 3$, so 6 is a perfect number. The next two perfect numbers are 28 and 496, and Euclid evidently noticed that

$$6 = 2 \cdot 3 = 2^1(2^2 - 1),$$

$$28 = 4 \cdot 7 = 2^2(2^3 - 1),$$

$$496 = 16 \cdot 31 = 2^4(2^5 - 1),$$

because he noticed that $2^{n-1}(2^n - 1)$ *is perfect when* $2^n - 1$ *is prime*. His proof is quite simple: if we write $p = 2^n - 1$ then the proper divisors of $2^{n-1} p$ are (because of unique prime factorization)

$$1, 2, 2^2, \ldots, 2^{n-1} \quad \text{and} \quad p, 2p, 2^2 p, \ldots, 2^{n-2} p.$$

The sum of the first group is $2^n - 1$ and the sum of the second group $(2^{n-1} - 1)p = (2^{n-1} - 1)(2^n - 1)$, so the sum of all proper divisors is

$$(2^n - 1)(1 + 2^{n-1} - 1) = 2^{n-1}(2^n - 1), \quad \text{as required.}$$

Since Euclid reported this discovery, the only real progress on perfect numbers has been a theorem of Euler that every *even* perfect number is of Euclid's form. We still do not have a clear description of the primes of the form $2^n - 1$. We do not even know whether there are infinitely many of them. And, finally, we do not know whether there are any *odd* perfect numbers at all!

Even less is known about the primes of the form $2^n + 1$, but they are worth mentioning because they play an unexpected role in an ancient geometric problem: *constructing regular m-gons by straightedge and compass*. Euclid gave constructions for $m = 3$ (equilateral triangle) and $m = 5$ (regular pentagon), and for values of m derived from these by taking their product and repeatedly doubling the number of sides (see sections 5.4 and 5.6). No further progress was made (or expected) until the nineteen-year-old Gauss discovered a construction of the regular 17-gon in 1796.

The key to Gauss's discovery is that 3, 5, and 17 are primes of the form $2^n + 1$, namely,

$$3 = 2^1 + 1, \quad 5 = 2^2 + 1, \quad 17 = 2^4 + 1.$$

Gauss in fact found that *the only constructible polygons with a prime number of sides are those for which the prime is of the form $2^n + 1$.* It can be shown quite easily that such primes are actually of the form $2^{2^k} + 1$, but *only five* of them are known:

$$3 = 2^{2^0} + 1,$$
$$5 = 2^{2^1} + 1,$$
$$17 = 2^{2^2} + 1,$$
$$257 = 2^{2^3} + 1,$$
$$65537 = 2^{2^4} + 1.$$

Thus, notwithstanding Euclid's proof that there are infinitely many primes, attempts to find infinitely many primes of a particular form, such as $2^n - 1$ or $2^n + 1$, have been a miserable failure. There is no doubt that primes are the most "simple yet difficult" concept in mathematics, so they are probably the concept that best epitomizes the nature of mathematics itself. We will see them reappear, time and time again, at places where mathematics is particularly interesting and difficult.

Descent

But to return to the method of infinite descent ... Further novel applications of this method were discovered by Fibonacci (1202) and Fermat (published after his death, in 1670). Fibonacci used the method to find what are now called *Egyptian fractions*. The ancient Egyptians had a curious way of dealing with fractions by writing each fraction between 0 and 1 as a sum of distinct fractions of the form $1/n$, called *unit fractions*. Examples are

$$\frac{3}{4} = \frac{1}{2} + \frac{1}{4},$$

$$\frac{2}{3} = \frac{1}{2} + \frac{1}{6},$$

$$\frac{5}{7} = \frac{1}{2} + \frac{1}{7} + \frac{1}{14}.$$

It is not hard to find Egyptian fractions like these by trial and error, but how can we be sure that we will always succeed? Fibonacci gave a method that can be proved to succeed: namely, *repeatedly remove the largest unit fraction*.

Fibonacci's method always works because, if a/b is a fraction in lowest terms and $1/n$ is the largest unit fraction less than a/b, then

$$\frac{a}{b} - \frac{1}{n} = \frac{na - b}{bn} = \frac{a'}{bn}$$

is such that $a' < a$. (If $na - b \geq a$ then $a/b > 1/(n-1)$, so $1/n$ is *not* the largest unit fraction less than a/b.) Thus the numerator a' of the remainder continually decreases until halting occurs (necessarily with value 1) in a finite number of steps. Here is how the method works with 5/7:

$$\frac{1}{2} = \text{largest unit fraction} < \frac{5}{7},$$

so consider

$$\frac{5}{7} - \frac{1}{2} = \frac{3}{14} \qquad \text{(notice } 3 < 5\text{)}.$$

Next,

$$\frac{1}{5} = \text{largest unit fraction} < \frac{3}{14},$$

so consider

$$\frac{3}{14} - \frac{1}{5} = \frac{1}{70} \qquad \text{(which gives termination)}.$$

Thus

$$\frac{5}{7} = \frac{1}{2} + \frac{1}{5} + \frac{1}{70}.$$

Fermat's result was considerably more sophisticated than Fibonacci's, but similar in its dependence on descent. He proved that *there are no positive integers x, y, z such that $x^4 + y^4 = z^2$* by showing that, for any supposed solution, there is a smaller solution. Since positive integers cannot decrease indefinitely, we have a contradiction. It was Fermat who introduced the term "descent" for this kind of proof. He may have had a special kind of "descent" in mind—which works only for a certain kind of equation—but it seems fair to apply the word "descent" to any proof that relies on the fact that infinite descent is impossible in the positive integers. As we will see in chapter 9, this applies to virtually all proofs in number theory.

Algebraic Number Theory

The term "algebraic number theory" is usually understood to mean "number theory using algebraic numbers" rather than "number theory using algebra," though of course one does algebra when working with algebraic numbers. Thus our examples using the algebraic numbers $\sqrt{-1}$, $\sqrt{-2}$, and $\sqrt{2}$ to solve problems about ordinary integers are part of algebraic number theory. Several of these examples depend on the miraculous identity

$$(ad - bc)^2 + (ac + bd)^2 = (a^2 + b^2)(c^2 + d^2),$$

expressing the fact that

$$|uv|^2 = |u|^2 |v|^2 \quad \text{when} \quad u = a + bi \quad \text{and} \quad v = c + di,$$

or "norm of a product equals product of the norms." A special case of this fact seems to have been observed by Diophantus. It is presumably behind his remark (*Arithmetica*, Book III, Problem 19) that

> 65 is naturally divided into two squares in two ways, namely into $7^2 + 4^2$ and $8^2 + 1^2$, which is due to the fact that 65 is the product of 13 and 5, each of which is the sum of two squares.

The "two ways" come about because there are in fact two identities, obtained by exchanging the $+$ and $-$ signs inside the two squares on the left. In this case

$$65 = 13 \cdot 5 = (2^2 + 3^2)(1^2 + 2^2)$$
$$= (2 \cdot 2 \mp 3 \cdot 1)^2 + (2 \cdot 1 \pm 3 \cdot 2)^2$$
$$= 1^2 + 8^2, \ 7^2 + 4^2.$$

The general form of the identity was stated by al-Khazin around 950 CE, in comments on Diophantus, and a proof by algebraic calculation was given by Fibonacci (1225).

So it could be said that a characteristic property of the complex numbers,

$$|uv| = |u||v|,$$

was known long before the complex numbers themselves! Complex numbers, written in terms of square roots of negative numbers, were first used in the 1500s in connection with the solution of cubic equations. But they were not completely understood until the 1800s (see section 4.11). The property $|uv| = |u||v|$ is important not only in number theory, but also in geometry and algebra, as we will see in section 10.3.

As we saw, the norm of u defined by $|u|^2$ also applies to integers of the form $u = a + b\sqrt{-2}$, and in fact the concept of norm extends much further than this. It is also useful to define the norm of an integer of the form

$$a + b\sqrt{2}, \quad \text{where } a \text{ and } b \text{ are ordinary integers,}$$

by $\text{norm}(a + b\sqrt{2}) = a^2 - 2b^2$, because again it can be checked that

$$\text{norm}(uv) = \text{norm}(u)\text{norm}(v).$$

As a result, questions about divisibility of the "integers" $a + b\sqrt{2}$ reduce to the corresponding questions about their norms, which are ordinary integers. So we can explore the concepts of gcd and primes for integers of the form $a + b\sqrt{2}$. It turns out that unique prime factorization holds "up to factors of norm 1." But the uniqueness is much "less unique" than for the Gaussian integers, because there are now *infinitely many* integers of norm 1, thanks to the infinitely many solutions of $a^2 - 2b^2 = 1$.

This example highlights the importance of integers of norm 1, which were called *units* and studied by Dirichlet in the 1840s. We can also see the need for a concept of *algebraic integer* to cover the various kinds of "integer" studied so far, and to lay a general foundation for the concept of unique prime factorization. This was done by Dedekind (1871b) in response to the discovery of Kummer in the 1840s that unique prime factorization *is lost* for certain algebraic integers, and that some kind of "ideal prime factors" must be invented to recover it. In doing so, Dedekind lifted algebra to a new level of abstraction, where there are "ideal" objects consisting of infinitely many ordinary objects. This level is beyond the elementary level of number theory we wish to consider in this book.

The Pell Equation

The equation $x^2 - my^2 = 1$, where m is a nonsquare positive integer, is called the *Pell equation*. The reason for the name is rather stupid—Euler once mistakenly attributed the equation to the seventeenth-century English mathematician John Pell— but the name has stuck. Actually, Pell equations are much older than that, and they seem to have arisen independently in Greece and India.

The Pell equation with $m = 2$, as we have said, appears to have been studied in ancient Greece in connection with the irrationality of $\sqrt{2}$. A much fancier example occurs in a problem posed by Archimedes called the *cattle problem*. This problem turns on the solution of the Pell equation

$$x^2 - 4729494y^2 = 1,$$

the smallest solution of which has 206545 decimal digits! It is inconceivable that Archimedes could have found such a number, considering the primitive state of Greek computation, but perhaps he understood Pell equations enough to know that their solutions can be huge in comparison with the number m. The first manageable solution of the cattle problem equation, using algebraic numbers rather than decimals, was given by Lenstra (2002).

Pell equations were rediscovered by Indian mathematicians, a few hundred years after the Greeks. They used algebra, which the Greeks did not have, with considerable success. For example, Brahmagupta (628) found that the least positive solution of $x^2 - 92y^2 = 1$ is $x = 1151$, $y = 120$, and Bhaskara II around 1150 gave a method which always finds a solution, though without proving that it works. He illustrated his method with the first really hard example, $x^2 - 61y^2 = 1$, for which the least positive solution is $x = 1766319049$, $y = 226153980$.

The latter example was rediscovered by Fermat (1657), who posed it as a challenge to a colleague. Fermat was unaware of the Indian discoveries, so he too must have known that this is the first really hard example. Whether he could prove that each Pell equation has a positive solution is not known: the first published proof is due to Lagrange (1768). Lagrange pointed out that solving the Pell equation $x^2 - my^2 = 1$ is essentially the same as finding the continued fraction for $\sqrt{m}$, and proving that it is *periodic*.

We can illustrate this connection with the continued fraction for $\sqrt{2}$. As we saw in section 2.2,

$$\sqrt{2} + 1 = 2 + \frac{1}{\sqrt{2} + 1},$$

so substituting $2 + \frac{1}{\sqrt{2}+1}$ for $\sqrt{2} + 1$ is a neverending, periodic process. In fact it makes sense to write

$$\sqrt{2} + 1 = 2 + \cfrac{1}{2 + \cfrac{1}{2 + \cfrac{1}{2 + \cfrac{1}{\ddots}}}}.$$

The expression on the right is called the *continued fraction* for $\sqrt{2}+1$, and it is "periodic" in the sense that denominator 2 recurs forever. Subtracting 1 from each side, we get the continued fraction for $\sqrt{2}$, which is "ultimately periodic" in the sense that 2 recurs forever after the initial 1:

$$\sqrt{2} = 1 + \cfrac{1}{2 + \cfrac{1}{2 + \cfrac{1}{2 + \cfrac{1}{\ddots}}}}.$$

If we now truncate this fraction at finite levels, we get fractions approximating $\sqrt{2}$, which are alternately quotients of solutions of $x^2 - 2y^2 = 1$ and $x^2 - 2y^2 = -1$. For example,

$$1 + \cfrac{1}{2} = \cfrac{3}{2}, \qquad 1 + \cfrac{1}{2 + \cfrac{1}{2}} = \cfrac{7}{5}, \qquad 1 + \cfrac{1}{2 + \cfrac{1}{2 + \cfrac{1}{2}}} = \cfrac{17}{12}, \qquad \dots\dots$$

Thus the continued fraction encodes not only $\sqrt{2}$ but also the solutions of the corresponding Pell equation!

2.10 Philosophical Remarks

The arguments in the early sections of this chapter, about the Euclidean algorithm and the infinitude of primes, are often described as *pure* or *elementary* number theory. They are pure in the sense that they involve only concepts that clearly belong to number theory: whole numbers, addition, and multiplication. They are considered elementary because of this and also because the arguments are quite simple (albeit surprising and ingenious). We will say more about elementary number theory when we discuss logic in chapter 9. There we explain why elementary number theory is in some sense the theoretical minimum of mathematical knowledge, which must be part of any account of what mathematics, and mathematical proof, really is.

The arguments in the later sections of the chapter are *impure*[2] in the sense that they bring in concepts from algebra and geometry. However, they are still elementary, in my opinion, because elementary mathematics includes algebra and geometry at the level used here. Indeed, it is the artful use of concepts from algebra and geometry that *makes* these arguments elementary. Without the introduction of geometry, for example, it would be much harder to see why the division property (and hence unique prime factorization) is true. And without the guiding hand of unique prime factorization we would not know how to carry out Euler's proof in section 2.7.

Euler's theorem is *deeper than* the infinitude of primes in the sense that it depends on a "higher level," or more abstract, concept (unique prime factorization for algebraic numbers) to become comprehensible. Clearly, the more abstract concepts we introduce, the more likely we are to enter *advanced* mathematics, though in my opinion Euler's theorem remains on the elementary side of the line. More convincing examples of advanced mathematics occur when unique prime factorization is lost, as it is with the numbers discovered by Kummer in the 1840s. As mentioned in the previous section, "ideal prime factors" must be introduced to recover unique prime factorization in this case. "Ideal factors" are not only more abstract than algebraic numbers—they are in fact *infinite sets* of algebraic numbers—but also it takes a lot more work to develop the theory needed to make them useful. So they are quite far from elementary.

Irrational and Imaginary Numbers

It might seem, given the irrationality of $\sqrt{2}$ proved in sections 2.2 and 2.3, that we are unwise to use $\sqrt{2}$ to study the Pell equation. It seems even more unwise to use $\sqrt{-2}$ to study $y^3 = x^2 + 2$. If we do not really know what $\sqrt{2}$ is, how can we trust it to give correct answers? The reason is that we do not have to know what $\sqrt{2}$ *is*—only how it *behaves*—and all we need to know about the behavior of $\sqrt{2}$ is that

---

[2] Philosophers use this term, though not in a derogatory sense I hope. Mathematicians greatly admire such proofs for their elements of surprise and creativity.

$(\sqrt{2})^2 = 2$. We can in fact replace the symbol $\sqrt{2}$ by a letter x, and calculate with x as in high school algebra, replacing x^2 by 2 where necessary. The usual laws of algebra apply to expressions involving x, and hence they cannot lead to incorrect conclusions. We will see precisely why this is so, and what the "usual laws" are, in chapter 4.

It is true that we can give general definitions of *real* and *complex* numbers that explain what $\sqrt{2}$ and $\sqrt{-2}$ really are. And we can also prove that the real and complex numbers obey the "usual laws" of calculation. We do this in chapter 9. However, this is deeper mathematics, involving infinity in a serious way. As we will see in chapter 4, using $\sqrt{2}$ and $\sqrt{-2}$ in calculations that otherwise involve only rational numbers is essentially just as finite as calculating with the rational numbers themselves. It may be "higher arithmetic," but it is still arithmetic.

Elementary School Arithmetic

To many people, the word "arithmetic" suggests something much *more* elementary than the material covered in this chapter, namely the facts about numbers one learns in elementary school. Indeed, mathematicians often call the material of this chapter "number theory" to distinguish it from facts about specific numbers such as

$$1 + 1 = 2,$$

or

$$2 + 3 = 3 + 2,$$

or (to take a more complicated example)

$$26 \cdot 11 = 286.$$

But even the world of addition and multiplication facts about specific numbers is quite complicated, and it takes several years of work to master it in elementary school. We will explore the reasons for this more deeply in the next chapter.

In the meantime, it may be worth pointing out that there is actually a very concise way to encapsulate all the addition and multiplication facts about specific numbers. By the inductive definitions given in section 1.8, they all unfold from the four equations below, in which $S(n)$

denotes the successor of n, $n + 1$.

$$m + 0 = m, \tag{1}$$
$$m + S(n) = S(m + n), \tag{2}$$
$$m \cdot 0 = 0, \tag{3}$$
$$m \cdot S(n) = m \cdot n + m. \tag{4}$$

Equation (1) defines $m + n$ for all m and $n = 0$. Equation (2) defines $m + n$ for $m = k + 1$, assuming it is already defined for $n = k$. Thus (1) and (2) are the base step and induction step of the definition of $m + n$ for all m and n. Similarly, once $+$ is defined, equations (3) and (4) define $m \cdot n$ for all natural numbers m and n.

So, in principle, equations (1) to (4) generate all facts about sums and products of specific numbers. The price we pay for this simplicity is that we have to work with the names 0, $S(0)$, $SS(0)$, $SSS(0)$, ... for the natural numbers $0, 1, 2, 3, \ldots$, so that the name of a number is as large as the number itself. For example, the equation $1 + 1 = 2$ has to be written in the form

$$S(0) + S(0) = SS(0).$$

Similarly, $2 + 3 = 3 + 2$ becomes $SS(0) + SSS(0) = SSS(0) + SS(0)$, and $26 \cdot 11 = 286$ becomes the following highly inconvenient formula, whose right-hand side has been written in 11 rows with 26 copies of the letter S:

$$SSSSSSSSSSSSSSSSSSSSSSSSSS(0) \cdot SSSSSSSSSS(0)$$
$$= SSSSSSSSSSSSSSSSSSSSSSSSSS$$
$$SSSSSSSSSSSSSSSSSSSSSSSSSS$$
$$SSSSSSSSSSSSSSSSSSSSSSSSSS$$
$$SSSSSSSSSSSSSSSSSSSSSSSSSS$$
$$SSSSSSSSSSSSSSSSSSSSSSSSSS$$
$$SSSSSSSSSSSSSSSSSSSSSSSSSS$$

$$SSSSSSSSSSSSSSSSSSSSSSSSSSS$$
$$SSSSSSSSSSSSSSSSSSSSSSSSSSS$$
$$SSSSSSSSSSSSSSSSSSSSSSSSSSS$$
$$SSSSSSSSSSSSSSSSSSSSSSSSSSS$$
$$SSSSSSSSSSSSSSSSSSSSSSSSSSSS(0).$$

Nevertheless, anyone with sufficient patience can *prove* all such formulas by appeal to the equations (1), (2), (3), and (4). In particular, the proof of $1 + 1 = 2$ takes the form

$$
\begin{aligned}
S(0) + S(0) &= S(S(0) + 0) &&\text{by equation (2),} \\
&= S(S(0)) &&\text{by equation (1),} \\
&= SS(0).
\end{aligned}
$$

Proofs of multiplication facts reduce to facts about repeated addition, since it follows from equation (4) that

$$m \cdot n = m + m + \cdots + m,$$

where there are n occurrences of m on the right-hand side.[3]

Thus to show that all addition and multiplication facts about specific numbers follow from equations (1) to (4) it suffices to derive just the addition facts. It suffices in turn to show this for all facts of the form

$$\text{sum} = \text{number},$$

where sum is a sum of numerals and number is a single numeral. This is because different sums can be proved equal by showing that they each equal the same number. Finally, we can show equations (1) to (4) imply all true equations of the form sum = number by induction on the number of plus signs in sum.

---

[3] Strictly speaking, the right-hand side should contain parentheses. If there are three terms, it should look like $(m + m) + m$; if there are four terms it should look like $((m + m) + m) + m$, and so on. However, the position of the parentheses is irrelevant to the argument that follows.

If there is one plus sign in sum, then the value number of sum is obtained as we obtained the value $SS(0)$ for the sum $S(0) + S(0)$ above. If there are $k + 1$ plus signs in sum then

$$\text{sum} = \text{sum}_1 + \text{sum}_2,$$

where sum_1 and sum_2 are sums with at most k plus signs. It follows by induction that equations (1) to (4) imply

$$\text{sum}_1 = \text{number}_1 \quad \text{and} \quad \text{sum}_2 = \text{number}_2,$$

and then it follows from the base step again that we can obtain the value number of sum as the value of $\text{number}_1 + \text{number}_2$.

This shows (in outline) why equations (1) to (4) capture all the addition and multiplication facts about specific numbers. We have yet to capture the *algebraic structure* of the numbers, which is expressed by rules such as $a + b = b + a$ and $a \cdot b = b \cdot a$. We will discuss algebraic structure in chapter 4, and show in chapter 9 that it too is intimately related to induction.

3

$\backsim$

Computation

PREVIEW

Computation has always been part of the *technique* of mathematics, but it did not become a mathematical *concept* until the twentieth century. The best-known definition of this concept, due to Turing (1936), is modeled on human computation with pencil and paper. To prepare for the definition, the first sections of this chapter review the system of decimal notation and the computations used in elementary arithmetic. This review has the dual purpose of breaking down computations into elementary steps a machine could carry out, and estimating the *number* of elementary steps as a function of the number of input digits.

We take this dual approach because there are two main questions about computation of interest today: whether a problem can be solved *at all* by a computer and, if so, whether the computation is *feasible*.

The first question arises for certain very general problems in mathematics and computation, such as deciding the truth of all mathematical statements, or all statements about computation. These problems are *not* solvable by computer, as we will show. It follows quite easily from Turing's definition.

The second question arises for certain problems which can clearly be solved by a finite computation, but for which the known methods of solution take an astronomical amount of time. Adding to the frustration of such problems, it is often the case that a correct solution can be *verified* quite quickly. An example is the problem of factorizing numbers with a large number of digits.

Problems whose answers are hard to find, *by any method*, but easy to verify are not yet proved to exist (for example, there is conceivably a fast method for factorizing large numbers which we have not discovered yet). However, because of their great interest, we discuss some candidates.

3.1 Numerals

As will be clear from the previous chapter, many interesting facts about numbers have nothing to do with the notation used to represent them. Proving that there are infinitely many primes, for example, does not depend on decimal numerals. But when it comes to computation—even simple operations such as addition and multiplication—the notation for numbers makes a big difference. In this section we will compare three notations for positive integers: the naive "base 1" or "tally" notation which represents each number by a symbol as large as itself, the common decimal or base 10 notation in which the symbol for a number is exponentially smaller than the number itself, and the binary or base 2 notation, which is somewhat longer than base 10, but still exponentially smaller than the number itself.

In base 1 notation a number n is represented by a string of n copies of the digit 1. Thus the base 1 numeral for n literally has *length n*, if we measure length by the number of digits. In this notation, arithmetic is very simple; in fact, it is more like geometry than arithmetic. For example, we decide whether m is less than n by seeing whether the numeral for m is *shorter* than the numeral for n. Thus it is clear that

$$11111 < 111111111$$

(which says that $5 < 9$, as we would write it in base 10 notation). Addition is equally trivial. We add two numerals as we would add lengths, by placing them end to end; for example:

$$11111 + 111111111 = 11111111111111$$

(which expresses the result we would write as $5 + 9 = 14$ in base 10). Subtraction is equally easy, and multiplication not much worse, though it points towards a looming problem with base 1 numerals: *it is impractical to deal with large numbers.*

Multiplication of m by n *magnifies* the length of m by n, so one quickly runs out of space to write down the result of repeated multiplications. We need a more compact notation, and this can be achieved by using two or more digits.

A base 10 numeral, as we all know, is a string of digits chosen from among 0, 1, 2, 3, 4, 5, 6, 7, 8, 9, and with the leftmost digit not equal to 0. There are exactly 10^n strings of n digits and *each of them represents a different number.* (For strings beginning with zeros, the initial zeros are ignored.) Thus base 10 notation is as compact as possible for notations with 10 symbols, and it is "exponentially more compact" than base 1 notation, which requires strings of length up to 10^n to represent the first 10^n numbers.

But there is a price to pay for compact notation. Comparison, addition, and subtraction are no longer as simple as they are for base 1 notation, and some operations are barely feasible at all. We discuss arithmetic operations and the meaning of "feasibility" below. Here we consider just comparison of numerals, since it shows how compact notation complicates even the simplest problems.

If one base 10 numeral is shorter than another, then of course the shorter numeral represents a smaller number. But suppose we have two base 10 numerals of equal length, say

$$54781230163846 \quad \text{and} \quad 54781231163845.$$

To find which is smaller we must search for the leftmost digit where they differ, and the smaller number is the one for which this digit is smaller. In the present case we find:

$$54781230163846 < 54781231163845.$$

This rule for ordering numerals is called *lexicographical* ordering, because it is like the rule for ordering words (of the same length) in the dictionary. Just as one needs to know alphabetical order to look up words in the dictionary, one needs to know "digit order" in order to compare base 10 numerals.

The simplest "alphabet" for writing numbers compactly is the two-letter alphabet of *binary*, or *base 2*, notation, consisting of the symbols

0 and 1. Base 2 is worth studying, not only because it is the simplest compact notation, but also because it helps us to become aware of the workings of base 10 notation, which most of us have forgotten (or never understood in the first place).

A base 2 numeral, say 101001, represents a number as a sum of powers of 2, in this case the number (with digits written bold so that they stand out)

$$n = \mathbf{1} \cdot 2^5 + \mathbf{0} \cdot 2^4 + \mathbf{1} \cdot 2^3 + \mathbf{0} \cdot 2^2 + \mathbf{0} \cdot 2^1 + \mathbf{1} \cdot 2^0.$$

Thus we are writing n as sum of descending powers of 2, namely $n = 2^5 + 2^3 + 2^0$. Every natural number m can be written uniquely in this way by removing the largest power of 2 from m, then repeating the process. Necessarily, a smaller power of 2 is removed each time, so the process expresses n as a sum of distinct powers of 2. For example,

$$
\begin{aligned}
37 &= 32 + 5 \\
&= 2^5 + 5 \\
&= 2^5 + 4 + 1 \\
&= 2^5 + 2^2 + 2^0 \\
&= \mathbf{1} \cdot 2^5 + \mathbf{0} \cdot 2^4 + \mathbf{0} \cdot 2^3 + \mathbf{1} \cdot 2^2 + \mathbf{0} \cdot 2^1 + \mathbf{1} \cdot 2^0,
\end{aligned}
$$

so the binary numeral for 37 is 100101.

The base 10 numeral for a number n is obtained by the similar process of removing, at each stage, the largest possible power of 10. The same power of 10 can be removed as many as nine times, so the coefficients of the powers can be any of the digits 0, 1, 2, 3, 4, 5, 6, 7, 8, 9. For example, the base 10 numeral 7901 denotes the number

$$\mathbf{7} \cdot 10^3 + \mathbf{9} \cdot 10^2 + \mathbf{0} \cdot 10^1 + \mathbf{1} \cdot 10^1.$$

Or, as we used to say in elementary school,

7 thousands, 9 hundreds, 0 tens, and 1 unit.

In school, we got used to thinking that this is a pretty simple idea, but look again. The formula

$$7901 = 7 \cdot 10^3 + 9 \cdot 10^2 + 0 \cdot 10^1 + 1 \cdot 10^0$$

involves addition, multiplication, and exponentiation. Thus we need a sophisticated set of concepts just to find convenient *names* for numbers. It is no wonder that arithmetic is a deep subject!

In fact, some of the most intractable questions in number theory involve the decimal or binary representation of numbers. For example: are there infinitely many primes whose binary digits are all 1? We do not know—this is the same as asking whether there are infinitely many primes of the form $2^n - 1$.

3.2 Addition

To understand what happens to base 10 numerals when we add the corresponding numbers, consider the example $7924 + 6803$. If we expand the numerals as sums of powers of 10, and write the coefficients in bold symbols as above, then

$$
\begin{aligned}
7924 + 6803 = \; & 7 \cdot 10^3 + 9 \cdot 10^2 + 2 \cdot 10 + 4 \\
& + 6 \cdot 10^3 + 8 \cdot 10^2 + 0 \cdot 10 + 3 \\
= \; & (7 + 6) \cdot 10^3 + (9 + 8) \cdot 10^2 + (2 + 0) \cdot 10 + (4 + 3) \\
= \; & (13) \cdot 10^3 + (17) \cdot 10^2 + (2) \cdot 10 + (7),
\end{aligned}
$$

collecting powers of 10. The last line does not immediately translate into a base 10 numeral, because the coefficients are not all less than 10. Those greater than 10, such as $13 = 10 + 3$, create an "overflow" of higher powers of 10 that must be "carried" to the left and added to the coefficients already present. In this case, we get an extra power 10^3 and an extra power 10^4, because

$$
\begin{aligned}
& (13) \cdot 10^3 + (17) \cdot 10^2 + (2) \cdot 10 + (7) \\
= \; & (10 + 3) \cdot 10^3 + (10 + 7) \cdot 10^2 + (2) \cdot 10 + (7) \\
= \; & (10 + 3 + 1) \cdot 10^3 + (7) \cdot 10^2 + (2) \cdot 10 + (7) \\
= \; & 1 \cdot 10^4 + 4 \cdot 10^3 + 7 \cdot 10^2 + 2 \cdot 10 + 7.
\end{aligned}
$$

This is why we mumble under our breath "9 plus 8 gives 7, carry 1," and so forth as we add 7924 to 6803 by hand. We are concisely describing

Figure 3.1: Written calculation and abacus calculation. Image courtesy of the Erwin Tomash Library on the History of Computing, http://www.cbi.umn.edu /hostedpublications/Tomash/. I thank Professor Michael Williams of the University of Calgary for granting permission.

what happens to the **17** in the first three lines of the calculation above: 7 remains as the coefficient of 10^2, while **10** contributes a coefficient **1** of 10^3. "Carrying" also occurs when addition is done using the abacus, the method generally used in Europe before the publication of Fibonacci's *Liber abaci* in 1202 (and also used extensively afterwards). There is really no difference between abacus addition and written base 10 addition, since the underlying processing of powers of 10 is the same.

Indeed, it was not immediately clear that written computation was superior to abacus computation, and rivalry between the two continued for centuries. Figure 3.1, from the *Margarita philosophica* of Gregor Reisch (1503), shows a duel between a calculator using written numerals and one using the abacus. (Strictly, he is using a counting board, but it is essentially an abacus.)

As long as we use a compact notation, such as base 10 or base 2, carrying is an unavoidable part of the addition process. Yet, surprisingly, there are hardly any theorems about carrying. It seems as though carrying is a necessary evil, without any interesting properties in its own right.

Apart from carrying, the only other knowledge needed for base 10 addition is the *addition table*, which gives the results of adding any two of the digits 0, 1, 2, 3, 4, 5, 6, 7, 8, 9. It consists of 100 facts, such as "5 plus 7 gives 2, carry 1," so it requires a certain amount of memory. But assuming the addition table is known, and the calculator does not suffer from fatigue, there will be a bound b on the time required to add any two digits (plus any 1 that may be carried over from adding the previous digits). It follows that the time needed to add two n-digit numbers will be bounded by bn.

This is a very desirable (but quite rare) state of affairs, where the time needed to produce an answer is roughly proportional to the length of the question. For most computational problems, the time required to produce an answer grows *faster* than the length of the question, sometimes dramatically so. About the only other nontrivial questions about base 10 or base 2 numerals that can be answered in time proportional to the length of the question are comparison (is $m < n$?) and subtraction (what is $m - n$?). We invite the reader to think about the usual methods of comparison and subtraction to see why this is so. (An example of a computationally trivial question is deciding whether n is even, since we need only look at the last digit of n.)

3.3 Multiplication

Though he had both esteem and admiration for the sensibility of the human race, he had little respect for their intelligence: man has always found it easier to sacrifice his life than to learn the multiplication table.

W. Somerset Maugham, "Mr. Harrington's Washing," in Maugham (2000), p. 270

Somerset Maugham knew the main difficulty that most people have with multiplication, but the difficulty goes deeper, as one can see by analyzing the multiplication of large numbers. The usual way to multiply numbers given as base 10 numerals, say 4227 and 935, is to work out the products of 4227 by the individual digits of 935:

$$4227 \times 9 = 38043$$

$$4227 \times 3 = 12681$$

$$4227 \times 5 = 21135.$$

From this we conclude immediately that

$$4227 \times 900 = 3804300$$

$$4227 \times 30 = 126810$$

$$4227 \times 5 = 21135,$$

from which we get 4227×935 by adding the three numerals on the right-hand side. The computation is usually set out in a concise tableau, something like:

$$
\begin{array}{r}
4227 \\
\times\, 935 \\
\hline
21135 \\
126810 \\
3804300 \\
\hline
3952245
\end{array}
$$

The number of digits in the tableau is a reasonable measure of the time needed for the computation, since the time needed to get each digit is bounded by a constant. The constant reflects the maximum amount of mental arithmetic needed to get each digit, by retrieving information from the multiplication and addition tables in our head. If we multiply an m-digit number M by an n-digit number N then the number of digits in the tableau is $< 2mn$, so it is fair to say that the time needed to compute MN is bounded by cmn, for some constant c.

For people who cannot remember the multiplication table, it is possible to do the single-digit multiplications by repeated addition, since

$$2M = M + M$$
$$3M = M + M + M$$
$$\vdots$$
$$9M = M + M + M + M + M + M + M + M + M.$$

This does not significantly change the time estimate, since the time required to add as many as nine m-digit numbers is still bounded by a constant times m. Thus the time to find each line in the tableau is bounded by a constant times m, so the time to compute n lines, and to add them, is bounded by dmn, for some constant d. People who do not know the multiplication table just have to put up with a constant d somewhat larger than c.

An interesting way to see directly that multiplication can be done without the multiplication table is to do it in base 2 numerals. In this case the only single-digit multiplications are by 1 or 0, and hence trivial. The resulting multiples then have to be shifted by attaching zeros at their right, and the results added. Thus the only real work is in adding the rows of the tableau. But again we see that the time to multiply an m-digit number by an n-digit number is bounded by emn, for some constant e.

There is a useful variant of base 2 multiplication which does not require the numbers to be written in base 2. It uses only the related fact, from section 3.1, that any positive integer can be written as a sum of distinct powers of 2. When a number M is written in this way, we can multiply by M using only addition and multiplication by 2. For example, to multiply a number N by 37 we write

$$37 = 1 + 2^2 + 2^5,$$

so that

$$37N = (1 + 2^2 + 2^5)N = N + 2^2N + 2^5N.$$

Thus we need only to write down N, double it twice, then double another three times, and add the results. This idea of "multiplication by

Figure 3.2: Visualizing quotient and remainder.

repeated doubling" has an important consequence—"exponentiation by repeated squaring"—which will be discussed in section 3.5.

3.4 Division

When a and b are positive integers, a is not usually divisible by b, so the general division operation is *division with remainder*. That is, for given a and b we wish to find the *quotient* $q \geq 0$ and *remainder* $r \geq 0$ such that

$$a = qb + r \quad \text{with} \quad r < b.$$

We saw why the existence of q and r is obvious from a picture in section 2.1. To refresh your memory we repeat the picture here (figure 3.2), with a lying on a line among the multiples of b. The greatest multiple of b not exceeding a is qb, and the remainder is $r = a - qb$, which is less than the distance b between the successive multiples qb and $(q+1)b$.

We were content with the existence of a remainder $r < b$ when we proved that the Euclidean algorithm produces the gcd in section 2.1. However, we also claimed that the division process is fast enough to compute the gcd for numbers with thousands of digits. We now make good on this claim by proving that division with remainder can be done in time similar to that required for multiplication. That is, the time is roughly proportional to mn when a and b have m and n digits, respectively.

To obtain the time estimate we use a division method quite like the "long division" process taught in schools, but somewhat simpler to describe and analyze. We illustrate the method by dividing 34781 by 26.

The idea is to scan the digits of 34781 one by one from left to right, at each step dividing by 26 and attaching the remainder on the left of the next digit, then dividing this by 26 in turn. Each quotient is a single digit number, so it is not really necessary to know how to divide by

26—only how to *multiply* 26 by each single digit number. Here are the five steps in the calculation for this example.

Divide 3 by 26, obtaining quotient 0 and remainder 3,
then divide 34 by 26, obtaining quotient 1 and remainder 8,
then divide 87 by 26, obtaining quotient 3 and remainder 9,
then divide 98 by 26, obtaining quotient 3 and remainder 20,
then divide 201 by 26, obtaining quotient 7 and remainder 19.

The quotient of 34781 by 26 is then the sequence of quotient digits, 1337, and the remainder is the last remainder, 19.

One can see how this works by interpreting the digits in 34781 as numbers of units, tens, hundreds, and so on. For example, 34 stands for 34 thousands, so its quotient by 26 is 1 thousand with remainder 8 thousand. Attaching this remainder to the next digit, 7, which stands for 7 hundreds, gives 87 hundreds. So its quotient by 26 is 3 hundreds with 9 hundreds remaining, and so on.

In general, if we use the method to divide an m-digit number a by an n-digit number b there are m steps, each of which multiplies the n-digit number b by a single digit and subtracts it from a number previously obtained so as to obtain a remainder $< b$. Since there are only 10 possible digits to be tried, the time for each step is bounded by a constant multiple of n. Hence the time for m steps is bounded by a constant multiple of mn, as claimed.

The Euclidean Algorithm, Extended

The Euclidean algorithm in section 2.1, when speeded up by using division with remainder, is now confirmed to be efficient. The argument about this algorithm in section 2.3, implying that $\gcd(a, b) = ma + nb$ for some integers m, n, can in fact be realized by extension of the Euclidean algorithm itself, thus providing an efficient determination of m and n. The idea is to do the same thing algebraically on the *letters a* and *b* that we do numerically on the *numbers a* and *b* to find their gcd. The advantage of algebra is that by computing with letters we can keep track of the coefficients of a and b.

Here are the two computations side by side in the case where $a = 5$ and $b = 8$:

$$\begin{aligned} \gcd(5,\, 8) &= \gcd(5,\, 8 - 5) & \gcd(a,\, b) &= \gcd(a,\, b - a) \\ &= \gcd(3,\, 5) & &= \gcd(b - a,\, a) \\ &= \gcd(3,\, 5 - 3) & &= \gcd(b - a,\, a - (b - a)) \\ &= \gcd(3,\, 2) & &= \gcd(b - a,\, 2a - b) \\ &= \gcd(2,\, 3 - 2) & &= \gcd(2a - b,\, b - a - (2a - b)) \\ &= \gcd(2,\, 1) & &= \gcd(2a - b,\, -3a + 2b). \end{aligned}$$

Comparing the numbers on the left with the letters on the right, we find that $1 = -3a + 2b$ equals $\gcd(5,\, 8) = \gcd(a,\, b)$.

Clearly, the algebraic computation takes the same number of steps (and a similar amount of time) as the numerical computation. Thus we also have an efficient algorithm for finding the m and n such that $\gcd(a,\, b) = ma + nb$. This is of interest in the next chapter, where we discuss the concept of the *inverse* of an integer modulo a prime p, or of a polynomial modulo an irreducible polynomial $p(x)$. In both cases the inverse arises from a Euclidean algorithm just as the number m does in $\gcd(a,\, b) = ma + nb$. Hence it is also possible to calculate inverses efficiently.

3.5 Exponentiation

The easiest case of exponentiation, raising 10 to the power N, gives a 1 followed by N zeros. For example, the base 10 numeral for $10^{1000000}$ is 1 followed by a million zeros. Thus, *just writing down the base 10 numeral for M^N takes exponentially longer than writing down the numerals for M and N*. The situation is similar with base 2 numerals, and it should not be a surprise: we already know from section 3.1 that the lengths of base 10 and base 2 numerals are exponentially *shorter* than the numbers they represent.

Thus, the fact that base 10 numerals can compactly represent very large numbers has a downside: exponentiation of short numerals can produce very long numerals. So exponentiation is not "feasible" to compute in the way that addition and multiplication are. Regardless of how cleverly we compute M^N (and in fact it can be done with relatively few multiplications), the time to write down the result generally prevents the computation from being completed.

Indeed, the similar problem of computing $M^N \bmod K$ (the remainder when M^N is divided by K) is perfectly feasible, simply because the computation can be done by multiplying numbers less than K. This is due to the fact, mentioned in section 2.4, that the remainder when AB is divided by K equals the remainder when $(A \bmod K)(B \bmod K)$ is divided by K. So the only numbers we ever have to multiply are remainders on division by K, which are necessarily less than K. We can therefore keep the computation time small by keeping the number of multiplications small; more precisely, to a number exponentially smaller than the exponent N.

The trick is to use the method of *exponentiation by repeated squaring*, which follows from the method of multiplication by repeated doubling, discussed in section 3.3.

We illustrate this method with the problem of finding

$$79^{37} \bmod 107,$$

that is, finding the remainder when 79^{37} is divided by 107. Since we never have to multiply numbers bigger than 107, the main problem is to find the 37th power using far less than 37 multiplications. To do this, we use the representation of 37 as a sum of powers of 2, from the previous section:

$$37 = 1 + 2^2 + 2^5.$$

It follows that

$$79^{37} = 79^{1+2^2+2^5} = 79 \cdot 79^{2^2} \cdot 79^{2^5}.$$

Now notice that

$$79^{2^2} = 79^{2 \cdot 2} = (79^2)^2,$$

which we obtain by squaring twice; that is, by two multiplications. After that three more squarings give us

$$((((79^2)^2)^2)^2)^2 = 79^{2^5}.$$

Thus with five multiplications (mod 107) we get 79^{2^2} and 79^{2^5}. Finally, two more multiplications give us

$$79^{37} = 79^1 \cdot 79^{2^2} \cdot 79^{2^5}.$$

So seven multiplications suffice to obtain the 37th power. And (as we have emphasized) each multiplication is of numbers less than 107, hence bounded by the time needed to multiply such numbers plus the time to reduce the product mod 107. Reducing mod 107 amounts to dividing by 107 and taking the remainder, and this takes much the same time as multiplication.

In general, the number of multiplications required to compute M^N is less than twice the length of the binary numeral for N. The worst case is when each of the n digits of N is 1. In that case we have to compute M to the exponents

$$1, \quad 2^1, \quad 2^2, \quad \ldots, \quad 2^{n-1},$$

by squaring $n - 1$ times, and then multiply the powers

$$M^1, \quad M^{2^1}, \quad M^{2^2}, \quad \ldots, \quad M^{2^{n-1}},$$

which takes $n - 1$ multiplications. Thus we have less than $2n$ multiplications. When all the multiplications are mod K, we can assume that the numbers being multiplied are less than K. So the multiplications take time bounded by some multiple of k^2, where k is the number of binary digits of K, as does the reduction of each product mod K.

Thus the time to find M^N mod K is bounded by ek^2n, for some fairly small constant e. This makes the computation "feasible," in the sense that a computer can easily carry it out for numbers M, N, K with hundreds of digits. This result is the key to one of the most important computations in the world today—the encryption of transactions on the internet. Many of these use the well-known RSA encryption method, which involves raising numbers to large powers mod K. Encryption is feasible because of the fast exponentiation method above. As far as we know (and hope) decryption is *not* feasible because it depends on factoring large numbers. We discuss the factoring problem in the next section.

3.6 P and NP Problems

Our study of addition, multiplication, and exponentiation has raised the issue of *feasible* computation. Until now we have been vague about what feasibility means, but there is a precise notion that seems to capture the concept pretty well: *polynomial time* computation. To define polynomial time computation properly we must first define computation, which we will do in the next section. Right now we will give some examples that illustrate the concept, and examples that illustrate the related concept of *nondeterministic* polynomial time computation. The examples will also help to explain why we allow time to be measured by general polynomial functions, rather than sticking to, say, polynomials of degree 1, 2, or 3.

In section 3.2 we observed how two n-digit numbers (in base 10 or base 2) can be added in time bounded by cn, for some constant c. We can therefore say that addition is a problem *solvable in linear time*. It is an ideal state of affairs when the time needed to answer a question is proportional to the time needed to read it—but not one that is very common. In section 3.3 we saw that multiplication of n-digit numbers takes time bounded by dn^2, if we multiply by the usual method. There are in fact methods that are faster, for very large numbers. But no linear time method of multiplication is yet known, for any realistic model of computation.

This brings us to the question of defining computation and computation time. Until now we have assumed that computation is done the way humans do it: mostly with pencil and paper and a limited amount of mental arithmetic (using facts in memory such as "7 plus 5 gives 2, carry 1"). This turns out to be basically the right idea! It is made precise by the *Turing machine* concept of computation, described in section 3.7, which covers all known computational devices. A Turing machine operates in a sequence of discrete steps, so it gives a precise measure of computation time, as the number of steps. In terms of the pencil-and-paper image, a step means moving the pencil from one symbol to the next, or replacing one symbol by another (which includes writing a symbol in a blank space).

Now it happens that, in the Turing machine model, symbols are written in a single row. For example, to add the numbers 58301 and 29946 one has to write something like

$$58301 + 29946$$

and to write the sum on the same line. This demands a lot of zigzagging back and forth. First find the unit digits and add them

$$7 = 5830\not{1} + 2994\not{6}$$

(crossing off the unit digits to avoid getting confused at the next stage). Then find the tens digits and add them:

$$47 = 583\not{0}\not{1} + 299\not{4}\not{6},$$

and so on, carrying when necessary. After five trips back and forth we are done:

$$88247 = \not{5}\not{8}\not{3}\not{0}\not{1} + \not{2}\not{9}\not{9}\not{4}\not{6}.$$

If the numbers being added have n digits then it takes n trips back and forth to add them. Since the trip itself traverses at least n symbols, doing addition by this method takes on the order of n^2 steps.

Thus, *if we vary the model of computation*, computation time can change from linear to quadratic. We similarly find that doing multiplication on a Turing machine, rather than in the usual two-dimensional format, changes the computation time from quadratic to cubic. To avoid worrying about such variations, we introduce the concept of *polynomial time* computation.

First we define a *problem* $\mathcal{P}$ to be a set of strings of symbols that we interpret as "instances" of the problem, or "questions." Then $\mathcal{P}$ is said to be *solvable in polynomial time* if there is a polynomial p and a Turing machine M such that, given an instance I of $\mathcal{P}$ with n symbols, M will compute the correct answer to I in time $T \leq p(n)$.

The class of problems solvable in polynomial time is called P.

Thus P includes the problems of addition and multiplication of base 10 numerals. It also includes the related problems of subtraction and division, and fancier problems such as computing n decimal digits of $\sqrt{2}$, which can be done using the algorithm for generating integer solutions of the equation $x^2 - 2y^2 = 1$ (section 2.8). A more remarkable example is the problem of *recognizing primes*. As long ago as 1801, Gauss drew attention to this problem, insisting that

> the dignity of the science itself seems to require that every possible means be explored for the solution of a problem so elegant and so celebrated.
>
> Gauss (1801), article 329

A polynomial time solution was not found until recently, by Agrawal et al. (2004). An improved version of their method, due to Lenstra and Pomerance in 2011, takes on the order of n^6 steps for an n-digit number.

The quotation from Gauss above is actually not entirely about recognizing primes. It is actually about

> the problem of distinguishing prime numbers from composite numbers and of resolving the latter into their prime factors.
>
> Gauss (1801), article 329

The latter part of the problem, finding the factors of composite numbers, is *not* solved by the Agrawal-Kayal-Saxena method, which in fact identifies composite numbers *without* finding factors. We still do not know a polynomial time method for finding factors of a number M, even though correct factors can be *verified* in polynomial time by multiplying and comparing the result with M. Notice that it is no good to try dividing M by all smaller numbers, because if M has m digits there are about 10^m numbers less than M. And 10^m, like any exponential function, grows faster than any polynomial function of m.

The situation encountered in the factorization problem—where the answer is hard to find but easy to verify—is surprisingly common. Another simple example is solving polynomial equations mod 2. Given a polynomial $p(x_1, x_2, \ldots, x_n)$, with integer coefficients, we ask whether the congruence

$$p(x_1, x_2, \ldots, x_n) \equiv 0 \pmod 2$$

has any solution. To find such a solution we seemingly have to try all assignments of the values 0 or 1 to the variables $x_1, x_2, \ldots, x_n$, and there are 2^n such assignments. Yet a correct assignment can be verified in polynomial time by substituting the values and computing mod 2 sums and products.

Problems whose answers are hard to find yet easy to verify are so ubiquitous that they too have a name: *nondeterministic* polynomial time, or NP, problems. The word "nondeterministic" refers to the fact that their polynomial time solution succeeds only with the help of nondeterministic steps (typically, guessing the answer).

The ultimate example of an NP problem is *finding proofs in mathematics*. Ideally, the steps in a proof can be simply and mechanically checked for correctness, so that correct proofs can be verified in polynomial time, relative to the number of symbols they contain. (This ideal has been substantially realized, as proofs of several difficult theorems have actually been checked by computers.) But finding proofs remains difficult. It has not been mechanized because the number of proofs with n symbols grows exponentially with n. For some reason we do not yet understand (because we have been unable to prove that NP is a larger class than P), exponential growth is keeping mathematicians in business.

3.7 Turing Machines

We may compare a man in the process of computing
a real number to a machine which is capable of only a
finite number of conditions $q_1, q_2, \ldots, q_R$.

A. M. Turing (1936), p. 231

As we have seen in the preceding sections, various types of computation have been used in arithmetic for thousands of years. However, the need for a general concept of computation was not felt until the early twentieth century. It arose from quite a different form of computation: *symbolic logic*. The idea of symbolic logic, which was a dream of Leibniz in the seventeenth century, was to turn *reasoning* into a form of calculation. It would then be possible, Leibniz thought, to

| | 1 | 0 | 1 | + | 1 | 1 | 0 | 1 | | |

Figure 3.3: A Turing machine tape.

resolve disputes simply by saying "let us calculate." As it happened, the dream was not even partially realized until the nineteenth century, and it remains problematic today, for reasons such as the mystery of P and NP. However, we did find out the meaning of symbolic logic and computation.

We will say more about symbolic logic in chapter 9. For now, the main point to grasp is that reducing all conceivable forms of reasoning to calculation is a very general task—so general that it ought to embrace all conceivable forms of *computation* as well. The first to grasp this point was Post in 1921. He proposed a *definition* of computation, arrived at by generalizing the systems of symbolic logic then in use. However, he did not publish his results at the time—even though they included some sensational discoveries—because he was in some doubt that his definition really covered all possible forms of computation.

Because of Post's doubt, the concept of computation remained unknown to mathematics until two other definitions were proposed in the 1930s. Credit for the first published definition goes to Church (1935), but the first definition that was convincing enough to catch on was the *Turing machine* concept, introduced later in the same year. As indicated by the quotation at the beginning of this section, Turing arrived at his concept of machine by analyzing how a human being computes.

At a minimum, human computation requires a pencil, paper, and a limited amount of mental input to guide the pencil. As we foreshadowed in the previous section, the "paper" is a tape divided into squares, each of which can carry a single symbol (figure 3.3). The "pencil," also called a *read/write head*, is a device that can recognize and write a finite number of symbols $\Box, S_1, S_2, \ldots, S_n$, where $\Box$ denotes the blank square.

Finally, the machine has a finite number of *internal states* $q_1, q_2, \ldots, q_n$, corresponding to the mental states needed for the given computation. As we have seen, computations such as addition can be carried out using only a finite number of mental states, and Turing argued that no computation could require infinitely many mental states, otherwise some of them would be so similar that they would be confused. For the same reason, a computation cannot involve infinitely many different symbols.

This is why a Turing machine has only finitely many q_i and S_j. How does the machine operate? It performs a series of steps: at each step the read/write head observes the symbol S_j currently in view and, depending on its current internal state q_i, it replaces S_j by a symbol S_k, moves one square to the left or right, and enters a state q_l. Thus a machine M is specified by a table of *quintuples*, listing the action to be performed for certain pairs q_i, S_j. (If no action is listed for q_i, S_j then M *halts* in state q_i when viewing S_j.) We write

$$q_i \quad S_j \quad S_k \quad R \quad q_l$$

if the movement is to the right, and

$$q_i \quad S_j \quad S_k \quad L \quad q_l$$

if the movement is to the left.

Here is an example: a machine M that adds 1 to a base 2 numeral. We assume that the read/write head starts in state q_1 on the rightmost digit of the input numeral. Next to each quintuple, we give its interpretation.

q_1	0	1	L	q_2	Replace 0 by 1, move left in passive state q_2
q_1	1	0	L	q_3	Replace 1 by 0, move left in "carry" state q_3
q_2	0	0	L	q_2	Replace 0 by 0, move left in passive state q_2
q_2	1	1	L	q_2	Replace 1 by 1, move left in passive state q_2
q_3	0	1	L	q_2	Replace 0 by 1, move left in passive state q_2
q_3	1	0	L	q_3	Replace 1 by 0, move left in "carry" state q_3
q_2	□	□	L	q_4	Replace □ by □, move left in halting state q_4
q_3	□	1	L	q_4	Replace □ by 1, move left in halting state q_4

Thus M crawls across the input numeral, updating its digits for as far as the "carry" propagates, then going into a passive state that changes

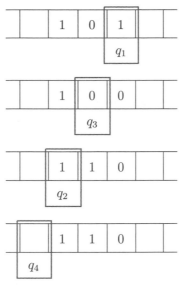

Figure 3.4: Snapshots of the computation of M.

nothing, before halting on a blank square to the left of the updated numeral. Figure 3.4 shows "snapshots" of successive steps in the work of M on the input numeral 101. The position of the read/write head is enclosed in a box labeled by the current state.

The machine M can be simplified a little by noticing that the initial state q_1 behaves exactly the same as the "carry" state q_3, so we could set $q_3 = q_1$ and get by with six quintuples instead of eight. With this identification, the "carry," passive, and halting states correspond rather naturally to the three mental states a human being would need in order to perform this computation.

Designing a Turing machine to perform a given computation, though tedious, is basically a matter of thinking how you would do it if allowed to view only one symbol at a time. With some practice, one becomes convinced that any computation is possible, so it is intuitively plausible that Turing machines can compute anything that is computable. Further evidence comes from the competing models of computation proposed by Post, Church, and many others. It has been checked that all the proposed models of computation can be simulated by Turing machines. Because of this, mathematicians have accepted the principle, called *Church's thesis* or the *Church-Turing thesis*, that the Turing machine concept captures the intuitive concept of computability.

3.8 *Unsolvable Problems

By defining the concept of computing machine we also gain definitions of *algorithm* and *solvable problem*. A *problem* P, as we said in section 3.6, can be viewed as a set of *instances* or *questions*, which are finite strings of symbols in some finite alphabet. For example, the problem of recognizing primes can be given as the set of strings of the form

$$\text{Is } n \; prime?$$

where n is a base 10 numeral. In fact we could take the instances of this problem to be just the numerals n, since the numeral is all the machine needs to answer the question.

An *algorithm* for a problem P—informally, a rule for obtaining answers to the questions in P—is precisely a Turing machine that takes each instance Q of P as input and, after a finite number of steps, halts with the correct answer written on its tape. To be even more precise, if Q is a yes/no question (such as "Is n prime?") we can demand that M signal yes by halting on 1 and no by halting on $\square$.

Finally, a problem P is *solvable* if there is an algorithm that solves it; that is, a Turing machine M that halts with the correct answer to each input Q from P. We can also elaborate this definition to give a precise definition of *solvability in polynomial time*, discussed less formally in section 3.6. And, by extending the definition of Turing machine to allow nondeterministic computation steps—that is, by allowing a situation $q_i S_j$ to have more than one outcome—we can give a definition of nondeterministic computation. This makes the definitions of the classes P and NP from section 3.6 completely precise.

However, P and NP problems are definitely solvable. In this section we wish to investigate *unsolvable* problems. It turns out that some of them are quite easy to describe, and the easiest to prove unsolvable are problems concerning Turing machines themselves.

Consider the so-called *halting problem*, introduced by Turing (1936). It can be taken to consist of the following questions:

$Q_{M,I}$. Given a Turing machine M and input I, decide whether M eventually halts on $\square$ after receiving input I. (To be completely specific, assume that M starts in state q_1 on the rightmost symbol of I, on an otherwise blank tape.)

To see why this could be a difficult problem, suppose that I is a base 10 numeral and that M searches for a $J > I$ such that $2^{2^J} + 1$ is prime. At present we do not know whether M halts for $I = 5$, so there are instances of the halting problem that are hard to answer. Turing (1936) showed that the halting problem is unsolvable by considering questions $Q_{M,I}$ about the computation of real numbers. A somewhat simpler approach is to study what happens when Turing machines have to investigate their own behavior.

A Turing machine has a description, as we know from the previous section, consisting of a finite list of quintuples. Although the description is finite, it may contain arbitrarily many of the symbols q_1, q_2, . . . and S_1, S_2, ... Thus no Turing machine can accept all such descriptions as inputs. However, it is easy to replace each description by one in the fixed finite alphabet $\{q, S, ', R, L, \square\}$, by replacing each q_i by the string $q''^{\cdots}$ with i strokes, and each S_j by the string $S''^{\cdots}$ with j strokes. Let $d(M)$ be the description of M, rewritten in this manner. It now makes sense to ask the following questions, which form a subproblem Q of the halting problem:

$Q_{M,d(M)}$. Does Turing machine M, given input $d(M)$, eventually halt on $\square$?

If T is a Turing machine that solves this problem, it is fair to assume that T is given the question $Q_{M,d(M)}$ in the form of the description $d(M)$, since this input accurately describes M. Also, we can assume that T follows the convention of halting on 1 for yes and on $\square$ for no.

But then it is impossible for T to correctly answer the question $Q_{T,d(T)}$! If T answers no, then T halts on $\square$, in which case the answer is incorrect. And if T does *not* halt on $\square$, then the answer is no, in which case T is supposed to halt on $\square$. Thus T does not solve the problem Q, because it fails to correctly answer one of the questions in Q. Thus the problem Q is unsolvable, and hence so is the halting problem.

With a little reflection it is, I hope, *obvious* that the halting problem is unsolvable. It is so easy to thwart the hypothetical solving machine by asking it about its own behavior. Such "self-reference" has been the stuff of paradox since ancient times. A nice example occurs in *Don Quixote*:

> Before anyone crosses this bridge, he must first state on oath where he is going and for what purpose. If he swears truly, he may be allowed to pass; but if he tells a lie, he shall suffer death by hanging on the gallows there. ... Now it happened that they once put a man on his oath, and he swore that he was going to die on the gallows.
>
> Miguel Cervantes, *Don Quixote*, Part II, Chapter LI
> Translated by J. M. Cohen

What is more surprising is that genuine mathematical problems turn out to be unsolvable for essentially the same reasons. A famous example is *Hilbert's tenth problem*, so named because it was tenth on a list of problems that Hilbert posed to the mathematical community in 1900. It consists of questions we may call $Q_{p(x_1,\ldots,x_n)}$, where $p(x_1, \ldots, x_n)$ is a polynomial with integer coefficients.

$Q_{p(x_1,\ldots,x_n)}$: Does $p(x_1, \ldots, x_n) = 0$ have a solution in integers $x_1, \ldots, x_n$?

This problem was proved unsolvable via a long series of transformations of the halting problem. The first step, in the 1930s, was to *arithmetize* the Turing machine concept: that is, to encode successive tape configurations by numbers, and steps of computation by arithmetic operations (initially using addition, multiplication, and exponentiation). The hardest part, eventually accomplished by Matiyasevich (1970), was to eliminate the use of exponentiation,[1] thereby reducing questions about computation to questions about polynomials.

With this discovery of Matiyasevich, we see unsolvability cast a shadow even over elementary mathematics, in the arithmetic of addition and multiplication.

---

[1] It is *not* possible to eliminate either one of addition or mulitplication. The theory of addition alone has an algorithm for deciding the truth of any sentence, and so does the theory of multiplication. Unsolvability springs from the marriage of addition *and* multiplication.

3.9 *Universal Machines

A Turing machine is essentially a computer program, written in a very simple programming language. This is clear in the formulation of Post (1936), which has instruction numbers instead of internal states. Leaving aside some minor differences between Post and Turing, Turing's quintuple

$$q_i S_j S_k L q_l$$

becomes Post's instruction

i. Replace S_j by S_k, move left, and go to instruction l

(and similarly when the quintuple contains a move to the right). Thus the heart of the Turing machine programming language is the "go to" command, a command that has largely been replaced by more "structured" commands in modern programming languages.

When Turing machines are viewed as programs, it is natural (for us) to seek a machine that can *run* all these programs, and indeed Turing (1936) designed a *universal Turing machine* that does exactly that. Its details are unimportant, because once one has accepted the Church-Turing thesis—that any computation can be performed by a Turing machine—then it is clear that a universal machine exists. One simply has to think how a human being can simulate the computation of any Turing machine M on any input I, which is quite easy to do.

The main difficulty is that the arbitrary machine M can have arbitrarily many states q_i and symbols S_j in its description, whereas the universal machine U (like any Turing machine) is constrained to have only finitely many states and symbols. The inevitable way round this difficulty is to encode each q_i and S_j as a *string* of symbols from a fixed alphabet, say the alphabet $\{q, S,'\}$. Then the single symbol q_i can be encoded by the string $q^{(i)}$ (q with i primes) and the single symbol S_j can be encoded by the string $S^{(j)}$ (S with j primes). Consequently, when U has to simulate replacement of S_j by S_k it has to replace the string $S^{(j)}$ by the string $S^{(k)}$. Of course, this slows down the running of U considerably in comparison with the running of the machine M it is simulating, but it makes the universality of U possible.

With the existence of a universal machine U, each unsolvable problem about all machines M becomes an unsolvable problem about the single machine U. For example, the halting problem becomes the problem of deciding, for a given input I, whether U eventually halts after starting on input I. This is because an input to U can encode both an arbitrary machine M and an input to M.

Since Turing (1936) first outlined the description of a universal machine there have been many attempts to design universal machines that are as simple as possible. It is known, for example, that there is a universal machine with only two symbols (including the blank square) and a universal machine with only two states. It is not known, however, what is the smallest combination of state number and symbol number. The current record holder is a machine with four states and six symbols, found by Rogozhin (1996). We have not yet learned anything striking about computation from these small Turing machines, perhaps because they are not simple enough. However, there are other models of computation with strikingly simple universal machines, such as Conway's "game of life," which is described in Berlekamp et al. (1982).

3.10 Historical Remarks

The discovery of compact notation for numbers, and methods for computing their sums and products, go back thousands of years. In Europe and the Far East they were initially implemented on the abacus. Written computation became practical after the invention of a symbol for zero, in India around the fifth century CE. The Indian notation for numbers spread to the Arab world (hence our term "Arabic numerals") and then to Europe with the Moors in Spain. Computation with written numerals in Europe began around the time of Fibonacci's *Liber abaci* in 1202. As the title of the book indicates, computation until that time was synonymous with the abacus.

Indeed, for centuries to come there was opposition to written computation from users of the abacus, and the abacus users had a point: written arithmetic is no faster than abacus arithmetic for basic tasks. (As recently as the 1970s my father-in-law used the abacus in his shop in Malaysia.) Written computation did not really advance mathematics until it was used for computations not conceivable on the abacus: in algebra in the sixteenth century and in calculus in the seventeenth century. These two fields ushered in a golden age of written computation, with eminent mathematicians such as Newton, Euler, and Gauss producing virtuoso displays of both numerical and symbolic computation.

As early as the 1660s Leibniz foresaw the possibility of a *calculus ratiocinator*, a symbolic language in which reasoning was done by computation. The first concrete step towards Leibniz's dream was taken by Boole (1847), who created an algebraic symbolism for what we now call propositional logic. Boole took + and · to stand for "or" and "and," and 0 and 1 to stand for "false" and "true." Then his + and · satisfy laws similar to those of ordinary algebra, and one can decide whether certain types of statements are true by algebraic computation. In fact, if $p+q$ is taken to mean "p or q but not both," then the algebraic rules of propositional logic become exactly the same as those of mod 2 arithmetic. This remarkable parallel between arithmetic and logic is explained in section 9.1.

Propositional logic is not the whole of logic by any means, and checking logical truth is not generally as easy as mod 2 arithmetic. But Boole's success in reducing basic logic to calculation inspired Frege (1879), Peano (1895), and Whitehead and Russell (1910) to develop comprehensive symbolic systems for logic and mathematics. The aim of these *formal systems*, as they were called, was to avoid errors or gaps in proofs due to unconscious assumptions or other kinds of human error. The steps of a formal proof can be followed without knowing the meaning of the symbols so, in principle, a formal proof can be checked by a machine. Indeed, formal proofs (and hence theorems) can in principle be *generated* by a machine, by combining a machine that generates all possible strings of symbols with one that checks whether a given string is a proof.

At the time when the first formal systems were developed, no such machines had been built, and computability was not imagined to be a mathematical concept. But the idea gradually dawned that formal systems include *all possible* computation processes. As mentioned in section 3.7, Post in 1921 considered the most general symbol manipulations conceivable in formal systems and broke them down into simple steps. His aim at first was to simplify logic to the point where truth or falsehood became mechanically decidable, as Leibniz had hoped. He got as far as finding some very simple systems that generate all the theorems of Russell and Whitehead's system. Then, to his surprise, he found it was hard to foresee the outcomes of some very simple processes.

One that he discovered is called *Post's tag system*. The system takes any string s of 0s and 1s and repeatedly applies the following rules.

1. If the leftmost symbol of s is 0, attach 00 to s on the right, then delete the three leftmost symbols of the resulting string.
2. If the leftmost symbol of s is 1, attach 1101 to s on the right, then delete the three leftmost symbols of the resulting string.

Thus if $s = 1010$ the rules successively produce

$$\cancel{1}\cancel{0}\cancel{1}01101,$$

$$\cancel{0}\cancel{1}\cancel{1}0100,$$

$$\cancel{0}\cancel{1}\cancel{0}000,$$

$$\cancel{0}\cancel{0}\cancel{0}00,$$

$$\cancel{0}\cancel{0}\cancel{0}0,$$

$$\cancel{0}\cancel{0}\cancel{0},$$

at which point the string becomes empty and the process halts. It is fun to see what happens with various strings s; the process can run for a very long time, and it can also become periodic. Post was unable to find an algorithm to decide which initial strings s eventually lead to the empty string, and in fact this particular "halting problem" remains unsolved to this day.

After reaching this impasse, Post's train of thought took a dramatic change of direction. Simple systems could simulate all possible computations, yes, but this did *not* mean that there was an algorithm to answer all questions about computation. On the contrary, it implied the existence of *unsolvable* algorithmic problems, by an argument like that used to prove the unsolvability of Turing's halting problem in section 3.8. Post recounted his anticipation of Turing's idea in Post (1941).

However, Post paused on the brink of this momentous discovery because he was concerned about having to make the assumption we now call Church's thesis. It seemed to him more like a law of nature, in need of eternal testing, than a mathematical definition of computation. It was only after Church himself had proposed the thesis, and Turing was working on the same idea, that Post (1936) published one of his systems for computation—coincidentally, one very similar to the Turing machine concept.

In the meantime, Church and Turing had found unsolvable problems independently, and Gödel (1931) had found a related and equally momentous result: *incompleteness* of axiom systems for mathematics. That is, for any sound axiom system A for mathematics, there are theorems that A does not prove. Actually Gödel proved stronger results than this, which I will come to in a moment.

First we should observe, as Post did in the 1920s, that any unsolvable problem implies an infinite amount of incompleteness. Take Turing's halting problem for example. Suppose we want to prove theorems about Turing machine computations, and that we have a sound formal system A for doing so. (That is, A proves only *true* theorems about Turing machines.) Since A is formal, we can mechanically generate all its theorems, by a Turing machine T if we like. But then if A is complete it will prove all true facts of the form

Machine M, on input $d(M)$, eventually halts on $\square$.

And for the remaining machines M' it will prove all the facts of the form

Machine M', on input $d(M')$, never halts on $\square$.

Consequently, by looking down the list of theorems of A, we will be able to answer all of the questions $Q_{M,d(M)}$, and hence solve the halting problem. Since the halting problem is not solvable, there must be theorems of the above form that A fails to prove (and in fact infinitely many, because if there were only finitely many missing theorems we could add them as axioms to A).

Gödel discovered incompleteness by a different argument, which does not assume Church's thesis. Also, he was able to prove incompleteness of formal systems for mainstream mathematics; namely, any system that contains basic number theory. Gödel's argument and Post's argument can today be seen as two aspects of the same phenomenon: the *arithmetization* of symbolic computation, or logic. As mentioned in section 3.8, this means that all the operations of a Turing machine can be simulated by operations on numbers, ultimately definable in terms of $+$ and $\cdot$. So, in a sense, all computation is abacus computation!

A more explicit way to express all computation in terms of $+$ and $\cdot$ is inherent in the negative solution of Hilbert's tenth problem, also mentioned in section 3.8. Matiyasevich (1970) actually showed that knowing the outcome of any Turing machine computation is equivalent to knowing whether a certain polynomial equation

$$p(x_1, \ldots, x_n) = 0 \qquad (*)$$

has a solution in integers $x_1, \ldots, x_n$. This amounts to knowing whether there are integers $x_1, \ldots, x_n$ that yield 0 by a certain sequence of additions and multiplications (which produce the polynomial p).

Now suppose we replace Hilbert's tenth problem by the corresponding problem in mod 2 arithmetic, by asking whether ($*$) has a solution with $x_1, \ldots, x_n$ equal to 0 or 1, and $+$ and $\cdot$ interpreted as the mod 2 sum and product. Then we collapse a problem about arbitrary computation to a problem in propositional logic, called the *satisfiability problem*. The latter problem is solvable, because when the x_i can take only the values 0 and 1 there are only finitely many solutions of ($*$) to try.

However, the satisfiability problem is still interesting because, as mentioned in section 3.6, it is in NP but not known to be in P. The obvious method of solution is to substitute all 2^n values of the sequence $(x_1, \ldots, x_n)$ in ($*$), and no known method of solution is substantially faster than this. Cook (1971) in fact showed that the satisfiability problem is as hard as any NP problem because, if it were solvable in polynomial time, *any* NP problem would be solvable in polynomial time. Thus all the difficulties of NP problems are condensed into this single problem of mod 2 arithmetic. (Such problems—and many of them are now known— are called NP-*complete*.)

It is striking that general computation and NP computation share a simple description in terms of polynomial equations, but so far this common description has failed to throw any light on the baffling question: is NP $\neq$ P?

3.11 Philosophical Remarks

In this chapter we seem to have moved seamlessly from third grade arithmetic to the depths of unsolvability and incompleteness—a sea-change into something rich and strange. Where, if anywhere, did we cross the line between elementary and advanced mathematics? In my opinion, it was *not* with the definition of Turing machine. Admittedly, Turing machines are not yet part of every mathematician's education, but I believe they deserve to be because:

1. Computation is now one of the fundamental concepts of mathematics.
2. The Turing machine concept is the simplest and most convincing model of computation.
3. The concept is really very simple—not much more complicated than the algorithms for addition and multiplication of decimal numerals.

If this much is granted then the advanced step in the theory of computation must be either Church's thesis—making computability a part of mathematics—or the "self-referential" trick used to prove the unsolvability of the halting problem.

Despite their simplicity, both Church's thesis and the self-referential trick can be considered as deep ideas, and hence part of advanced mathematics. They are deep in the sense that they were not uncovered in the thousands of years of previous mathematical history, and also in the sense that they underlie and support the massive edifice of mathematical logic and set theory that has been built since their discovery. Let me expand a little on these claims.

For thousands of years mathematicians have done computations, so there has always been a *concept* of computation, albeit a vague one. Mathematicians have also engaged in the formalization of more or less vague concepts. This started with Euclid's formalization of the concept of "geometry" in the *Elements*, and it gathered pace around 1900, when Peano (1889) formalized "arithmetic" with axioms for the natural numbers and Zermelo (1908) formalized "set theory" with axioms for sets. But formalization received a severe setback when Gödel (1931) proved that all axiom systems for number theory and set theory are incomplete—showing that formalization of arithmetic is not entirely possible, after all. Gödel thought that his argument would show that the concept of computability cannot be formalized either. He became convinced that he was wrong when he saw the Turing machine concept, and later declared, in the paper Gödel (1946), that it is a "kind of miracle" that formalization of computability is possible.

Thus, if Church's thesis is correct, computability is actually a more precise and absolute concept than the concept of arithmetic! This is surely a deep and wonderful discovery.

Now let us turn to the "self-reference" trick, whereby we prove that no Turing machine T can solve the halting problem by confronting the hypothetical machine T with its own description $d(T)$. As I mentioned in section 3.8, a similar idea turned up centuries ago in philosophy and literature (in *Don Quixote*, for example) but only as a paradox—amusing and thought-provoking, yes, but not of much consequence. The mathematical versions have consequences of epic proportions; namely, the completely unexpected presence of unsolvability and incompleteness in elementary mathematics.

A less paradoxical kind of self-reference is known as the *diagonal argument* or *diagonal construction*. I will discuss the diagonal construction more fully in section 9.7, but it is worth giving a preview of it here, since it was actually Turing's starting point.

Given a list of real numbers $x_1, x_2, x_3, \ldots$, displayed as infinite decimals, the diagonal construction computes a number x *different from each x_n*, by ensuring that x differs from x_n in the nth decimal place. The construction is called "diagonal" because the digits x has to *avoid* are those lying on the diagonal of the array of digits of $x_1, x_2, x_3, \ldots$ (shown in bold type in figure 3.5). Now, if $x_1, x_2, x_3, \ldots$ is a *computable list of computable numbers* then x is also computable (by using a specific rule for avoiding the diagonal digits; say, using 2 to avoid 1 and using 1 to avoid any digit other than 2). Thus the list $x_1, x_2, x_3, \ldots$ does not include all computable numbers.

To see why this argument leads to the halting problem, we relate computable numbers to Turing machines. As the title "On computable numbers ..." of Turing (1936) suggests, the first application of his machines was to define computable real numbers. Turing defined a real

$$x_1 \quad 0.1374\ldots$$
$$x_2 \quad 0.9461\ldots$$
$$x_3 \quad 0.2222\ldots$$
$$x_4 \quad 0.3456\ldots$$
$$\vdots$$

Figure 3.5: The diagonal of digits to avoid.

number x to be *computable* if there is a machine that prints the decimal digits of x successively, and without later erasure, on alternate squares of the machine's tape. He chose this convention so as to leave infinitely many other squares available for the computations that produce the digits, but obviously other conventions are possible. The important parts of the definition are:

1. For each n, the nth digit of x is eventually printed.
2. Once printed the digit is never altered.

We can then define a *list* of computable numbers $x_1, x_2, x_3, \ldots$ to be computable if there is a Turing machine that computes a list of machine descriptions such that the nth machine on the list defines x_n. The diagonal construction now gives a computable number x different from the numbers $x_1, x_2, x_3, \ldots$ in any computable list. So it would seem, to anyone not convinced that there was a complete notion of computability, that here is a *proof* that the notion is incomplete.

But Turing was convinced that he had a complete notion of computable numbers, so he drew a different conclusion: *not all Turing machines define computable numbers, and it is impossible to compute the list of those that do.* The hard part is to confirm the two conditions above, which depend on knowing the machine's entire future behavior. By considering what is needed to pick out the machines that define computable numbers, Turing found the underlying difficulty: the halting problem. This is how he proved that the halting problem is unsolvable. Its unsolvability was later confirmed by more direct proofs, like the one given in section 3.8.[2]

---

[2] The first proof in this style that I saw was the one in Hermes (1965), §22, p. 145.

4

⌐

Algebra

PREVIEW

Classical algebra was what Newton called "Universal Arithmetick"; that is, calculations involving symbols for unknowns, but subject to the same rules as calculation with numbers. This view still holds for elementary algebra today, but attention has shifted to the *rules themselves*, and the different mathematical structures that satisfy them.

The reason for this shift was the failure of algebra to reach its original goal, which was to solve polynomial equations by the operations of $+$, $-$, $\cdot$, $\div$, and the so-called *radicals* $\sqrt{\ }$, $\sqrt[3]{\ }$, $\sqrt[4]{\ }$, ... In 1831 Galois changed the direction of algebra by proving that equations of degree 5 and higher are not generally solvable by radicals. To prove such a result, one needs to develop a general theory of algebraic operations and the rules they satisfy. The Galois theory of algebraic operations is beyond the scope of elementary algebra, but parts of it, such as the theory of *fields*, are not.

Fields are systems with operations $+$, $-$, $\cdot$, $\div$ obeying the ordinary rules of arithmetic. So they, and the related systems called *rings* (which lack the operation of $\div$), involve calculations and concepts already familiar from arithmetic. In fact, arithmetic inspires the development of field theory with the concepts of division with remainder and congruence, which prove to be useful far beyond their intended range of application. It might truly be said that field theory is the new "Universal Arithmetick."

More surprisingly, field theory also builds a bridge between algebra and geometry. The two are linked by the concept of *vector space* and the technique of *linear algebra*. We develop the algebraic side of vector spaces in this chapter, and the geometric side in the next.

4.1 Classical Algebra

The word "algebra" comes from the Arabic word *al-jabr*, meaning "restoring." It was once commonly used in Spanish, Italian, and English to mean the resetting of broken bones. The mathematical sense of the word comes from the book *Al-jabr w'al mûqabala* of al-Khwārizmi of 850, where it roughly means manipulation of equations. Al-Khwārizmi's algebra goes no further than the solution of quadratic equations, which was already known to the cultures of ancient Greece, the Middle East, and India. Nevertheless, al-Khwārizmi's algebra was the one to take hold in medieval Europe (his name also gave us the term "algorithm"). There it evolved into the symbolic arithmetic we know as "high school algebra" today.

A typical manipulation in al-Khwārizmi's algebra is the trick of "completing the square" to solve quadratic equations. Given the equation

$$x^2 + 10x = 39,$$

for example, we notice that $x^2 + 10x$ can be completed to the square

$$(x+5)^2 = x^2 + 10x + 25$$

by adding 25. We therefore add 25 to both sides of the original equation and get

$$x^2 + 10x + 25 = 39 + 25 = 64.$$

It follows that

$$(x+5)^2 = 64 = 8^2,$$

so

$$x + 5 = \pm 8$$

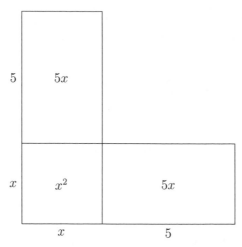

Figure 4.1: The incomplete square $x^2 + 10x$.

and therefore

$$x = -13 \quad \text{or} \quad x = 3.$$

Actually, al-Khwārizmi did not recognize the negative solution, because he justified his manipulations by a geometric argument in which x represents a length. This style of argument persisted for several centuries in the Muslim world and Europe, such was the enduring authority of Euclid. In the present example al-Khwārizmi began by interpreting $x^2 + 10x$ as a square of side x with two rectangles of area $5x$ attached as shown in figure 4.1.

Then he literally completed the square by filling the gap with a square of side 5 (and hence area 25), as in figure 4.2.

Since we had $x^2 + 10x = 39$, the completed square has area $39 + 25 = 64$, so it is the square of side 8. This makes $x = 3$.

Negative solutions had to wait for the acceptance of negative numbers, which happened rather slowly in the Middle East and Europe. In India, negative numbers were already accepted by Brahmagupta (628), who used them to obtain the full solution of the general quadratic equation $ax^2 + bx + c = 0$. Namely,

$$x = \frac{-b \pm \sqrt{b^2 - 4ac}}{2a}.$$

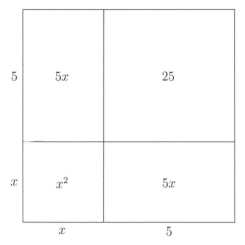

Figure 4.2: The completed square $x^2 + 10x + 25 = (x+5)^2$.

The first real advance in algebra after Brahmagupta and al-Khwārizmi was the solution of *cubic* equations by the sixteenth-century Italian mathematicians del Ferro and Tartaglia. In particular, they found that the solution of

$$x^3 = px + q$$

is

$$x = \sqrt[3]{\frac{q}{2} + \sqrt{\left(\frac{q}{2}\right)^2 - \left(\frac{p}{3}\right)^3}} + \sqrt[3]{\frac{q}{2} - \sqrt{\left(\frac{q}{2}\right)^2 - \left(\frac{p}{3}\right)^3}}.$$

This is known as the *Cardano formula*, due to its publication in the book *Ars magna* of Cardano (1545). This remarkable formula greatly influenced the development of mathematics. In particular, it forced mathematicians to consider "imaginary numbers," such as $\sqrt{-1}$. As Bombelli (1572) observed, such numbers arise when innocuous equations like

$$x^3 = 15x + 4$$

are solved by the Cardano formula. This particular equation has the obvious solution $x = 4$, but according to the Cardano formula its solution is

$$x = \sqrt[3]{2 + \sqrt{2^2 - 5^3}} + \sqrt[3]{2 - \sqrt{2^2 - 5^3}}$$

$$= \sqrt[3]{2 + \sqrt{-121}} + \sqrt[3]{2 - \sqrt{-121}}$$

$$= \sqrt[3]{2 + 11\sqrt{-1}} + \sqrt[3]{2 - 11\sqrt{-1}}.$$

Bombelli was able to reconcile the obvious solution with the formula solution by assuming that the "imaginary" $i = \sqrt{-1}$ obeys the same rules of calculation as the ordinary "real" numbers. If it does, then one can check that

$$2 + 11i = (2+i)^3 \quad \text{and} \quad 2 - 11i = (2-i)^3,$$

so

$$\sqrt[3]{2 + 11i} = 2+i \quad \text{and} \quad \sqrt[3]{2 - 11i} = 2-i,$$

and therefore

$$\sqrt[3]{2 + 11i} + \sqrt[3]{2 - 11i} = (2+i) + (2-i) = 4. \qquad \text{QED!}$$

This raises the question: exactly what are the "rules of calculation" obeyed by the real and imaginary numbers? Answering this question leads us towards *algebraic structure*, the more abstract approach to algebra developed in the nineteenth century. We say more about this in the sections that follow. But first we should explain the problem that finally brought algebraic structure into sharp focus.

Solution by Radicals

The formulas above for solving quadratic and cubic equations express the roots of an equation in terms of its coefficients, using the operations of $+$, $-$, $\cdot$, $\div$, and the so-called radicals $\sqrt{}$ and $\sqrt[3]{}$. Indeed, the word "radical" comes from the Latin word *radix* meaning "root." Quite soon after the solution of the cubic became known, Cardano's student Ferrari found that the general *quartic* (fourth degree) equation could also be solved in terms of its coefficients by means of $+$, $-$, $\cdot$, $\div$, and the same radicals.

This sparked a search for similar solutions of the *quintic* (fifth degree) equation and equations of higher degree, admitting the possible need for the radicals $\sqrt[4]{}$, $\sqrt[5]{}$, and so on. The goal of this search was called *solution by radicals*.

However, no solution by radicals of the general quintic equation

$$ax^5 + bx^4 + cx^3 + dx^2 + ex + f = 0$$

was ever found. By 1800, suspicion was growing that the general quintic is *not* solvable by radicals, and Ruffini (1799) had made an attempt to prove it. Ruffini's 300-page attempt was too long and unclear to convince his contemporaries, but it foreshadowed a valid proof by Abel (1826). Ironically, Abel's much shorter proof was so terse that few of his contemporaries understood it either. Doubts about Abel's unsolvability proof were put to rest only by the exposition of Hamilton (1839).

In the meantime, a dramatically new and elegant unsolvability proof had appeared in 1831. It too was not understood at first, and it was published only in 1846. This was the proof of Galois, who threw new light on the nature of solution by radicals by identifying the algebraic structures now called *fields* and *groups*.

Examples of fields were already well known in mathematics. The most important example is the system $\mathbb{Q}$ of rational numbers with the operations of addition and multiplication, introduced in section 1.3. $\mathbb{Q}$ can be viewed as the system resulting from the positive integers by applying the operations of $+$, $-$, $\cdot$, and $\div$. One obtains larger fields by throwing irrational numbers into the mix. The result $\mathbb{Q}(\sqrt{2})$ of throwing in (or "adjoining," which is the technical term) $\sqrt{2}$ is the field of numbers of the form $a + b\sqrt{2}$, where a and b are rational. When irrational numbers are adjoined to $\mathbb{Q}$ the resulting fields may have *symmetries*. For example, each element $a + b\sqrt{2}$ of $\mathbb{Q}(\sqrt{2})$ has a "conjugate" $a - b\sqrt{2}$ which behaves in the same way—any equation involving members of $\mathbb{Q}(\sqrt{2})$ remains true when each number in the equation is replaced by its conjugate.

Galois introduced the algebraic concept of *group* to study the symmetries arising when radicals are adjoined to $\mathbb{Q}$. He was able to show that solvability of an equation by radicals corresponds to a "solvability" property of the group attached to the equation. This property made it possible to show in one fell swoop that many equations, including the general quintic, are not solvable by radicals, because their groups lack the necessary "solvability" property.

The group concept became one of the most important concepts in mathematics, because it is useful in any situation where there is a notion of symmetry. Nevertheless, I am reluctant to call it an elementary concept. Its most impressive applications, such as the Galois theory of equations, depend on a substantial amount of accompanying *group theory*. I think, rather, that the general group concept is one of the keys to advanced mathematics, and hence that it lies just outside elementary mathematics.

The field concept, on the other hand, seems to lie inside elementary mathematics. Important fields, such as $\mathbb{Q}$, occur at an elementary level. Other important elementary concepts, such as *vector space*, are based on it (see section 4.6). And some ancient problems from elementary geometry can be quite easily solved with its help (see section 5.9). In the next two sections we approach the concept of field by reflecting on the basic operations of arithmetic—addition, subtraction, multiplication, and division—and their rules for calculation.

4.2 Rings

As mentioned in the previous section, we arrive at the system $\mathbb{Q}$ of rational numbers by applying the operations $+$, $-$, $\cdot$, and $\div$ to the positive integers. Before going this far, however, it is instructive to study the effect of applying the operations $+$, $-$, and $\cdot$ alone. These produce the *integers*

$$\ldots, -3, -2, -1, 0, 1, 2, 3, \ldots,$$

which already form an interesting system. It is called $\mathbb{Z}$ (for "Zahlen," as we explained in section 1.3).

$\mathbb{Z}$ is the smallest collection of numbers that includes the positive integers and is such that the sum, difference, and product of any two members is a member. The rules for calculating with members of $\mathbb{Z}$ can be distilled down to the following eight, called the *ring axioms*.

$$a+b=b+a \qquad\qquad ab=ba \qquad \text{(commutative laws)}$$

$$a+(b+c)=(a+b)+c \quad a(bc)=(ab)c \qquad \text{(associative laws)}$$

$$a+(-a)=0 \qquad\qquad\qquad\qquad \text{(inverse law)}$$

$$a+0=a \qquad\qquad a\cdot 1=a \qquad \text{(identity laws)}$$

$$a(b+c)=ab+ac \qquad\qquad\qquad \text{(distributive law)}$$

This set of rules is the result of much thought to minimize their number, and it is not entirely obvious that all the commonly used rules follow from them. Take a moment to think why the following facts are consequences of the eight above.

1. For each a there is exactly one a' such that $a+a'=0$; namely, $a'=-a$.
2. $-(-a)=a$.
3. $a\cdot 0=0$.
4. $a\cdot(-1)=-a$.
5. $(-1)\cdot(-1)=1$.

While you are thinking, I will point out another feature of the ring axioms. They involve only the operations of addition, multiplication, and *negation* (forming the negative, or *additive inverse*, $-a$, of a). This is another feat of distillation, whereby the operation of subtraction is eliminated in favor of a combination of addition and negation; namely, we make the definition

$$a-b=a+(-b).$$

Now let's return to the five facts enumerated above. The first follows by solving the equation $a+a'=0$, which we can do by adding $-a$ to both sides. This sounds obvious, but it actually takes several steps if we strictly observe the rules. Certainly the right-hand side becomes $-a$ when we add $-a$ on its left, by the identity law. The left-hand side becomes

$$(-a)+(a+a') = [(-a)+a]+a' \qquad \text{by the associative law,}$$
$$= [a+(-a)]+a' \qquad \text{by the commutative law,}$$
$$= 0+a' \qquad \text{by the inverse law,}$$
$$= a'+0 \qquad \text{by the commutative law,}$$
$$= a' \qquad \text{by the identity law.}$$

Equating the two sides, we finally get $a' = -a$.

Now to prove $-(-a) = a$ we observe that $a' = -(-a)$ is a solution of the equation $(-a)+a' = 0$ by the identity law. But $a' = a$ is also a solution of this equation because

$$(-a)+a = a+(-a) = 0 \qquad \text{by the commutative and inverse laws.}$$

There is only one solution, by Fact 1, so $-(-a) = a$.

To prove the third fact, $a \cdot 0 = 0$, consider

$$a \cdot 1 = a \cdot (1+0) \qquad \text{by the identity law,}$$
$$= a \cdot 1 + a \cdot 0 \qquad \text{by the distributive law.}$$

Now add $-a \cdot 1$ to both sides, obtaining 0 on the left (by the inverse law) and $a \cdot 0$ on the right (by the associative, commutative, and inverse laws). Thus $a.0 = 0$.

Next we find $a \cdot (-1)$ by considering

$$0 = a \cdot 0 \qquad \text{by Fact 3,}$$
$$= a \cdot [1+(-1)] \qquad \text{by the inverse law,}$$
$$= a \cdot 1 + a \cdot (-1) \qquad \text{by the distributive law,}$$
$$= a + a \cdot (-1) \qquad \text{by the identity law.}$$

Then $a \cdot (-1) = -a$ by Fact 1.

Finally, to find $(-1) \cdot (-1)$, we have

$$(-1) \cdot (-1) = -(-1) \qquad \text{by Fact 4,}$$
$$= 1 \qquad \text{by Fact 2.}$$

Proving these well-known facts about $\mathbb{Z}$ is not a waste of time because the proofs apply to *any other* system that satisfies the eight rules. These systems are very numerous, and they can be quite different from $\mathbb{Z}$. For example, they can be finite. In fact, it is probably from the finite examples (see figure 4.3) that such systems got their name: *rings*. From now on we will call the eight ring axioms the *ring properties*.

Finite Rings

The most important finite rings arise from the notion of congruence mod n, introduced in section 2.4. The idea of doing arithmetic mod n, mentioned there, amounts to working in a ring whose members are $0, 1, 2, \ldots, n - 1$ and whose $+$ and $\cdot$ operations are addition and multiplications mod n. It is equivalent, and more convenient, to view the members of this ring as the *congruence classes* $[0], [1], [2], \ldots, [n - 1]$ mod n, where

$$[a] = \{\ldots, a - 2n, a - n, a, a + n, a + 2n, \ldots\}$$

consists of all the numbers congruent to a, mod n. We can then define addition and multiplication mod n very simply by

$$[a] + [b] = [a + b] \quad \text{and} \quad [a] \cdot [b] = [a \cdot b].$$

The only subtle point is to check that these operations are actually *well* defined, in other words, that they do not depend on the numbers a and b chosen to represent the congruence classes. If, instead, we choose a' and b', then

$$a' = a + cn \quad \text{and} \quad b' = b + dn \quad \text{for some integers } c \text{ and } d.$$

Then if we use a' and b' to define sum and product we indeed get the same congruence classes, because

$$[a' + b'] = [a + cn + b + dn] = [a + b + (c + d)n] = [a + b]$$

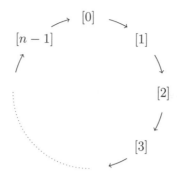

Figure 4.3: A "ring" in arithmetic mod n.

and

$$[a' \cdot b'] = [(a + cn) \cdot (b + dn)]$$
$$= [a \cdot b + adn + bcn + cdn^2]$$
$$= [a \cdot b + (ad + bc + cdn)n] = [a \cdot b].$$

This definition has two great advantages.

- We can use the ordinary $+$ and $\cdot$ symbols without ambiguity. Applied to congruence classes, they mean addition and multiplication mod n; applied to numbers (that is, inside the square brackets) they mean ordinary addition and multiplication.

- The ring properties for congruence classes mod n are "inherited" from $\mathbb{Z}$ immediately. For example, $[a] + [b] = [b] + [a]$ because

$$[a] + [b] = [a + b] \qquad \text{by definition,}$$
$$= [b + a] \qquad \text{by the commutative law in } \mathbb{Z},$$
$$= [b] + [a] \qquad \text{by definition.}$$

Thus the congruence classes $[0], [1], [2], \ldots, [n-1]$ form a ring under the operations of addition and multiplication mod n. This finite ring is indeed "ring-shaped" in the sense that its members are naturally arranged in a circle (figure 4.3), in which each element results from the one before by adding $[1]$, mod n.

Addition mod n by any element $[a]$ fits easily into this picture: every element advances a places around the circle. However, multiplication mod n is more complicated.

One important way in which finite rings differ from $\mathbb{Z}$ is in the presence of *multiplicative inverses*. Ring elements x and y are said to be multiplicative inverses of each other if $x \cdot y = 1$. Thus, in $\mathbb{Z}$ (and every other ring) 1 is inverse to itself and -1 is inverse to itself, but in $\mathbb{Z}$ no other element has a multiplicative inverse. On the other hand, in arithmetic mod n, multiplicative inverses are quite common. Sometimes every nonzero element has a multiplicative inverse. For example, in arithmetic mod 5 we have

$$[1] \cdot [1] = [1],$$
$$[2] \cdot [3] = [6] = [1],$$
$$[3] \cdot [2] = [6] = [1],$$
$$[4] \cdot [4] = [16] = [1],$$

so each of $[1]$, $[2]$, $[3]$, $[4]$ has a multiplicative inverse. In the next section we will find precise conditions for the existence of multiplicative inverses.

4.3 Fields

The system $\mathbb{Q}$ of rational numbers comes from the ring $\mathbb{Z}$ when we include the quotient m/n of any two integers m and $n \neq 0$. Or, more economically, when we include the *multiplicative inverse $n^{-1} = 1/n$* of each integer $n \neq 0$. This is because, when inversion is combined with the operation of multiplication already present, we get all the fractions m/n. The multiplicative inverse of the fraction $\frac{m}{n} \neq 0$ is literally the inverted fraction $\frac{n}{m}$.

It follows that $\mathbb{Q}$ has the following nine properties; eight of them being ring properties inherited from $\mathbb{Z}$ and the ninth being the existence of a multiplicative inverse a^{-1} for each $a \neq 0$.

$$a + b = b + a \qquad ab = ba \qquad \text{(commutative laws)}$$
$$a + (b + c) = (a + b) + c \quad a(bc) = (ab)c \qquad \text{(associative laws)}$$

$$a + (-a) = 0 \quad a \cdot a^{-1} = 1 \text{ for } a \neq 0 \quad \text{(inverse laws)}$$

$$a + 0 = a \quad\quad a \cdot 1 = a \quad\quad \text{(identity laws)}$$

$$a(b + c) = ab + ac \quad\quad\quad \text{(distributive law)}$$

Of course, the operations of $+$ and $\cdot$ must first be extended to the fractions m/n, and this is known to cause headaches in elementary school (which sometimes persist into later life). However, if one accepts that

$$\frac{m}{p} + \frac{n}{q} = \frac{mq + np}{pq},$$

$$\frac{m}{p} \cdot \frac{n}{q} = \frac{mn}{pq},$$

then it is routine to check that the ring properties for fractions follow from those for integers, and that n/m is the multiplicative inverse of m/n.

It should perhaps be added that fractions are not strictly the same thing as rational numbers, because many fractions represent the *same* rational number, for example,

$$\frac{1}{2} = \frac{2}{4} = \frac{3}{6} = \frac{4}{8} = \cdots .$$

So the idea of treating a *class* of objects as a single entity, as we did with congruence classes in the previous section, is actually an idea you met in elementary school—as soon as you saw that a whole class of fractions represent the same number. We tend to identify a rational number with just one of the fractions that represent it, but it is sometimes important to be flexible. In particular, we use

$$\frac{m}{p} = \frac{mq}{pq} \quad \text{and} \quad \frac{n}{q} = \frac{np}{pq},$$

in order to find a "common denominator" and hence explain the sum formula above:

$$\frac{m}{p} + \frac{n}{q} = \frac{mq}{pq} + \frac{np}{pq} = \frac{mq + np}{pq}.$$

As we did with $\mathbb{Z}$ and rings, we take the laws of calculation for $\mathbb{Q}$ as the defining properties of a more general concept, called a *field*. We call the nine laws the *field properties* or *field axioms*.

There are many examples of fields, some of which are among the finite rings studied in the previous section. One of them consists of the congruence classes [0], [1], [2], [3], [4] under addition and multiplication mod 5. We saw that these classes form a ring, and we also saw that each nonzero class has a multiplicative inverse, so the classes form a field. We call this field $\mathbb{F}_5$. By investigating the conditions for multiplicative inverses to exist, we find infinitely many similar finite fields $\mathbb{F}_p$. There is one for each prime p.

The Finite Fields $\mathbb{F}_p$

In arithmetic mod n, the congruence class $[a]$ has a multiplicative inverse $[b]$ if

$$[a] \cdot [b] = [1].$$

(We also say, more loosely, that b is an inverse of a, mod n.) This means that the class $[a] \cdot [b] = [ab]$ is the same as the class

$$[1] = \{\ldots, 1 - 2n, 1 - n, 1, 1 + n, 1 + 2n, \ldots\},$$

which means in turn that

$$ab = 1 + kn \quad \text{for some integer } k,$$

or finally that

$$ab - kn = 1 \quad \text{for some integer } k.$$

This is clearly *not* possible if a and n have a common divisor $d > 1$, because d then divides $ab - kn$ and hence 1. But if $\gcd(a, n) = 1$ we can find a suitable b, because we know from section 2.3 that

$$1 = \gcd(a, n) = Ma + Nn \quad \text{for some integers } M \text{ and } N.$$

This gives us integers $b = M$ and $k = -N$ as required.

Thus *a has a multiplicative inverse*, mod *n, just in case* gcd(a, n) = 1. (And we can find the inverse, M, by the extended Euclidean algorithm described in section 3.4.) When n is a prime number, p, this is true for each $a \not\equiv 0$, mod n, so in this case *the congruence classes* [0], [1], [2], . . . , [$p - 1$] *form a field*, which we call $\mathbb{F}_p$.

The simplest example is the field $\mathbb{F}_2$ of arithmetic mod 2 that we studied in section 2.4. This field also plays an important role in *propositional logic*, as we will see in chapter 9. Indeed, $\mathbb{F}_2$ *is* propositional logic in algebraic form.

4.4 Two Theorems Involving Inverses

The concept of inverse is an old idea in ordinary arithmetic,[1]where we know that subtraction reverses the process of addition and division reverses the process of multiplication. However, the generality of the "inverse" concept was first realized by Galois, around 1830, and it was he who discovered finite fields. When other nineteenth-century mathematicians reflected on this discovery, they realized that inverses greatly clarify some theorems of number theory found in earlier centuries. Here are two celebrated examples.

Fermat's Little Theorem

In 1640 Fermat discovered an interesting theorem about prime powers. In the language of congruence mod p, it reads

If $a \not\equiv 0$ (mod p) and p is prime then $a^{p-1} \equiv 1$ (mod p).

Fermat's proof used the binomial theorem of section 1.6, but there is a more transparent proof that uses inverses mod p.

The condition $a \not\equiv 0$ (mod p) means that [a] is one of the nonzero congruence classes [1], [2], . . . , [$p - 1$]. If we multiply each of these

---

[1] One of my most vivid memories from elementary school was the day our sixth grade teacher showed us what "inversion" meant. Announcing that he was going to show us how to divide by a fraction, he called on the smallest boy in the front row to step forward. Then, without any warning, he lifted the boy off the ground and *turned him upside-down*. That, the teacher said, was how to divide by a fraction: "turn it upside-down and multiply."

classes by $[a]$ we get the nonzero classes

$$[a][1], \quad [a][2], \quad \ldots, \quad [a][p-1].$$

These classes are distinct, because we can recover the distinct classes

$$[1] \quad [2], \quad \ldots, \quad [p-1]$$

by multiplying by the inverse of $[a]$. This means that

$$[a][1], \quad [a][2], \quad \ldots, \quad [a][p-1]$$

are in fact the *same* classes as

$$[1], \quad [2], \quad \ldots, \quad [p-1]$$

(only, perhaps, in a different order).

It follows that both sets of classes have the same product, namely,

$$[a]^{p-1} \cdot [1] \cdot [2] \cdot \cdots \cdot [p-1] = [1] \cdot [2] \cdot \cdots \cdot [p-1].$$

Multiplying both sides of this equation by the inverses of $[1], [2], \ldots, [p-1]$ we get

$$[a]^{p-1} = [1], \quad \text{in other words,} \quad a^{p-1} \equiv 1 \pmod{p}.$$

This is what we call *Fermat's little theorem* today (because it is not as big as his "last" theorem about sums of nth powers). The version found by Fermat himself,

$$a^p \equiv a \pmod{p},$$

follows by multiplying both sides by a.

Euler's Theorem

Around 1750, Euler gave a proof of Fermat's little theorem quite close to the one above, and he generalized it to a theorem about congruence mod n, where n is not necessarily prime. In the language of congruence classes and inverses, the proof goes as follows (using the term *invertible* to mean "having an inverse").

Suppose that there are m invertible congruence classes mod n, say

$$[a_1], \quad [a_2], \quad \ldots, \quad [a_m].$$

If we multiply each of these by an invertible congruence class $[a]$ we get the classes

$$[a][a_1], \quad [a][a_2], \quad \ldots, \quad [a][a_m],$$

which are again invertible, and distinct, since we can recover the original list of invertible classes by multiplying by the inverse of $[a]$, mod n. Thus the second set of congruence classes is the same as the first, and hence they both have the same product:

$$[a]^m \cdot [a_1] \cdot [a_2] \cdot \cdots \cdot [a_m] = [a_1] \cdot [a_2] \cdot \cdots \cdot [a_m].$$

Multiplying each side by the inverses of $[a_1], [a_2], \ldots, [a_m]$, we get

$$[a]^m = [1], \quad \text{in other words} \quad a^m \equiv 1 \pmod{n},$$

where m is the number of invertible congruence classes, mod n.

Now from the previous section we know that $[a]$ is invertible just in case $\gcd(a, n) = 1$. So m is how many numbers a among $1, 2, \ldots,$ $n - 1$ have gcd 1 with n. Such numbers are also called *relatively prime* to n. The number of them is denoted by $\varphi(n)$, and φ is called the *Euler phi function*. Euler's theorem is usually stated in terms of the phi function, namely:

If $\gcd(a, n) = 1$ *then* $a^{\varphi(n)} \equiv 1 \pmod{n}$.

As an illustration of Euler's theorem, consider the numbers relatively prime to 8. They are $1, 3, 5, 7$, and hence there are four invertible congruence classes mod 8,

$$[1], \quad [3], \quad [5], \quad [7].$$

The argument above tells us that

$$a^4 \equiv 1 \pmod{8}$$

for any a that is relatively prime to 8. For example, if $a = 3$ we get $a^4 = 81$, which is indeed $\equiv 1 \pmod{8}$.

In recent years, Euler's theorem has become one of the most commonly used theorems in mathematics, because it is a crucial part of the RSA cryptosystem used in many internet transactions. I will not discuss RSA in any more detail, since it may now be found in almost any introduction to number theory.

4.5 Vector Spaces

The concept of a vector space was not articulated until the twentieth century, so you might think that it is very abstract and sophisticated. In a way it is, because it builds on the already abstract concept of field. But in another way it isn't, because it is the foundation of a very elementary topic—*linear algebra*—which goes back over 2000 years. It is one of the quirks of mathematical history that linear algebra was, for a long time, considered *too* elementary to be worth studying in its own right.

In our brief survey of polynomial equations in section 4.1 we skipped over the linear equation

$$ax + b = 0,$$

because it is indeed too simple to be worth discussing. Solving several linear equations in several unknowns is not quite so simple, even though it was done over 2000 years ago, using a method very similar to what we use today. The only thing one needs to recall here about this method is that it involves the operations of $+$, $-$, $\cdot$, and $\div$. So, if the coefficients of the equations belong to a certain field, then the solutions belong to the very same field.

I hope that these remarks suffice to show that the *field* concept is implicit in the machinery of linear algebra. For beginners in linear algebra, the field is usually assumed to be $\mathbb{R}$, the system of *real* numbers that we will discuss further in chapter 6. But the field could also be $\mathbb{Q}$, $\mathbb{F}_p$, or any other field.

When the field is $\mathbb{R}$, linear equations in two variables represent lines in the plane $\mathbb{R}^2$. This, after all, is why such equations are called "linear." Linear equations in three variables represent planes in the three-dimensional space $\mathbb{R}^3$, and so on. Indeed, this is where the word "space" in "vector space" comes from. $\mathbb{R}$, $\mathbb{R}^2$, and $\mathbb{R}^3$ are all examples

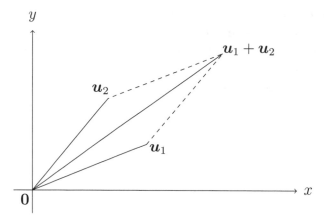

Figure 4.4: Vector addition in the plane.

of vector spaces. The *vectors* in these spaces are their points, and we have rules for adding points and multiplying them by numbers. In $\mathbb{R}$, these operations are ordinary addition and multiplication; in the higher-dimensional spaces they are called *vector addition* and *scalar multiplication*.

Figure 4.4 shows what vector addition looks like in the plane $\mathbb{R}^2$. Points of $\mathbb{R}^2$ are ordered pairs (x, y) and they are added "coordinate-wise" by the rule:

$$(x_1, y_1) + (x_2, y_2) = (x_1 + x_2, y_1 + y_2).$$

Thus the sum of points $u_1 = (x_1, y_1)$ and $u_2 = (x_2, y_2)$ is the fourth vertex of a parallelogram whose other vertices are $\mathbf{0} = (0, 0)$, u_1, and u_2.

This picture suggests that the concept of vector addition has some geometric content. So does the concept of scalar magnification,

$$a(x, y) = (ax, ay),$$

which *magnifies* the line from $\mathbf{0}$ to (x, y) by a factor of a. Indeed, we will see that the concept of vector space is a kind of halfway house between algebra and geometry. On the one hand, algebra reduces many geometric problems to routine computation; on the other hand, geometric intuition suggests concepts like *dimension*, which can guide algebraic arguments. We study the concept of dimension in the next section. Now it is time to give the definition of vector space.

Given a field $\mathbb{F}$ and a set V of objects with an operation $+$, we say that V is a *vector space over* $\mathbb{F}$ if the following conditions are satisfied. First $+$ must have the usual properties of addition. For any u, v, w in V:

$$u + v = v + u,$$

$$u + (v + w) = (u + v) + w,$$

$$u + 0 = u,$$

$$u + (-u) = 0.$$

Thus, in particular, there is a zero vector 0 and a negative $-u$ of each vector u. Second, there must be a multiplication of vectors by members of $\mathbb{F}$ (the "scalars") which is compatible with the multiplication in $\mathbb{F}$ and which makes addition in $\mathbb{F}$ compatible with vector addition. That is, for any a, b in $\mathbb{F}$ and u, v in V:

$$a(bu) = (ab)u,$$

$$1u = u,$$

$$a(u + v) = au + av,$$

$$(a + b)u = au + bu.$$

Altogether, a vector space is defined by eight properties, on top of the nine properties it takes to define a field. This is a lot, admittedly. However, all the properties resemble the rules we already use for ordinary calculation with numbers, so we can do vector space calculations almost without thinking.

It is easy to check, by such calculations, that the vector space addition and scalar multiplication defined above for $\mathbb{R}$ satisfy the eight conditions, so $\mathbb{R}^2$ is a vector space over $\mathbb{R}$. Analogous definitions of vector addition and scalar multiplication show that $\mathbb{R}^n$, the space of n-tuples of real numbers, is also a vector space over $\mathbb{R}$.

Another interesting example is the set

$$\mathbb{Q}(\sqrt{2}) = \{a + b\sqrt{2} : a, b \text{ in } \mathbb{Q}\}.$$

This set, which happens to be a field itself, is a vector space over the field $\mathbb{Q}$. The vector addition operation is ordinary addition of numbers, and the scalar multiplication is ordinary multiplication—by rational numbers. If we view the rational numbers a and b as the "coordinates" of the number $a + b\sqrt{2}$, then vector addition is "coordinatewise"—

$$(a_1 + b_1\sqrt{2}) + (a_2 + b_2\sqrt{2}) = (a_1 + a_2) + (b_1 + b_2)\sqrt{2}$$

—just like vector addition in $\mathbb{R}^2$. In fact, we will see in the next section that $\mathbb{Q}(\sqrt{2})$ is *two-dimensional* as a vector space over $\mathbb{Q}$. We will also see that it is no coincidence that the dimension number is the same as the *degree* of $\sqrt{2}$; that is, the degree of the polynomial equation $x^2 - 2 = 0$ that defines it.

This example nicely shows how the vector space concept mediates between algebra and geometry.

4.6 Linear Dependence, Basis, and Dimension

We can see the dimension 2 of $\mathbb{R}^2$ in its two coordinate axes, or in the two directions of the unit vectors $i = (1, 0)$ and $j = (0, 1)$. These two vectors *span* $\mathbb{R}^2$ in the sense that any vector is a *linear combination* of them:

$$(x, y) = xi + yj.$$

Also, i and j are *linearly independent* in the sense that neither is a multiple of the other or, in other words, the linear combination

$$ai + bj \neq 0 \quad \text{unless } a = b = 0.$$

These two properties—spanning and linear independence—make the pair i, j what we call a *basis* of $\mathbb{R}^2$ (over $\mathbb{R}$).

This train of thought leads us to define a *basis of a vector space V over* $\mathbb{F}$. A set $\{v_1, v_2, \ldots, v_n\}$ is called a basis of V over $\mathbb{F}$ if the following conditions hold.

1. The vectors $v_1, v_2, \ldots, v_n$ span V; that is, for each v in V,

$$v = a_1v_1 + a_2v_2 + \cdots + a_nv_n, \quad \text{where } a_1, a_2, \ldots, a_n \text{ are in } \mathbb{F}.$$

2. The vectors $v_1, v_2, \ldots, v_n$ are linearly independent; that is

$$a_1 v_1 + a_2 v_2 + \cdots + a_n v_n = 0, \quad \text{with } a_1, a_2, \ldots, a_n \text{ in } \mathbb{F},$$

only if $a_1 = a_2 = \cdots = a_n = 0$.

A space with such a basis is called *finite-dimensional*. We will see how "dimension" comes into play shortly. Not all vector spaces are finite-dimensional. It may depend on the field we view them over; for example, $\mathbb{R}$ has infinite dimension over $\mathbb{Q}$. However, it seems fair to say that infinite-dimensional spaces belong to advanced mathematics. So in elementary mathematics we should consider just finite-dimensional spaces. For them we can prove the important property that *all bases have the same size*.

To do this we use the following simple but clever result. It is named after Steinitz because of its appearance in Steinitz (1913), but it is actually due to Grassmann (1862), Chapter 1, Section 2.

Steinitz exchange lemma. If n vectors span V, then no $n+1$ vectors are independent. *(So, if there is a basis with n vectors, there cannot be a basis with more than n vectors.)*

Proof. Suppose (for the sake of contradiction) that $u_1, u_2, \ldots, u_n$ span V and that $v_1, v_2, \ldots, v_{n+1}$ are linearly independent in V. The plan is to replace one of the u_i by v_1, while ensuring that the remaining u_i, together with v_1, still span V. We then replace one of the remaining u_i by v_2, retaining the spanning property, and continue in this way until all of the u_i are replaced by $v_1, v_2, \ldots, v_n$. This means that $v_1, v_2, \ldots, v_n$ is a spanning set, so v_{n+1} is a linear combination of them, contrary to the assumption that $v_1, v_2, \ldots, v_{n+1}$ are independent.

To see why the plan succeeds, suppose that we have successfully replaced $m-1$ of the u_i by $v_1, v_2, \ldots, v_{m-1}$, so now we want to replace another of the u_i by v_m. (So when $m = 1$ we have done nothing yet.) Our success so far means that $v_1, v_2, \ldots, v_{m-1}$ and the remaining u_i span V, so

$$v_m = a_1 v_1 + a_2 v_2 + \cdots + a_{m-1} v_{m-1} + \text{terms } b_i u_i,$$

where the coefficients $a_1, \ldots, a_{m-1}$ and b_i are in $\mathbb{F}$. Since $\boldsymbol{v}_1, \boldsymbol{v}_2, \ldots,$ $\boldsymbol{v}_m$ are linearly independent, we must have some $b_j \neq 0$. But then, dividing by b_j, we obtain $\boldsymbol{u}_j$ as a linear combination $\boldsymbol{v}_1, \boldsymbol{v}_2, \ldots, \boldsymbol{v}_m$ and the remaining $\boldsymbol{u}_i$, so $\boldsymbol{u}_j$ can be replaced by $\boldsymbol{v}_m$ and we still have a spanning set.

Thus we can continue the exchange process until all of $\boldsymbol{u}_1, \boldsymbol{u}_2, \ldots, \boldsymbol{u}_n$ are replaced by $\boldsymbol{v}_1, \boldsymbol{v}_2, \ldots, \boldsymbol{v}_n$, leading to the contradiction foreshadowed above. This proves the Steinitz exchange lemma, and hence all bases of a finite-dimensional vector space have the same size. $\qquad\qquad\square$

This result enables us to define dimension.

Definition. The *dimension* of a finite-dimensional vector space V is the number of elements in a basis of V.

The concepts of basis and dimension are nicely illustrated by the vector space $\mathbb{Q}(\sqrt{2})$ over $\mathbb{Q}$. The numbers 1 and $\sqrt{2}$ span $\mathbb{Q}(\sqrt{2})$, because every $a + b\sqrt{2}$ in $\mathbb{Q}(\sqrt{2})$ is a linear combination of 1 and $\sqrt{2}$ with coefficients a and b in $\mathbb{Q}$.

Also, 1 and $\sqrt{2}$ are linearly independent over $\mathbb{Q}$, because if

$$a + b\sqrt{2} = 0 \quad \text{with } a, b \text{ rational and not both zero,}$$

then both a, b must be nonzero, in which case

$$\sqrt{2} = -a/b,$$

contradicting the fact that $\sqrt{2}$ is irrational.

Thus 1 and $\sqrt{2}$ comprise a basis for $\mathbb{Q}(\sqrt{2})$ over $\mathbb{Q}$, and therefore $\mathbb{Q}(\sqrt{2})$ has dimension 2 over $\mathbb{Q}$.

4.7 Rings of Polynomials

In high school algebra we get used to calculating with *polynomials*— objects such as $x^2 - x + 1$ or $x^3 + 3$ which can be added, subtracted, and multiplied according to the same rules as for numbers. Indeed, it is because the symbol x for the "unknown" or "indeterminate" behaves exactly like a number that Newton called this kind of algebra "Universal

Arithmetick." Strictly speaking, we should check that it is *consistent* to assume that x behaves like a number. This amounts to checking that the rules for adding, subtracting, and multiplying polynomials satisfy the ring properties listed in section 4.2. However, this is a routine, if tedious, task.

What is more interesting is how closely polynomials resemble integers in ways other than their ring properties. In particular, there is a concept of "prime" polynomial, a Euclidean algorithm, and a unique prime factorization theorem. These facts show that algebra imitates arithmetic even more than Newton realized, and they suggest that it will be interesting to see what happens when we import other ideas from number theory into the algebra of polynomials.

To put this idea on a sound foundation we begin by specifying the polynomials we are most interested in. They are the ring $\mathbb{Q}[x]$ of *polynomials in x with rational coefficients*. Thus any member of $\mathbb{Q}[x]$ has the form

$$p(x) = a_0 + a_1 x + \cdots + a_n x^n \quad \text{with } a_0, a_1, \ldots, a_n \text{ in } \mathbb{Q}.$$

For a nonzero polynomial we can assume that $a_n \neq 0$, in which case n is called the *degree*, deg p, of $p(x)$. Any constant polynomial has degree zero. The degree serves to measure the "size" of a polynomial. In particular it motivates the idea that a "prime" polynomial is one that is not the product of polynomials in $\mathbb{Q}[x]$ of smaller "size." More formally, we have:

Definition. A polynomial $p(x)$ in $\mathbb{Q}[x]$ is called *irreducible* if $p(x)$ is not the product of two polynomials in $\mathbb{Q}[x]$ of lower degree.

For example, $x^2 - 2$ is irreducible because any factors of $x^2 - 2$ of lower degree in $\mathbb{Q}[x]$ are necessarily constant multiples of $x - a$ and $x + a$, for some rational a. But then we get $a^2 = 2$, which contradicts the irrationality of $\sqrt{2}$. On the other hand, $x^2 - 1$ is *reducible* because it splits into the factors $x - 1$ and $x + 1$, which are of lower degree and in $\mathbb{Q}[x]$.

To obtain a Euclidean algorithm for polynomials it suffices, as for positive integers, to perform *division with remainder*. Recall from section 2.1 that, for positive integers a and $b \neq 0$, division with remainder gives integers q and r ("quotient" and "remainder") such that

$$a = qb + r \quad \text{with } |r| < |b|.$$

For polynomials $a(x)$ and $b(x) \neq 0$ in $\mathbb{Q}[x]$ we seek polynomials $q(x)$ and $r(x)$ in $\mathbb{Q}[x]$ such that

$$a(x) = q(x)b(x) + r(x) \quad \text{with } \deg(r) < \deg(b).$$

The polynomials $q(x)$ and $r(x)$ are exactly what one obtains from the "long division" process for polynomials sometimes taught in high school. It goes back to Stevin (1585b), who also observed that it gives a Euclidean algorithm for polynomials.

We illustrate long division with $a(x) = 2x^4 + 1$ and $b(x) = x^2 + x + 1$. The idea is to subtract or add multiples of $b(x)$ by suitable constant multiples of powers of x from $a(x)$, successively removing the highest powers of $a(x)$ until what remains has degree less than that of $b(x)$. First, subtract $2x^2 b(x)$ from $a(x)$ to remove the x^4 term:

$$a(x) - 2x^2 b(x) = 2x^4 + 1 - 2x^2(x^2 + x + 1) = -2x^3 - 2x^2 + 1.$$

Next, add $2xb(x)$ to remove the x^3 term (which also removes the x^2 term):

$$a(x) - 2x^2 b(x) + 2xb(x) = -2x^3 - 2x^2 + 1 + 2x(x^2 + x + 1) = 2x + 1.$$

That is,

$$a(x) - (2x^2 - 2x)b(x) = 2x + 1,$$

so we have

$$a(x) = (2x^2 - 2x)b(x) + (2x + 1).$$

This gives $q(x) = 2x^2 - 2x$ and $r(x) = 2x + 1$, so $\deg(r) < \deg(b)$ as required.

Once we have division with remainder we can sail through the remaining steps to unique prime factorization in $\mathbb{Q}[x]$.

- The Euclidean algorithm gives the *greatest common divisor* of two polynomials $a(x)$, $b(x)$ (meaning, common divisor of highest degree) in the form

$$\gcd(a(x), b(x)) = m(x)a(x) + n(x)b(x)$$

$$\text{with } m(x), n(x) \text{ in } \mathbb{Q}[x].$$

- This representation of the gcd gives the "prime divisor property": if $p(x)$ is an irreducible polynomial that divides a product $a(x)b(x)$ then $p(x)$ divides $a(x)$ or $p(x)$ divides $b(x)$.

- Unique factorization into irreducible polynomials follows, though uniqueness is only "up to constant (rational number) factors."

As in the case of primality testing for positive integers, it is not always easy to tell whether a polynomial is irreducible. However, we will be able to do this in some interesting cases of low degree. We are also interested to see what happens when we import the idea of "congruence modulo a prime" into the world of polynomials. We know from section 4.3 that congruence of integers modulo a prime p leads to the finite field $\mathbb{F}_p$. In the next section we will see that congruence in $\mathbb{Q}[x]$ modulo an irreducible $p(x)$ also leads to a field "of finite degree."

The Rings $\mathbb{R}[x]$ and $\mathbb{C}[x]$

Prime factorization is difficult in $\mathbb{Q}[x]$ because we insist that the factors have rational coefficients. For example, the factorization

$$x^2 - 2 = (x + \sqrt{2})(x - \sqrt{2})$$

does not exist in $\mathbb{Q}[x]$ because $\sqrt{2}$ is irrational. More generally, $x^n - 2$ is irreducible in $\mathbb{Q}[x]$, so the primes of $\mathbb{Q}[x]$ can have arbitrarily high degree.

We can simplify the class of irreducible polynomials by enlarging the class of numbers allowed as coefficients. With coefficients in the *real* numbers $\mathbb{R}$, the irreducible polynomials include some quadratics, such as $x^2 + 1$, but they all have degree ≤ 2. With coefficients in the *complex* numbers $\mathbb{C}$ the irreducible polynomials are linear. These results follow from the so-called *fundamental theorem of algebra*, according to which every polynomial equation $p(x) = 0$ with coefficients in $\mathbb{C}$ has a solution in $\mathbb{C}$. We discuss this theorem further in the Philosophical Remarks to this chapter, because it is not entirely a theorem of algebra.

Factorization into linear factors follows from the fundamental theorem by a simple application of division with remainder. Given that $p(x) = 0$ has a solution $x = c$, use division with remainder to write

$$p(x) = q(x)(x - c) + r(x) \quad \text{with } \deg(r) < \deg(x - c).$$

Substituting $x = c$ we find that $r(c)$ is necessarily zero and hence $x - c$ is a factor of $p(x)$. Also, $q(x)$ has degree one less than that of $p(x)$, so we can repeat the argument a finite number of times to get a factorization of $p(x)$ into linear factors.

Thus in the ring $\mathbb{C}[x]$ of polynomials with complex coefficients we obtain the simplest possible factorization: every polynomial splits into *linear* factors. In the ring $\mathbb{R}[x]$ of polynomials with real coefficients we cannot go so far, because examples such as $x^2 + 1$ do not factorize further. However, we can obtain factors of degree *at most* 2 because of the following convenient fact, due to Euler (1751).

If the coefficients of $p(x)$ are real, any nonreal solution $x = a + ib$ of $p(x) = 0$ is accompanied by the complex conjugate *solution $\overline{x} = a - ib$.* The reason why nonreal solutions occur in conjugate pairs lies in the "symmetry" properties (which are like those mentioned in section 4.1)

$$\overline{c_1 + c_2} = \overline{c_1} + \overline{c_2} \quad \text{and} \quad \overline{c_1 \cdot c_2} = \overline{c_1} \cdot \overline{c_2}$$

of the complex conjugation operation. These properties are easily checked from its definition $\overline{a + ib} = a - ib$.

It follows from these properties that if

$$p(x) = a_0 + a_1 x + \cdots + a_n x^n$$

then

$$\overline{p(x)} = \overline{a_0} + \overline{a_1 x} + \cdots + \overline{a_n x^n}.$$

So if $a_0, a_1, \ldots, a_n$ are real, and hence equal to their own conjugates, we have

$$\overline{p(x)} = a_0 + a_1 \overline{x} + \cdots + a_n \overline{x}^n = p(\overline{x}).$$

Therefore, if $p(x) = 0$ then $\overline{p(x)} = \overline{0} = 0 = p(\overline{x})$, as claimed.

The conjugate solutions $x = a + ib$ and $a - ib$ correspond to factors $x - a - ib$ and $x - a + ib$, which give a real quadratic factor because

$$(x - a - ib)(x - a + ib) = (x - a)^2 - (ib)^2 = x - 2ax + a^2 + b^2.$$

Thus any $p(x)$ in $\mathbb{R}[x]$ splits into real linear and quadratic factors, so the irreducibles of $\mathbb{R}[x]$ all have degree ≤ 2, as claimed.

4.8 Algebraic Number Fields

If we take the quotients $q(x)/r(x)$ of all the $q(x)$ and nonzero $r(x)$ in $\mathbb{Q}[x]$ we get a field $\mathbb{Q}(x)$ called the *field of rational functions* with coefficients in $\mathbb{Q}$. Then if we replace x by a number α we get a field of numbers, denoted by $\mathbb{Q}(\alpha)$ and called the result of *adjoining* α to $\mathbb{Q}$. We have already used this notation in the special case $\alpha = \sqrt{2}$ when we mentioned $\mathbb{Q}(\sqrt{2})$ in sections 4.5 and 4.6.

The field $\mathbb{Q}(\alpha)$ is of special interest when α is an *algebraic number*; that is, the solution of an equation $p(x) = 0$ for some $p(x)$ in $\mathbb{Q}[x]$. In this case $\mathbb{Q}(\alpha)$ is called an *algebraic number field*. For example, $\mathbb{Q}(\sqrt{2})$ is an algebraic number field because $\sqrt{2}$ is a solution of the equation $x^2 - 2 = 0$. Since $\sqrt{2}$ is irrational, $x^2 - 2$ is in fact a polynomial of minimal degree satisfied by $\sqrt{2}$.

Notice that we cannot simply substitute the algebraic number α for x in the quotients $q(x)/r(x)$ that define $\mathbb{Q}(x)$, because $r(\alpha)$ is zero when $r(x)$ contains the factor $p(x)$. We avoid this problem by using division with remainder for polynomials. This gives us a field in much the same way that we obtained the finite fields $\mathbb{F}_p$ in section 4.3. The role of the prime p is now played by an irreducible polynomial $p(x)$.

Congruence Modulo a Polynomial

If α is any algebraic number there is a polynomial of minimal degree such that $p(\alpha) = 0$. This *minimal polynomial* is unique (up to multiples by nonzero rational numbers) because if $q(x)$ is a polynomial of the same degree as $p(x)$, satisfied by α, we can assume (mutiplying by some rational) that $p(x)$ and $q(x)$ have the same coefficient for the highest power of x. Then if $p(x)$ and $q(x)$ are not identical, $p(x) - q(x)$ is a polynomial of lower degree satisfied by α, contrary to the minimality of $p(x)$.

Likewise, the minimal polynomial $p(x)$ for α is irreducible, because if $p(x) = q(x)r(x)$, with $q(x)$, $r(x)$ in $\mathbb{Q}[x]$ and of lower degree, then $p(\alpha) = 0$ implies $q(\alpha) = 0$ or $r(\alpha) = 0$—again contradicting the minimality of $p(x)$. The irreducible polynomial $p(x)$ for α gives a different, and more enlightening, way to obtain the number field $\mathbb{Q}(\alpha)$. Namely, take *the congruence classes of polynomials in $\mathbb{Q}[x]$ under "congruence modulo $p(x)$."*

Polynomials $a(x)$, $b(x)$ in $\mathbb{Q}[x]$ are *congruent modulo $p(x)$*, written

$$a(x) \equiv b(x) \ (\mathrm{mod} \ p(x)),$$

if $p(x)$ divides $a(x) - b(x)$ in $\mathbb{Q}[x]$. Thus the *congruence class $[a(x)]$* of $a(x)$ consists of all the polynomials that differ from $a(x)$ by a polynomial $n(x)p(x)$, where $n(x)$ is in $\mathbb{Q}[x]$. (Informally, the polynomials in this class are those that have to equal $a(x)$ when we interpret x as α.)

It follows, just as for integers modulo a prime p, that the congruence classes of polynomials modulo an irreducible $p(x)$ form a field. In particular, each polynomial $a(x)$ has an *inverse* mod $p(x)$; that is, a polynomial $a^*(x)$ in $\mathbb{Q}[x]$ such that

$$a(x)a^*(x) \equiv 1 \ (\mathrm{mod} \ p(x)).$$

What is this field? Well, if α is a solution of the polynomial equation $p(x) = 0$, the field is none other than $\mathbb{Q}(\alpha)$! Or, at least, it is a field with the "same structure" as $\mathbb{Q}(\alpha)$ in the following sense.

Number field construction. *If α is an algebraic number with minimal polynomial $p(x)$, then the members of $\mathbb{Q}(\alpha)$ are in one-to-one correspondence with the congruence classes of polynomials in $\mathbb{Q}[x]$ modulo $p(x)$. Sums and products also correspond.*

Proof. To establish the correspondence, it suffices to show that the values $a(\alpha)$ of *polynomials $a(x)$* in $\mathbb{Q}[x]$ are in one-to-one correspondence with the congruence classes $[a(x)]$ modulo $p(x)$, since the inverse of a value, $1/a(\alpha)$, corresponds to the congruence class $[a^*(x)]$ inverse to $[a(x)]$.

Thus we have to prove, for any polynomials $a(x)$, $b(x)$ in $\mathbb{Q}[x]$, that

$$a(\alpha) = b(\alpha) \quad \text{if and only if} \quad a(x) \equiv b(x) \; (\text{mod } p(x)),$$

or, letting $c(x) = a(x) - b(x)$, that

$$c(\alpha) = 0 \quad \text{if and only if} \quad c(x) \equiv 0 \; (\text{mod } p(x)).$$

Certainly, if $c(x) \equiv 0 \; (\text{mod } p(x))$ then $c(x) = d(x)p(x)$ for some $d(x)$ in $\mathbb{Q}[x]$. Hence

$$c(\alpha) = d(\alpha)p(\alpha) = 0 \quad \text{because} \quad p(\alpha) = 0.$$

Conversely, if $c(\alpha) = 0$ we divide $c(x)$ by $p(x)$ with remainder, obtaining quotient and remainder polynomials $q(x)$ and $r(x)$ in $\mathbb{Q}[x]$ such that

$$c(x) = q(x)p(x) + r(x) \quad \text{with} \quad \deg(r) < \deg(p).$$

Since $c(\alpha) = 0 = p(\alpha)$, it follows that $r(\alpha) = 0$ too. If $r(x)$ is nonzero this contradicts the minimality of $p(x)$. So in fact

$$c(x) = q(x)p(x) \quad \text{and therefore} \quad c(x) \equiv 0 \; (\text{mod } p(x)).$$

Thus there is a one-to-one correspondence between the values and the congruence classes of polynomials in $\mathbb{Q}[x]$, as required.

Finally, sums of values correspond to sums of congruence classes because $[a(x)] + [b(x)] = [a(x) + b(x)]$. Namely, the values $a(\alpha)$ and $b(\alpha)$ correspond to $[a(x)]$ and $[b(x)]$ and $a(\alpha) + b(\alpha)$ corresponds to $[a(x) + b(x)]$, which is the sum of $[a(x)]$ and $[b(x)]$. Products of values similarly correspond to products of congruence classes, because $[a(x)] \cdot [b(x)] = [a(x) \cdot b(x)]$. $\qquad\square$

The one-to-one correspondence, preserving sums and products, is what we mean by saying that the field of congruence classes has the "same structure" as the number field $\mathbb{Q}(\alpha)$. This structural correspondence between fields is an example of an *isomorphism* (from the Greek for "same form"). The isomorphism concept is everywhere in advanced algebra, so one might take its appearance in this theorem as a sign that we are approaching the boundary of elementary algebra. Nevertheless, I prefer to view the proof above as elementary, since it hinges on another simple application of division with remainder.

The analogy with the construction of fields from congruence classes of integers modulo a prime is also too nice to neglect. Where the result in the integer case is a finite field, the result in the polynomial case is also finite, in a sense. In this case "finiteness" is in the degree, which also equals the dimension of the field as a vector space over $\mathbb{Q}$.

4.9 Number Fields as Vector Spaces

The view of $\mathbb{Q}(\alpha)$ as the field of congruence classes of polynomials mod $p(x)$, where $p(x)$ is the minimal polynomial for α, is enlightening because it gives us a natural basis for $\mathbb{Q}(\alpha)$ as a vector space over $\mathbb{Q}$.

Basis for $\mathbb{Q}(\alpha)$. *If the minimal polynomial for α has degree n, then the numbers $1, \alpha, \alpha^2, \ldots, \alpha^{n-1}$ form a basis for $\mathbb{Q}(\alpha)$ over $\mathbb{Q}$.*

Proof. Suppose that $p(x) = a_0 + a_1 x + \cdots + a_n x^n$. Then

1. The congruence classes $[1], [x], \ldots, [x^{n-1}]$ are linearly independent over $\mathbb{Q}$, because if

$$b_0[1] + b_1[x] + \cdots + b_{n-1}[x^{n-1}] = [0]$$

   with $b_0, b_1, \ldots, b_{n-1}$ in $\mathbb{Q}$, not all zero,

   we have, by the correspondence in the number field construction,

$$b_0 + b_1 \alpha + \cdots + b_{n-1} \alpha^{n-1} = 0,$$

   contrary to the minimality of $p(x)$.

2. The congruence classes $[1], [x], \ldots, [x^{n-1}]$ span the field of all congruence classes mod $p(x)$. This field is certainly spanned by the infinitely many congruence classes $[1], [x], [x^2], \ldots$, but the subset of $[1], [x], \ldots, [x^{n-1}]$ also spans, because their linear combinations include

$$-[x^n] = \frac{1}{a_n}(a_0[1] + a_1[x] + \cdots + a_{n-1}[x^{n-1}]),$$

which gives us in turn

$$-[x^{n+1}] = \frac{1}{a_n}(a_0[x] + a_1[x^2] + \cdots + a_{n-1}[x^n]),$$

and so on.

Thus $[1], [x], \ldots, [x^{n-1}]$ is a basis for the field of conguence classes mod $p(x)$ over $\mathbb{Q}$, and hence the corresponding numbers $1, \alpha, \alpha^2, \ldots, \alpha^{n-1}$ form a basis for $\mathbb{Q}(\alpha)$ over $\mathbb{Q}$. $\quad\square$

It follows in particular that the vector space $\mathbb{Q}(\alpha)$ has *finite* dimension over $\mathbb{Q}$ for any algebraic number α. This is not easy to prove directly. It is a nice exercise to prove by direct calculation of sums, products, and inverses that

$$\mathbb{Q}(\sqrt{2}) = \{a + b\sqrt{2} : a, b \text{ in } \mathbb{Q}\},$$

thus showing that $\mathbb{Q}(\sqrt{2})$ has dimension 2 over $\mathbb{Q}$. But try proving, say, that $\mathbb{Q}(2^{1/5})$ has dimension 5 over $\mathbb{Q}$. Just try expressing

$$\frac{1}{2^{1/5} + 7 \cdot 2^{3/5} - 2^{4/5}} \text{ as a linear combination of } 1, 2^{1/5}, 2^{2/5}, 2^{3/5}, 2^{4/5}.$$

There is a converse to the fact that $\mathbb{Q}(\alpha)$ has *finite* dimension over $\mathbb{Q}$ for any algebraic number α. However, we will prove only an easier theorem, which is still surprising. In this proof we denote the vector space by $\mathbb{F}$, because it is in fact a field.

Finite-dimensional vector spaces over $\mathbb{Q}$. *If $\mathbb{F}$ is a vector space of dimension n over $\mathbb{Q}$, then each member of $\mathbb{F}$ is algebraic of degree $\leq n$.*

Proof. Let α be a member of $\mathbb{F}$ and consider the $n+1$ elements $1, \alpha, \alpha^2, \ldots, \alpha^n$. Since $\mathbb{F}$ has dimension n over $\mathbb{Q}$ these elements are linearly dependent, which means

$$a_0 + a_1\alpha + \cdots + a_n\alpha^n = 0 \quad \text{for some } a_0, a_1, \ldots, a_n \text{ in } \mathbb{Q}, \text{ not all zero.}$$

But this says that α is an algebraic number of degree $\leq n$. $\quad\square$

Since $\mathbb{F}$ has a basis of n elements, we can in fact say that $\mathbb{F}$ results from $\mathbb{Q}$ by adjoining n algebraic numbers of degree $\leq n$. There is a theorem that in this case there is a *single* number α of degree n such that $\mathbb{F} = \mathbb{Q}(\alpha)$; α is called a *primitive element*. We will not go so far as to prove the primitive element theorem.[2] However, we will be interested in the effect that successive adjunctions have on the dimension of a number field. For this we have the following theorem about "relative dimension," pointed out by Dedekind (1894), p. 473.

Dedekind product theorem. *If $\mathbb{E} \subseteq \mathbb{F} \subseteq \mathbb{G}$ are fields such that $\mathbb{F}$ has dimension m over $\mathbb{E}$ and $\mathbb{G}$ has dimension n over $\mathbb{F}$, then $\mathbb{G}$ has dimension mn over $\mathbb{E}$.*

Proof. Let $u_1, u_2, \ldots, u_m$ be a basis for $\mathbb{F}$ over $\mathbb{E}$, so each f in $\mathbb{F}$ can be written

$$f = e_1 u_1 + \cdots + e_m u_m \quad \text{with } e_1, \ldots, e_m \text{ in } \mathbb{E}. \tag{*}$$

Also let $v_1, v_2, \ldots, v_n$ be a basis for $\mathbb{G}$ over $\mathbb{F}$, so each g in $\mathbb{G}$ can be written

$$g = f_1 v_1 + \cdots + f_n v_n \quad \text{with } f_1, \ldots, f_n \text{ in } \mathbb{F},$$
$$= (e_{11} u_1 + \cdots + e_{1m} u_m) v_1 + \cdots + (e_{n1} u_1 + \cdots + e_{nm} u_m) v_n,$$

rewriting each f_i as a linear combination given by (*).

Thus each g in $\mathbb{G}$ can be written as a linear combination of the elements $u_i v_j$ with coefficients e_{ji} from $\mathbb{E}$. This means the mn elements $u_i v_j$ *span* $\mathbb{G}$ over $\mathbb{E}$.

Also, these elements are linearly independent over $\mathbb{E}$. If we have a linear combination of the elements $u_i v_j$ with coefficients e_{ji} from $\mathbb{E}$ that is equal to 0 then, retracing our steps, we see that

$$0 = (e_{11} u_1 + \cdots + e_{1m} u_m) v_1 + \cdots + (e_{n1} u_1 + \cdots + e_{nm} u_m) v_n.$$

---

[2] An example that makes a good exercise is the following. Adjoining $\sqrt{2}$ and $\sqrt{3}$ to $\mathbb{Q}$ is equivalent to adjoining the single number $\sqrt{2} + \sqrt{3}$.

Since $v_1, \ldots, v_n$ are linearly independent over $\mathbb{F}$, each of the coefficients

$$(e_{11}u_1 + \cdots + e_{1m}u_m), \quad \ldots, \quad (e_{n1}u_1 + \cdots + e_{nm}u_m)$$

must be zero. But then, since $u_1, \ldots, u_m$ are linearly independent over $\mathbb{E}$, each of the coefficients e_{ji} must be zero.

Thus the mn elements $u_i v_j$ form a basis for $\mathbb{G}$ over $\mathbb{E}$, so $\mathbb{G}$ has dimension mn over $\mathbb{E}$. $\qquad\qquad\square$

The theorems above are useful because an algebraic number α is often best studied by immersing it in the field $\mathbb{Q}(\alpha)$. For example, there is an ancient question in geometry ("duplication of the cube") that amounts to asking: can $\sqrt[3]{2}$ be obtained from rational numbers by means of square roots? In the next chapter we will show that the answer is NO by comparing the dimension of the field $\mathbb{Q}(\sqrt[3]{2})$ with the dimensions of fields obtained from $\mathbb{Q}$ by adjoining square roots. The latter dimensions are easily obtained from the Dedekind product theorem.

4.10 Historical Remarks

Like geometry and number theory, algebra has been known for thousands of years. Quadratic equations were solved by the Babylonians about 4000 years ago, and systems of linear equations in several unknowns were solved by the Chinese about 2000 years ago. However, algebra was slow to reach the level of generality and abstractness attained by geometry and number theory in Euclid's *Elements*. Perhaps, this was due to the split in Greek mathematics between number theory and geometry. Number theory dealt with whole numbers and geometry dealt with the other magnitudes, including irrational lengths, which were not thought to be capable of multiplication in the way that numbers were. Perhaps, too, it was because suitable notation was lacking. What today would be expressed by a one-line equation, might be expressed by the Greeks as a page of prose, making it hard to conceive the very idea of equations, let alone to compute with them fluently.

Whatever the reason, algebra did not really flourish until the sixteenth century, when Italian mathematicians discovered how to solve cubic and quartic (3rd and 4th degree) equations. The methods of solution were published in the *Ars Magna* (Great Art) of Cardano (1545), and for a moment it looked as though anything was possible:

> Since this art surpasses all human subtlety and the perspicuity of mortal talent and is a truly celestial gift and a very clear test of the capacity of men's minds, whoever applies himself to it will believe that there is nothing he cannot understand.
>
> Cardano (1545), p. 8

As it turned out, the Italian art of solving equations had almost reached its limit with the solution of the quartic, and with Bombelli's discovery of the algebra of complex numbers. However, this was enough to keep mathematicians busy for the next 300 years. Algebra was recruited to help out in geometry by Fermat and Descartes in the 1620s, and from there it spread to calculus in the 1660s (Newton, and later Leibniz, the Bernoullis, and Euler in the eighteenth century). The mere existence of written symbolic computation made all this possible, without much conceptual input from algebra.

Calculus in fact surpassed algebra, to the extent that it was able to solve the main algebraic problem of the time: proving the fundamental theorem of algebra. The solution was not what had been hoped for in the sixteenth century—a formula giving the roots of any equation in terms of the coefficients—but a new kind of proof: a *proof of existence*. We say more about this theorem in the next section, and we give a proof in section 10.3, but the idea of an existence proof can be illustrated with the cubic equation.

Any cubic polynomial, say $x^3 - x - 1$, corresponds to a cubic curve, in this case $y = x^3 - x - 1$, whose graph for real values is shown in figure 4.5. This picture confirms something that can also be seen algebraically, that the polynomial function $x^3 - x - 1$ has large positive values for large positive x and large negative values for large negative x. We can also see something that is more subtle to explain algebraically, that the curve is *continuous*, and hence that *it meets the x axis somewhere*, at an x value that is a root of the equation $x^3 - x - 1 = 0$.

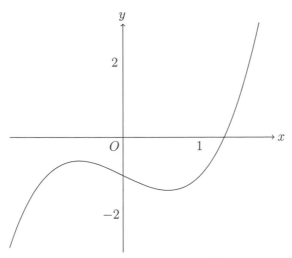

Figure 4.5: Graph of the curve $y = x^3 - x - 1$.

Thus the fundamental theorem of algebra turns out to involve a concept from *outside* algebra:[3] the concept of *continuous function*. As we will see in chapter 6, continuity is a fundamental concept of calculus, though its importance was not properly understood until the nineteenth century. For many reasons, continuity is an advanced concept, despite its apparent simplicity in cases like the graph of $y = x^3 - x - 1$.

One might hope, perhaps, for a proof of the fundamental theorem of algebra that does not involve the continuity concept. However, there is another reason why the theorem falls short of what the Italians were hoping for. As explained in section 4.1, the Italians sought solution "by radicals." They found them for equations of degree ≤ 4, but they were bound to fail for degree 5, as Abel (1826) and Galois (1831) showed, since the general equation of degree 5 does not have a solution by radicals. As mentioned in section 4.1, Galois introduced the concepts of field and group to explain why solution by radicals fails. We know from section 4.3 what a field is, so why do I not complete the story of solution by radicals by explaining what a group is? Unfortunately, there

---

[3] It is worth mentioning that some of the *motivation* for the fundamental theorem also came from outside algebra; namely, from the calculus problem of *integrating rational functions*. To solve this problem it is important to be able to factorize polynomials, and it turns out to be sufficient to obtain real linear and quadratic factors, which the fundamental theorem supplies.

is more to it than just defining the group concept.

1. The group concept is more remote from elementary mathematical experience than the concepts of ring and field. They encapsulate familiar calculations with numbers, while the group concept encapsulates the concept of "symmetry," where calculations are not immediately evident.
2. Moreover, it is not enough to know the concept of group. One also needs to develop a considerable amount of group *theory* before the group concept can be applied to solution by radicals. Most of group theory is remote from elementary mathematics because the "multiplication" of group elements is generally not commutative.

For further discussion why fields can be considered elementary, but groups probably cannot, see the next section.

The results of Galois were not understood by his contemporaries, but in the second half of the nineteenth century it became clear to a few mathematicians (particular the number theorists Richard Dedekind and Leopold Kronecker) that the abstract concepts of ring, field, and group were needed to properly understand the behavior of numbers and equations. As Dedekind said, in 1877:

> It is preferable, as in the modern theory of functions, to seek proofs based immediately on fundamental characteristics, rather than on calculation, and indeed to construct the theory in such a way that it is able to predict the results of calculation.
>
> Dedekind (1877), p. 37

In the 1920s, Emmy Noether became the first algebraist to fully embrace the abstract point of view, though she modestly used to say that "es steht schon bei Dedekind" (it's already in Dedekind). Her viewpoint was picked up by her students Emil Artin and B. L. van der Waerden, the latter of whom brought it into the mathematical mainstream with his book *Moderne Algebra* of 1930.

Today, "abstract algebra" is a standard topic in the undergraduate curriculum, generally at the upper level. But does abstraction belong at the elementary level? I believe so, but just how much abstraction belongs there is a delicate matter. The next section explores this question further.

4.11 Philosophical Remarks

Irrational and Imaginary Numbers

The construction of the algebraic number field $\mathbb{Q}(\alpha)$ gives a concrete, constructive way to deal with numbers such as $\sqrt{2}$, $\sqrt{-1}$, and $\sqrt{-2}$, whose "arithmetic" we studied in chapter 2. As we claimed in section 2.10, we do not have to view them as infinite decimals, or points in the plane, but merely as symbols that obey certain rules. Those rules are the field axioms and the defining equation (minimal polynomial equals zero) of the number in question. For example, we can take $\sqrt{-1}$ to be a symbol z that satisfies the field axioms and $z^2 = -1$. It follows that the calculations with the numbers of the forms $a + b\sqrt{-1}$ and $a + b\sqrt{-2}$ for rational a and b by no means assume a theory of all real and complex numbers. One need not believe in the infinite decimal for $\sqrt{2}$, for example, in order to justify the use of the symbol $\sqrt{2}$ in section 2.8 to find solutions of the Pell equation $x^2 - 2y^2 = 1$.

More generally, if we want to use any algebraic irrational number α, with minimal polynomial $p(x)$, we can work instead with polynomials in x and their congruence classes mod $p(x)$. This is what Kronecker (1887) called "general arithmetic," and why he is rumored to have said "irrational numbers do not exist." If he said it, he did not mean it literally, but rather that algebraic irrational numbers can be treated symbolically just like rational numbers when we use them in calculations with actual rational numbers. In effect, the algebraic number field construction "rationalizes" calculations with algebraic numbers.

However, not all numbers are algebraic, as we will see in section 9.8. If we wish to deal with numbers such as π or e, and particularly with the *totality* of real numbers, we cannot avoid using concepts that go beyond arithmetic. The need for the totality of real numbers—the number *line*—will become clear in chapter 6, and we will see how to "realize" the line in chapter 9.

* *The Fundamental Theorem of Algebra*

As mentioned in the previous section, the search for solutions of polynomial equations changed direction with the so-called fundamental theorem of algebra. According to this theorem, every polynomial equation with real coefficients has a root in the set $\mathbb{C}$ of complex numbers. We also showed, in section 4.8, that every polynomial equation $p(x) = 0$, where p is irreducible with rational coefficients, has a root in the field of polynomials modulo $p(x)$.

The latter result, due to Kronecker (1887) might be called the "algebraist's fundamental theorem of algebra," because it shows that an algebraically given polynomial equation has a root in an algebraically defined field. Some mathematicians prefer the algebraist's fundamental theorem to the classical version, since $\mathbb{R}$ and $\mathbb{C}$ are not obtainable from $\mathbb{Q}$ "algebraically": their construction involves infinite processes typical of analysis rather than algebra (see chapter 6). On the other hand, the construction of the number field of an irreducible $p(x)$ in section 4.8 is truly algebraic and in a certain sense it gives a "solution" of $p(x) = 0$— as the congruence class $[x]$ in the field.

A few mathematicians also prefer the algebraist's version because it is more *constructive*. Roughly speaking, a proof is called constructive if it provides an explicit construction of all objects claimed to exist. The construction can be infinite but, if so, only *potentially* infinite in the sense of section 1.10. That is, we allow an object to be constructed step by step, with each part of it being obtained at some finite step. The field of polynomials modulo $p(x)$ is "constructed" in this sense, because one can constructively list the rational numbers, and with them the polynomials with rational coefficients, and so on.

A proof is *non*constructive, or a "pure existence" proof, if it provides no construction of the object proved to exist. Typically, this happens when the proof involves objects dependent on an *actual* infinity, because such objects themselves cannot be constructed in step-by-step fashion. The sets $\mathbb{R}$ and $\mathbb{C}$ are actual infinities, as we will see in chapter 9, so the classical fundamental theorem of algebra is nonconstructive. One need not object to nonconstructive proofs (and I don't) to take the point that there is something problematic about them. A proof that depends on $\mathbb{R}$ is almost certainly advanced in some sense. We will see in chapter 6 that there is a whole family of theorems about continuous functions, including the fundamental theorem of algebra, that depend in a very similar way on properties of $\mathbb{R}$. These theorems mark out a substantial section of the boundary between elementary and advanced mathematics.

The constructive approach to the fundamental theorem of algebra was initiated by Kronecker (1887). He proposed replacing it with what he called the "fundamental theorem of general arithmetic," a special case of which is indeed the "algebraist's fundamental theorem of algebra." In a letter, Kronecker (1886) spoke in a dogmatic style that became popular with later constructivists:

> My treatment of algebra ... is in all subjects in which one makes use of common divisors the only one possible. The place of the so-called fundamental theorem of algebra, which is not applicable in such subjects, is taken by my new "fundamental theorem of general arithmetic."

For more on Kronecker's view of the fundamental theorem of algebra, see Edwards (2007).

*Group Theory

A group is defined to be a structure satisfying the following axioms, which should look rather familiar.

$$a(bc) = (ab)c \qquad \text{(associativity)}$$

$$a \cdot 1 = a \qquad \text{(identity)}$$

$$a \cdot a^{-1} = 1 \qquad \text{(inverse)}$$

Of course, we have seen these axioms as properties of the nonzero elements of a field but, in the absence of the other field axioms, they have vastly greater scope. In particular, the *identity element* **1** need not be the number 1.

The *group operation*, which combines the group elements a and b into the element written ab here, could in fact be addition on $\mathbb{Z}$, in which case the combination of a and b would be written $a+b$, the inverse of a would be written $-a$, and the identity element is the number 0. The associativity axiom in this case is $a+(b+c) = (a+b)+c$, which we know to be true in $\mathbb{Z}$. Thus we can say that $\mathbb{Z}$ is a group under the $+$ operation. The same is true of any vector space under the operation of vector sum (with the zero vector as the identity element).

Another example, in which we can revert to the "product" notation, is the collection of invertible functions on any set, under the operation of function composition ("taking function of a function"). If f, g, h, ... are functions, we take fg to be the function defined by $fg(x) = f(g(x))$, and it is easy to see that

$$f(gh)(x) = (fg)h(x) = f(g(h(x))),$$

so this "product" of functions is associative. The identity element $\mathbf{1}$ is the *identity* function defined by $\mathbf{1}(x) = x$, from which it easily follows that $f \cdot \mathbf{1} = f$. Finally, the inverse f^{-1} of function f is its *inverse function*, defined by the condition that

$$f(x) = y \text{ if and only if } f^{-1}(y) = x.$$

From this it easily follows that $f \cdot f^{-1} = \mathbf{1}$.

It might be thought that the concept of group is simpler than that of ring or field, because it requires fewer axioms, but in fact the opposite is true. The example of the group of invertible functions should already ring some alarm bells: how often have you taken function of a function of a function? Often enough to realize that it is associative? There are underlying mathematical reasons for the difficulty of the group concept, but first let us consider how groups emerged historically.

In the previous section I claimed that the group concept is more remote from elementary mathematics than the concepts of ring or field. The reason is that representative rings and fields, namely $\mathbb{Z}$ and $\mathbb{Q}$, have been familiar since ancient times, and the ring and field axioms describe fundamental properties of $\mathbb{Z}$ and $\mathbb{Q}$. The fact that the same properties hold for certain other structures is all to the good, because it means we can use familiar methods of calculation in other places.

The situation is quite different with the group concept. Before the concept was identified by Galois (and perhaps glimpsed by Lagrange a generation earlier), the only familiar groups were quite atypical, with a *commutative* group operation, such as $\mathbb{Z}$ under addition. The most important groups, such as those arising from general polynomial equations, are not commutative. So the group axioms had to *omit* a statement of commutativity, and mathematicians had to get used to noncommutative multiplication. Before the time of Galois, there was little experience with such computation, and indeed we now know that it is inherently difficult: we can prove it!

The reason that noncommutative multiplication is difficult is that it is quite close to the operations of a Turing machine. If the elements of a group are denoted by a, b, c, d, e, ... then their products are "words," such as cat and dog. Noncommutativity means that the order of letters generally cannot be changed, so that a "word" maintains its integrity, in some sense, though it can be disturbed somewhat by the insertion of adjacent inverse letters. For example,

$$cat = cabb^{-1}t.$$

Because of this, it can be shown that the operations of a Turing machine can be simulated in a group by a finite set of equations between words, allowing replacement of certain strings of symbols by other strings. The idea is to encode the initial configuration of the machine—consisting of the input, position of the reading head, and initial state—by a certain word, and to obtain the encodings of subsequent configurations by replacing substrings. It takes great ingenuity to avoid too much disturbance by inverse letters, and it was a great achievement when P. S. Novikov (1955) first succeeded.[4]

However, once it is known that it can be done, it follows from the unsolvability of the halting problem that various problems about computation in groups are also unsolvable. In particular, one can give a finite set of equations between words, for which the problem of deciding whether a given word equals 1 is unsolvable. Since this "word problem" is almost the simplest problem one can imagine about noncommutative multiplication, I take it that groups are officially hard, and hence not as elementary as rings and fields. (The corresponding problem becomes solvable when letters are allowed to commute.)

---

[4] Without the inverse letters (in which case one has what is called a *semi*group) the encoding of configurations is not so difficult, and we carry it out in section 10.2. This shows quite simply and directly why noncommutative multiplication leads to unsolvability.

5

◠

Geometry

PREVIEW

Geometry was the first mathematical discipline to be developed in detail, and a large swath of elementary geometry is already in Euclid's *Elements*. The early sections of this chapter sample some of the highlights of Euclid's approach to geometry, which is an attractive blend of visualization and logic. In fact, for many people it is still the most compelling example of mathematical reasoning.

Euclid's geometry is literally "hands on," because it uses hand tools: the *straightedge* and the *compass*. These tools determine the subject matter of Euclid's geometry—straight lines and circles—but this subject matter embraces the measurement of length, area, and angles (and the sometimes unexpected connections between them, such as the Pythagorean theorem).

More surprisingly, Euclid's geometry has a rich *algebraic* content. This was unknown to Euclid, because the Greeks avoided numbers in geometry, so algebra was not even contemplated. It came to light only around 1630, when Fermat and Descartes introduced numbers and equations. The middle of this chapter describes the relationship, discovered by Descartes, between straightedge and compass constructions and *constructible numbers*—those obtainable from rational numbers by the operations of $+$, $-$, $\cdot$, $\div$, and $\sqrt{}$.

Finally, we discuss Euclid's modern incarnation: the *Euclidean geometry* of vector spaces with an *inner product*. Vector spaces capture the *linearity* of Euclid's geometry, traceable back to the straightedge, while the inner product captures the concept of length compatible with the Pythagorean theorem. Perhaps Euclid's geometry is so durable because it fits into both the ancient and modern worlds. Its ancient basis in hand tools—straightedge and compass—is exactly equivalent to its modern basis in vector spaces and inner products.

5.1 Numbers and Geometry

In modern mathematics, elementary geometry is *arithmetized* by re-ducing it to linear algebra over the real numbers. We will see how this is done in sections 5.7 and 5.8. Advanced geometry is arithmetized, too, by means of algebraic or differentiable functions of real or complex numbers. Yet somehow it is still easier for most humans to approach geometry visually, so visual objects re-emerge from the primordial soup of numbers and we find ways to simulate visual operations by opera-tions on numbers. We represent points as ordered pairs (x, y) of real numbers, lines by linear equations $ax + by + c = 0$, circles by quadratic equations $(x - a)^2 + (y - b)^2 = r^2$; we find their intersection points by solving pairs of equations; we move objects by linear transformations of the set $\mathbb{R}^2$ of pairs (x, y); and so on. In this way we can model Euclid's geometry by higher-order concepts in the world of numbers.

This process has been recapitulated, in the last few decades, in the evolution of computers. Numbers are of course the native language of computers and, in their early decades of existence, computers were used mainly for processing numerical data. Then, in the 1970s, we saw the first crude printouts of images, typically made by assembling large arrays of symbols in a pattern printed on a long strip of paper. In the 1980s we started to program images, pixel by pixel, to be shown on computer screens. At the resolution then available (320×200) the results were usually hideous, as can be confirmed by looking at certain mathematics books of the period. When graphics commands were added to programming languages, it was a serious problem to program the best-looking "straight line" between two points. Any sloping line looked more or less like a staircase.

But resolution gradually improved and, by the 1990s, lines looked straight, curves looked smooth, and most pictures could be reproduced quite faithfully. This opened the floodgates of visual computing, leading

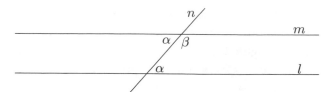

Figure 5.1: Angles involved in the parallel axiom.

to the world today in which users (particularly of phones and tablets) interact mainly with pictures on their screens: tapping them, translating and rotating them, magnifying them, and flipping them over by direct touch. It seems as though we *want* computers to be geometric, even if they have to be built on numbers, so programmers have to develop high-level programming concepts to model geometric operations.

In the present chapter we will study the interplay between numbers and geometry. There are reasons for doing geometry *without* a general concept of number, as we will see, and indeed Euclid did remarkable things without it. In the next two sections we review some of the big successes of his geometry: the theory of angles and the theory of area. But on balance geometry is easier with a suitable concept of number, and in the rest of the chapter we develop geometry on that basis.

5.2 Euclid's Theory of Angles

We are used to measuring angles by numbers: initially by degrees and at a higher level by radians, which involve the specific number π. But Euclid got by with the *right angle* as the unit of angle measure, and the concept of angle *equality*. His main conceptual tools for determining equality of angles were *congruence of triangles* and the *parallel axiom*. Indeed, the parallel axiom is the characteristic feature of Euclidean geometry, so we begin there. For convenience we use a slight variant of the parallel axiom, but it is equivalent to Euclid's.

Parallel Axiom. If lines l and m are parallel (that is, do not meet), and line n crosses them, then the angles α and β made by n with l and m, respectively, on the same side of n, have sum $\alpha + \beta$ equal to two right angles (figure 5.1).

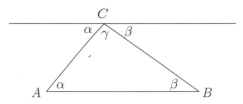

Figure 5.2: Angle sum of a triangle.

It follows that the angle adjacent to β is also α, since its sum with β is a straight angle, which equals two right angles. This immediately gives:

Angle sum of a triangle. *The angles α, β, γ of any triangle have sum $\alpha + \beta + \gamma$ equal to two right angles.*

Proof. Given a triangle ABC, with angles α, β, γ at the vertices A, B, C, respectively, consider the parallel to AB through the vertex C (figure 5.2).

Then it follows from the parallel axiom that angles α and β are adjacent to angle γ at C. Together, the angles at C below the parallel make a straight angle, so

$$\alpha + \beta + \gamma = \text{two right angles.} \qquad \square$$

Now we come to the use of congruence in establishing equality of angles. The relevant congruence axiom is the one today called SSS (for "side side side"): *if triangles ABC and $A'B'C'$ have corresponding sides equal then their corresponding angles are equal.* A famous consequence of SSS is a theorem about *isosceles* triangles (triangles with two equal sides).

Isosceles triangle theorem. *If ABC is a triangle with $AB = AC$, then the angle at B equals the angle at C.*

Proof. A stunning proof of this theorem (different from Euclid's) was given by Pappus, a Greek mathematician who lived some centuries later than Euclid.

Pappus noticed that triangle ABC is congruent to triangle ACB (yes, the same triangle, but "flipped," as it were) by SSS, because the

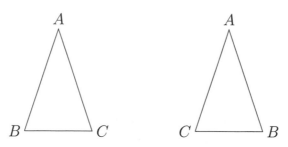

Figure 5.3: An isosceles triangle is congruent to itself.

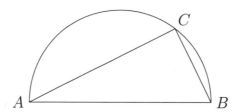

Figure 5.4: Angle in a semicircle.

corresponding sides

$$AB \text{ and } AC, \quad AC \text{ and } AB, \quad BC \text{ and } CB$$

are equal (see figure 5.3).

So the corresponding angles, at B and C, are also equal.   □

Now we combine these two theorems about angles to obtain:

Angle in a semicircle. *If AB is the diameter of a circle, and C is another point on the circle, then the angle at C in triangle ABC is a right angle (figure 5.4).*

Proof. Let O be the center of the circle, and draw the line segment OC. This creates two isosceles triangles, OAC and OBC, because $OA = OB = OC$ (as radii of the circle). Then it follows from the isosceles triangle theorem that we have equal angles as shown in figure 5.5.

Also, since the angle sum of the triangle ABC is two right angles, we have

$$\alpha + (\alpha + \beta) + \beta = \text{two right angles.}$$

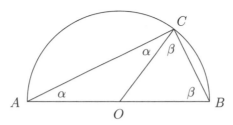

Figure 5.5: Isosceles triangles in a semicircle.

Therefore,

$$\alpha + \beta = \text{angle at } C = \text{right angle}. \qquad \square$$

5.3 Euclid's Theory of Area

According to legend, around 500 BCE the Pythagoreans discovered the irrationality of $\sqrt{2}$, or in their terms the *incommensurability* of the side and diagonal of the square. That is, there is no common unit of measure, *u*, such that the side and diagonal of the square are both integer multiples of *u*. Thus, the concept of positive integer—the *only* number concept the Pythagoreans had—is *inadequate* for describing all the lengths occurring in geometry. This momentous discovery led to a distinctive feature of Greek mathematics: the concept of *length* takes the place of number, but the arithmetic of lengths is quite limited. Lengths can be added and subtracted, but not multiplied. This means that area and volume have to be treated as separate kinds of magnitude, with even less capacity for addition and subtraction. In particular, the Greek concept of *equality* for areas and volumes is quite complicated.

Nevertheless, the Greeks were able to develop all the theory of area and volume needed for elementary mathematics. Here is how the theory of area gets off the ground.

Given lengths *a* and *b*, we modern mathematicians would like to say that the "product of *a* and *b*" is the area of the rectangle with adjacent sides *a* and *b*, which we may call the *rectangle of a* and *b*.

But what is area? It is certainly not a length, so from the Greek point of view there is no such thing as the area of the rectangle, other than the rectangle *itself* (figure 5.6). However there is a notion of *equality* for

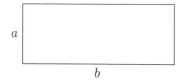

a

b

Figure 5.6: The rectangle of *a* and *b*.

rectangles which turns out to agree with the modern notion of equality of area, not only for rectangles but for all polygons. This notion of equality is based on the following principles, which are called "common notions" in Book I of Euclid's *Elements*.

1. Things which are equal to the same thing are also equal to each other.
2. If equals be added to equals, the wholes are equal.
3. If equals be subtracted from equals, the remainders are equal.
4. Things which coincide with one another are equal to one another.
5. The whole is greater than the part.

Notions 1, 2, 3, and 4 say essentially that objects are "equal" if one can be converted to the other by finitely often adding or subtracting identical ("coinciding") pieces. We already used these notions to prove the Pythagorean theorem in section 1.4. Notion 5 is not needed at this stage, because it turns out that Notions 1, 2, 3, and 4 suffice to prove that any polygons of equal area (in the modern sense) are "equal" in Euclid's sense.

The fact that "equal in area" is the same as "equal in Euclid's sense" (for polygons) is an elementary theorem. It was discovered in the nineteenth century by F. Bolyai and Gerwien, and is rather tedious to prove. Here we are content to show how the idea applies to the most important areas in elementary geometry—those of parallelograms and triangles. From now on we will say "equal in area" rather than "equal in Euclid's sense."

First, we see that the rectangle R of a and b is equal in area to any parallelogram P of height a and base b, because we can convert R to P by adding a triangle T to one side of R and then subtracting an identical triangle from the other side of the resulting shape (figure 5.7).

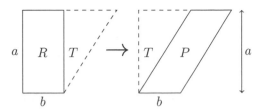

Figure 5.7: Converting a rectangle to a parallelogram.

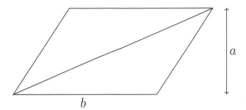

Figure 5.8: Making a parallelogram from two copies of a triangle.

Thus any parallelogram is equal in area to a rectangle with the same base and height. It follows that a triangle with height a and base b is equal in area to the rectangle with base b and half the height, $a/2$, because two copies of the triangle make a parallelogram of base b and height a (figure 5.8).

Since rectangles of equal area are actually "equal" in Euclid's sense we can *define* the product ab of lengths a and b to be the rectangle of a and b, and call products "equal" if the corresponding rectangles are "equal" in Euclid's sense. Then if $ab = cd$, we can *prove* this fact by converting the rectangle of c and d to the rectangle of a and b by adding and subtracting identical pieces. Thus, $ab = cd$ is actually *equivalent* to "equality" (in Euclid's sense) of the rectangle of a and b with the rectangle of c and d.

Even without knowing this general fact (which depends on the theorem of Bolyai and Gerwein) the Greeks could have proved "equality" of some interesting irrational rectangles, though it is not known whether they ever did so.

Geometric Proof that $\sqrt{2} \cdot \sqrt{3} = \sqrt{6}$

One of the supposed advantages of the theory of real numbers, introduced by Dedekind (1872), is that it permits rigorous proofs of

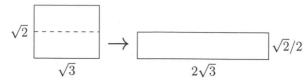

Figure 5.9: Converting the $\sqrt{2}$ by $\sqrt{3}$ rectangle to half height.

Figure 5.10: Converting rectangle to parallelogram.

results such as $\sqrt{2} \cdot \sqrt{3} = \sqrt{6}$. Indeed, Dedekind speculated that, with his definition of real number,

> we arrive at real proofs of theorems (as, e.g., $\sqrt{2} \cdot \sqrt{3} = \sqrt{6}$) that have never been established before.
>
> Dedekind (1901), p. 22

It may be true that $\sqrt{2} \cdot \sqrt{3} = \sqrt{6}$ had never really been proved before, but there was nothing to stop the Greeks from proving an equivalent proposition. Gardiner (2002), pp. 181–183 has pointed out that we can interpret and prove this equation quite easily in terms the Greeks would have accepted. Here is one such proof (a variation on Gardiner's), which is surely elementary.

By the Pythagorean theorem, $\sqrt{2}$ is the diagonal of the unit square and $\sqrt{3}$ is a side of the right-angled triangle with hypotenuse 2 and other side 1. Thus we can interpret $\sqrt{2}$ and $\sqrt{3}$ as lengths, and it is meaningful to speak of the rectangle of $\sqrt{2}$ and $\sqrt{3}$. We now transform this rectangle, by adding and subtracting equal pieces, into a rectangle of base 1.

First, cut the rectangle in two equal parts by a line parallel to the base $\sqrt{3}$, and then combine these two parts into a rectangle of height $\sqrt{2}/2$ and base $2\sqrt{3}$ (figure 5.9).

These rectangles are "equal" in Euclid's sense.

Next, we convert the new rectangle into a parallelogram by cutting off a right-angled triangle with equal sides $\sqrt{2}/2$ (and hence hypotenuse 1 by the Pythagorean theorem) from the left end and attaching it to the right (figure 5.10).

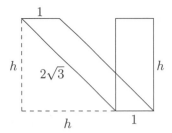

Figure 5.11: Converting parallelogram to rectangle with base 1.

The result is a parallelogram of the same width but sloping sides (at angle 45°) of length 1, and of course the same area again.

Finally, viewing the side 1 as the base of the parallelogram, we convert it to a rectangle with the same base and height, and hence the same area, by adding and subtracting a triangle (figure 5.11). It remains to find the height h of the rectangle.

We notice from figure 5.11 that h is the side of the isosceles right-angled triangle with hypotenuse $2\sqrt{3}$. Hence, by the Pythagorean theorem again,

$$h^2 + h^2 = (2\sqrt{3})^2 = 4 \times 3 = 12, \quad \text{so} \quad h^2 = 6.$$

This completes the proof that the rectangle with sides $\sqrt{2}$ and $\sqrt{3}$ is equal in area to the rectangle with sides $\sqrt{6}$ and 1. Hence (in modern terms) $\sqrt{2} \cdot \sqrt{3} = \sqrt{6}$.

The Concept of Volume

Euclid had a theory of volume that begins in the same way as his theory of area. The basic solid object is the *box* of a, b, and c; that is, with height a and base the rectangle of b and c, and all faces rectangular. "Equality" is defined by adding and subtracting identical pieces, and one finds that a *parallelepiped*[1] (a "squashed box" whose faces are parallelograms) is "equal" to the box with the same height, width, and depth (figure 5.12).

Next, by cutting a parallelepiped in half one finds that a *prism* is "equal" to the box with the same base but half the height. But that is

---

[1] This word is often misunderstood, misspelled, and mispronounced. It helps to read its parts as parallel-epi-ped, literally "parallel upon the foot," meaning that the top is parallel to the bottom, no matter which way is up.

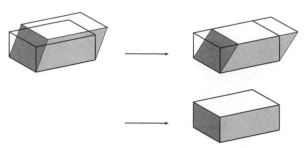

Figure 5.12: Transforming a squashed box to a box by adding and subtracting.

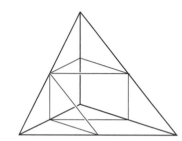

Figure 5.13: Two prisms in the tetrahedron.

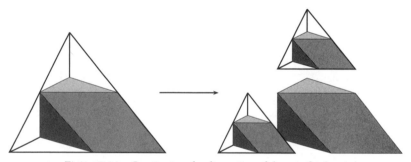

Figure 5.14: Continuing the dissection of the tetrahedron.

as far as one can go using finite numbers of pieces. In the case of the *tetrahedron*, Euclid was able to find an "equal" box only by cutting the tetrahedron into an *infinite* number of prisms. Figure 5.13 shows the first two prisms inside the tetrahedron, and figure 5.14 shows how to obtain further prisms, inside the tetrahedra that remain after the first two prisms are removed.

Much later, Dehn (1900) showed that the infinity is unavoidable: it is not possible to convert a regular tetrahedron to a cube by adding and subtracting finitely many pieces. This surprising discovery shows that volume is a deeper concept than area. If we suppose that volume belongs to elementary mathematics (as we presumably do), then we must accept that infinite processes have a place in elementary mathematics too. We pursue this train of thought further in the next chapter.

5.4 Straightedge and Compass Constructions

Many of the propositions in Euclid's *Elements* are *constructions*; that is, claims that certain figures in the plane can be constructed by drawing straight lines and circles. These constructions rest upon the ability to do two things that Euclid includes in his "postulates" (which we call axioms):

1. To draw a straight line from a given point to any other given point.
2. To draw a circle with given center and radius.

(Euclid actually splits the first axiom into two—drawing the line *segment* between given points and extending the segment arbitrarily far—because he does not admit infinite lines.) We now call the instruments for drawing lines and circles the "straightedge" and "compass," respectively,[2] so Euclid's constructions are called *straightedge and compass constructions*.

Euclid's very first proposition is a straightedge and compass construction: *drawing an equilateral triangle whose base is a given line segment AB.* It nicely illustrates how the compass is not only an instrument for drawing circles, but also for "storing" and "copying" lengths, because it allows a length to be carried from one place to another.

The third vertex C of the triangle is found at the intersection of the circles with centers A and B, respectively, and radius AB (figure 5.15). The triangle is then completed by drawing the line segments from A to C and from B to C.

---

[2] The straightedge is often called a ruler, but this wrongly suggests that there are marks on its edge that could be used for measurement. The compass used to be called a "pair of compasses." This term is falling into disuse, but I mention it because you may still see it in older books.

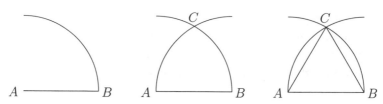

Figure 5.15: Constructing the equilateral triangle.

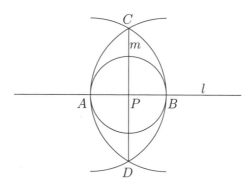

Figure 5.16: Constructing a perpendicular.

To prove that the construction is correct we observe that

$$AC = AB \quad \text{because both are radii of the first circle,}$$
$$BC = AB \quad \text{because both are radii of the second circle,}$$
$$\text{so} \quad AB = AC = BC,$$

because "things equal to the same thing are equal to one another."

Many other constructions in elementary geometry are variations on this one.

For example, the other intersection D of the two circles above gives a line CD *perpendicular* to AB and passing through its midpoint P. Conversely, by drawing a circle centered on an arbitrary point P on a line l we can make P the midpoint of a segment AB, and hence construct a line m perpendicular to l through P (figure 5.16).

Then, by constructing a perpendicular to m, we get a *parallel n* to the original line l. Parallels are a fundamental concept of Euclidean geometry, and they are the key to many constructions, because of the following theorem, which refers to figure 5.17.

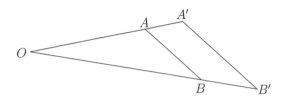

Figure 5.17: Setting of Thales' theorem.

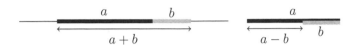

Figure 5.18: Forming the sum and difference of lengths.

Theorem of Thales. *If OAB and $OA'B'$ are triangles such that OAA' and OBB' are straight lines and AB is parallel to $A'B'$, then*

$$OA/OA' = OB/OB'.$$

Euclid's proof of this theorem is quite subtle, because his avoidance of numbers makes it hard to interpret the ratios OA/OA' and OB/OB'. When we accept that lengths are numbers we can argue much more simply (see section 5.7). This paves the way for straightedge and compass constructions that simulate all the arithmetic operations $+, -, \cdot, \div$ on lengths, as well as the $\sqrt{}$ operation. We carry out the most interesting parts of this program in the next section. But first we dispose of the $+$ and $-$ operations, which are simple even from Euclid's point of view and do not involve the theorem of Thales.

Given line segments a and b, we form $a + b$ by first extending the line segment a sufficiently far, then copying the line segment b with the compass and placing it alongside a in the same line. Similarly, if b is no greater than a, we form $a - b$ by carrying b to the interior of a with one end of b at the end of a (figure 5.18).

5.5 Geometric Realization of Algebraic Operations

As we saw at the end of the previous section, it is easy to obtain the sum and difference of lengths by constructions within the line. Product and quotient are also easy to obtain, but the construction makes use of the

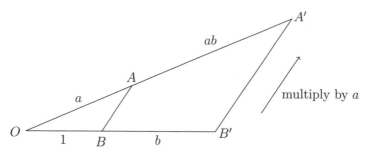

Figure 5.19: Constructing the product of lengths.

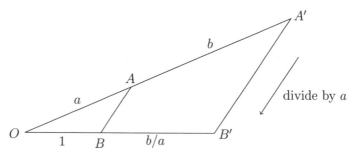

Figure 5.20: Constructing the quotient of lengths.

plane, and parallel lines. We also need to fix a *unit length*. But then, if we are given a length a, we can multiply any length b by a using the construction shown in figure 5.19.

The construction uses two lines through O. On one of them we mark the unit length $1 = OB$; on the other, the length $a = OA$. Then on the first line we add the length $b = BB'$. Finally, we draw AB and its parallel $A'B'$ through B'. By the theorem of Thales it follows that $AA' = ab$.

In effect, the parallels achieve "magnification by a," mapping the segments of length 1, b onto segments of length a, ab. The reverse direction gives "division by a," and hence we can obtain b/a for any lengths a, b as shown in figure 5.20.

Thus we now have geometric realizations of the operations $+, -, \cdot, \div$. To realize the operation $\sqrt{}$ is also surprisingly easy, but first we need the following consequence of the theorem of Thales.

Proportionality of Similar Triangles. *If ABC and $A'B'C'$ are triangles whose corresponding angles are equal (that is, angle at A equals angle*

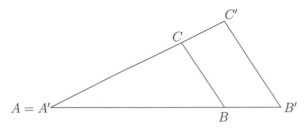

Figure 5.21: Similar triangles.

at A′ and so on) then their corresponding sides are proportional. That is

$$\frac{AB}{A'B'} = \frac{BC}{B'C'} = \frac{CA}{C'A'}.$$

Proof. Move the triangle $A'B'C'$ to the position where $A = A'$, A, B, B' are in the same line, and A, C, C' are in the same line. We assume A, C, C' are in that order, in which case we have the situation shown in figure 5.21. (The argument is similar if the order is A, C', C.)

Since the angles at C and C' are equal, the lines BC and $B'C'$ are parallel. Then it follows from the theorem of Thales that the upper and lower sides are divided proportionally; that is, $AB/A'B' = AC/A'C'$. Similarly, by moving the triangles so that B coincides with B', we find that $BC/B'C' = BA/B'A' = AB/A'B'$. So in fact all pairs of corresponding sides are proportional. □

Now to construct $\sqrt{l}$ for any length l we make the construction shown in figure 5.22. This is Euclid's construction in his Book VI, Proposition 13.

As we saw in section 5.2, the angles α and β at A and B, respectively, have sum $\alpha + \beta$ equal to a right angle. It follows, since the angle sum of each triangle is two right angles, that the angles at C are also α and β as shown. Thus triangles ADC and CDB are similar. Comparing corresponding sides we obtain the equal ratios

$$1/h = h/l,$$

from which it follows that

$$h^2 = l, \quad \text{and therefore} \quad h = \sqrt{l}.$$

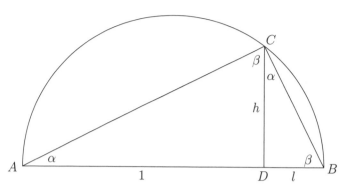

Figure 5.22: Constructing the square root of a length.

Moreover, we can *construct* $\sqrt{l}$ in the following steps.

1. Add l to 1 in the same line to form the line segment AB.
2. Bisect AB to find the center of the circle with diameter AB.
3. Draw this circle.
4. Construct the perpendicular to AB at the point D where the segments 1 and l meet.
5. Find the intersection C of this perpendicular with the circle of diameter AB and draw CD.

5.6 Algebraic Realization of Geometric Constructions

In the previous section we saw that straightedge and compass constructions include the algebraic operations of $+$, $-$, $\cdot$, $\div$, and $\sqrt{}$. In this section we will show that, when straightedge and compass operations are interpreted in the plane $\mathbb{R}^2$, *all* constructible lengths are obtainable from the unit length by the above operations. Thus we have an equivalence between geometric and algebraic concepts, and with it the opportunity to prove existence or nonexistence of geometric constructions by algebra. We will give an example of existence at the end of this section, and an example of nonexistence in section 5.9.

We begin by using coordinates to interpret the two basic constructions: drawing the line through two given points, and drawing the circle with given center and radius.

- Given points (a_1, b_1) and (a_2, b_2), the line between has slope $\frac{b_2-b_1}{a_2-a_1}$, so if (x, y) is any point on the line we have

$$\frac{y - b_1}{x - a_1} = \frac{b_2 - b_1}{a_2 - a_1}.$$

If we rewrite this equation in the standard form for the equation of a line, $ax + by + c = 0$, it is clear that a, b, and c come from the given coordinates a_1, b_1, a_2, b_2 by the operations $+$, $-$, $\cdot$, $\div$.

- Given a point (a, b) and radius r, the circle with center (a, b) and radius r has equation

$$(x - a)^2 + (y - b)^2 = r^2.$$

So, again, the coefficients are obtained from the given data by $+$, $-$, $\cdot$, $\div$.

New points are constructed as the intersections of previously constructed lines and circles. We now show that the coordinates of all such points arise from the coefficients of the lines and circles by the operations $+$, $-$, $\cdot$, $\div$, and $\sqrt{}$.

- Given lines $a_1x + b_1y + c_1 = 0$ and $a_2x + b_2y + c_2 = 0$, their intersection is found by solving for x and y. This can always be done with the operations $+$, $-$, $\cdot$, $\div$ alone, hence the coordinates of the intersection are expressible in terms of a_1, b_1, c_1, a_2, b_2, c_2 by the operations $+$, $-$, $\cdot$, $\div$.

- Given a line $a_1x + b_1y + c_1 = 0$ and a circle $(x - a_2)^2 + (y - b_2)^2 = r^2$, we can find their intersection by solving the first equation for y (if $b_1 \neq 0$; else solve for x instead) and substituting the resulting expression for y in the second equation. This gives a quadratic equation for x whose coefficients arise from a_1, b_1, c_1, a_2, b_2, r by the operations $+$, $-$, $\cdot$, $\div$. The quadratic formula solves this equation for x in terms of the coefficients and the operations $+$, $-$, $\cdot$, $\div$, and $\sqrt{}$. (This is where the $\sqrt{}$ operation becomes necessary.) Finally, the

y coordinates of the intersection come from the solutions for x and the first equation, again using $+, -, \cdot, \div$.

- Given two circles $(x - a_1)^2 + (y - b_1)^2 = r_1^2$ and $(x - a_2)^2 + (y - b_2)^2 = r_2^2$, we expand these two equations as

$$x^2 - 2a_1 x + a_1^2 + y^2 - 2b_1 y + b_1^2 = r_1^2,$$
$$x^2 - 2a_2 x + a_2^2 + y^2 - 2b_2 y + b_2^2 = r_2^2.$$

Subtraction cancels the x^2 and y^2 terms, leaving a linear equation whose common solutions with either of the circle equations can be found as in the previous case.

To sum up: the coordinates of all new points, and hence the coefficients in the equations of all new lines and circles, are obtainable from the coordinates of the initially given points by the operations $+, -, \cdot, \div$, and $\sqrt{\ }$. Therefore, the coordinates of all points constructible by straightedge and compass are obtainable from 1 by these operations. Conversely, all points whose coordinates are numbers obtainable from 1 by the operations $+, -, \cdot, \div$, and $\sqrt{\ }$ (which we may call *constructible numbers*) are constructible by straightedge and compass.

Now here is the promised example: using algebra to prove constructibility.

The Regular Pentagon

Figure 5.23 shows the regular pentagon with sides of length 1 and a diagonal whose length x is sought. The other two diagonals are drawn to create some similar triangles. The symmetry of the regular pentagon ensures that each diagonal is parallel to the opposite side, and the various parallels imply that the triangles ABC, $A'BC$, and $A'B'C'$ have the equal angles α shown, and hence also equal angles β.

Thus $A'BC$ is similar to ABC and, since they have the common side BC, they are actually *congruent*. This gives the equal sides of length 1 shown in figure 5.24, hence also the sides of length $x - 1$, since all diagonals have length x.

Then similarity of triangles ABC and $A'B'C'$ gives

$$\frac{1}{x} = \frac{x - 1}{1}$$

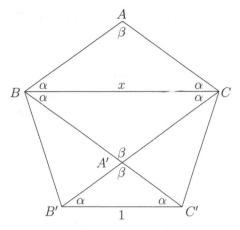

Figure 5.23: Angles in the regular pentagon.

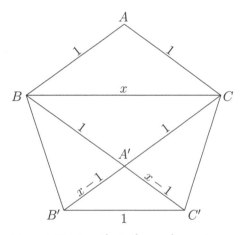

Figure 5.24: Lengths in the regular pentagon.

from the proportionality of corresponding sides. This is the quadratic equation $x^2 - x - 1 = 0$, which has the solutions $x = \frac{1 \pm \sqrt{5}}{2}$. The positive member of this pair is the length of the pentagon diagonal. It is a constructible number, so we can construct the diagonal by straightedge and compass. It is then easy to construct the triangle ABC, and to join a series of such triangles together to form the regular pentagon.

5.7 Vector Space Geometry

In this section we view the plane $\mathbb{R}^2$ purely as a vector space over $\mathbb{R}$ and we ask: how much of Euclidean geometry can we recover? Certainly not everything, because the concept of length is not available. However, there is a concept of "relative length" which allows us to compare the lengths of line segments with the same direction. This enables us to prove some important theorems that involve ratios of lengths, such as the theorem of Thales.

Before we state a vector version of the theorem of Thales, we need to express some key concepts in terms of vectors in $\mathbb{R}^2$. As in traditional geometry, we will denote a line segment by writing its endpoints side by side. Thus st will denote the line segment with endpoints s and t.

Triangles. Without loss of generality we can take one vertex of the triangle to be **0**, and the other two to be nonzero vectors u and v. But also, for the triangle to be nondegenerate, u and v must lie in different directions from **0**. That is, they must be *linearly independent*: $au + bv = 0$ only if $a = b = 0$.

Parallels. If s and t are points in $\mathbb{R}^2$, the vector $t - s$ represents the position of t relative to s; in particular, it gives the direction of the line segment st. If u and v are another two points, then uv is *parallel* to st if

$$t - s = c(v - u) \quad \text{for some real } c \neq 0.$$

Relative length. For any nonzero real number a, the segments from **0** to au and from **0** to u are in the same direction, and we call a the *length* of the segment from **0** to au *relative to the length* of the segment from **0** to u.

With these concepts expressed in the language of vectors, the proof is now quite easy.

Vector Thales theorem. *If* **0**, *u, v are the vertices of a triangle, s lies on* **0**u, *t lies on* **0**v, *and if st is parallel to* uv, *then the length of* **0**s *relative to the length of* **0**u *equals the length of* **0**t *relative to the length of* **0**v.

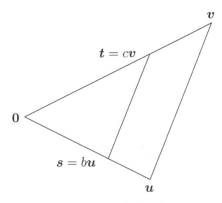

Figure 5.25: Setup for the theorem.

Proof. Figure 5.25 summarizes the hypotheses of the theorem. Since s lies on $\mathbf{0u}$ we have

$$s = bu \quad \text{for some real } b \neq 0,$$

and similarly

$$t = cv \quad \text{for some real } c \neq 0.$$

Also, since st is parallel to uv we have

$$t - s = cv - bu = a(v - u) \quad \text{for some real } a \neq 0,$$

hence

$$(b - a)u + (a - c)v = \mathbf{0}.$$

The linear independence of u and v then implies that

$$b - a = a - c = 0, \quad \text{so} \quad a = b = c.$$

Thus, in fact, $s = au$ and $t = av$. So s and t divide the segments $\mathbf{0u}$ and $\mathbf{0v}$ in the same ratio, namely a. $\qquad\square$

Remark. It is even easier to prove the converse theorem: *if s and t divide the segments $\mathbf{0u}$ and $\mathbf{0v}$ in the same ratio, a, then st is parallel to uv.* Because in this case $s = au$ and $t = av$, so $t - s = a(v - u)$, which says st is parallel to uv.

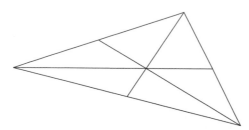

Figure 5.26: The centroid as the common point of the medians.

Other Theorems of Vector Geometry

Among the points on the segment from $\mathbf{0}$ to $\boldsymbol{u}$ is its *midpoint* $\frac{1}{2}\boldsymbol{u}$. More generally, the midpoint of the segment from $\boldsymbol{u}$ to $\boldsymbol{v}$ is $\frac{1}{2}(\boldsymbol{u}+\boldsymbol{v})$ because

$$\frac{1}{2}(\boldsymbol{u}+\boldsymbol{v}) = \boldsymbol{u} + \frac{1}{2}(\boldsymbol{v}-\boldsymbol{u}).$$

("Go to $\boldsymbol{u}$, then halfway towards $\boldsymbol{v}$.") Midpoints feature in several classical theorems of geometry. Here is one, normally proved using congruence, but which has a very simple proof using vector concepts.

Diagonals of a parallelogram. *The diagonals bisect each other.*

Proof. Without loss of generality, we can take three vertices of the parallelogram to be $\mathbf{0}$, $\boldsymbol{u}$, $\boldsymbol{v}$. Then the fourth vertex is $\boldsymbol{u}+\boldsymbol{v}$, because this point makes opposite sides parallel.

The diagonal from $\mathbf{0}$ to $\boldsymbol{u}+\boldsymbol{v}$ has midpoint $\frac{1}{2}(\boldsymbol{u}+\boldsymbol{v})$, which is also the midpoint of the diagonal from $\boldsymbol{u}$ to $\boldsymbol{v}$. Thus the two diagonals bisect other, since their midpoints coincide. □

Our second example is a theorem of Archimedes, who interpreted it as a theorem about the *center of mass* of a triangle (also known as the *barycenter* or *centroid* of the triangle; see figure 5.26).

Concurrence of medians. *In any triangle, the lines from each vertex to the midpoint of the opposite side (the* medians *of the triangle) have a common point.*

Proof. Let the triangle have vertices $\boldsymbol{u}$, $\boldsymbol{v}$, $\boldsymbol{w}$, so one median is the line segment from $\boldsymbol{u}$ to $\frac{1}{2}(\boldsymbol{v}+\boldsymbol{w})$. Making a guess (inspired perhaps by figure 5.26), we consider the point 2/3 of the way from $\boldsymbol{u}$ to $\frac{1}{2}(\boldsymbol{v}+\boldsymbol{w})$,

$$\boldsymbol{u}+\frac{2}{3}\left[\frac{1}{2}(\boldsymbol{v}+\boldsymbol{w})-\boldsymbol{u}\right],$$

which equals

$$\boldsymbol{u}+\frac{1}{3}(\boldsymbol{v}+\boldsymbol{w}-2\boldsymbol{u})=\frac{1}{3}(\boldsymbol{u}+\boldsymbol{v}+\boldsymbol{w}).$$

. This point involves the symbols $\boldsymbol{u}$, $\boldsymbol{v}$, $\boldsymbol{w}$ symmetrically, so we will get the *same* point if we travel 2/3 of the way along any other median, either from $\boldsymbol{v}$ to $\frac{1}{2}(\boldsymbol{u}+\boldsymbol{w})$ or from $\boldsymbol{w}$ to $\frac{1}{2}(\boldsymbol{u}+\boldsymbol{v})$. Thus the point $\frac{1}{3}(\boldsymbol{u}+\boldsymbol{v}+\boldsymbol{w})$ is on all three medians. □

5.8 Introducing Length via the Inner Product

We define the *inner product* $\boldsymbol{u}\cdot\boldsymbol{v}$ of vectors $\boldsymbol{u}=(a,b)$ and $\boldsymbol{v}=(c,d)$ in $\mathbb{R}^2$ by

$$\boldsymbol{u}\cdot\boldsymbol{v}=ac+bd.$$

It follows in particular that

$$\boldsymbol{u}\cdot\boldsymbol{u}=a^2+b^2,$$

which is the square of the "hypotenuse" $|\boldsymbol{u}|$ of the right-angled triangle whose width a and height b are the coordinates of $\boldsymbol{u}$. So it is natural to take $|\boldsymbol{u}|$ as the *length* of $\boldsymbol{u}$.

More generally, it is natural to define the *distance* from $\boldsymbol{u}_1=(x_1,y_1)$ to $\boldsymbol{u}_2=(x_2,y_2)$ as

$$|\boldsymbol{u}_2-\boldsymbol{u}_1|=\sqrt{(x_2-x_1)^2+(y_2-y_1)^2},$$

because this is the distance given by the Pythagorean theorem for the triangle with width x_2-x_1 and height y_2-y_1 (figure 5.27).

Thus the inner product gives us the Euclidean concept of length if we define the length $|\boldsymbol{u}|$ of $\boldsymbol{u}$ by

$$|\boldsymbol{u}|=\sqrt{\boldsymbol{u}\cdot\boldsymbol{u}}.$$

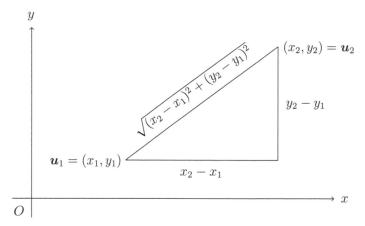

Figure 5.27: Distance given by the Pythagorean theorem.

For this reason, the vector space $\mathbb{R}^2$ with the inner product is called the *Euclidean plane*.[3]

The concept of length determines the concept of *angle*, in principle, because the angles of a triangle are determined by the lengths of its sides. However, the concept of angle is not completely simple—in the next chapter we will find that calculus is needed to answer some basic questions about it—so, like Euclid, we single out the *right angle* as the most important angle.

Vectors $\boldsymbol{u}$ and $\boldsymbol{v}$ of equal length make a right angle, or are perpendicular, precisely when $\boldsymbol{u} = (a, b)$ and $\boldsymbol{v} = (-b, a)$, as is clear from figure 5.28. From this we see that any *vectors $\boldsymbol{u}$ and $\boldsymbol{v}$ are perpendicular if and only if*

$$\boldsymbol{u} \cdot \boldsymbol{v} = 0,$$

because replacing either vector by a scalar multiple does not change the value 0.

This criterion makes it possible to prove many theorems about perpendicularity by simple algebraic calculations. These calculations involve the following rules for the inner product, which are easily

---

[3] It is straightforward to define *Euclidean space* of any dimension n by making the natural extension of the inner product to $\mathbb{R}^n$: $(a_1, a_2, \ldots, a_n) \cdot (b_1, b_2, \ldots, b_n) = a_1 b_1 + a_2 b_2 + \cdots + a_n b_n$. There is also a natural extension of the Pythagorean theorem to $\mathbb{R}^n$ which gives the same concept of length as this inner product.

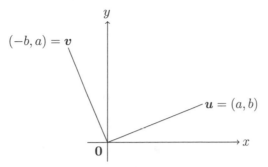

Figure 5.28: Perpendicular vectors of equal length.

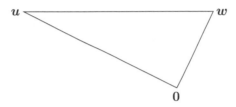

Figure 5.29: Right-angled triangle.

verified from its definition. (Notice, again, that the rules look like rules of ordinary arithmetic. So one can calculate almost without thinking, even though the · symbol now applies to vectors, not numbers.)

$$u \cdot v = v \cdot u,$$

$$u \cdot (v + w) = u \cdot v + u \cdot w,$$

$$(au) \cdot v = u \cdot (av) = a(u \cdot v).$$

First, let us verify that the Pythagorean theorem holds in complete generality in the Euclidean plane.

Vector Pythagorean theorem. *If u, v, w are the vertices of a right-angled triangle, with the right angle at v, then*

$$|v - u|^2 + |w - v|^2 = |u - w|^2.$$

Proof. Without loss of generality we can assume that $v = 0$, in which case $u \cdot w = 0$ by perpendicularity, and we wish to show (figure 5.29) that

$$|u|^2 + |w|^2 = |u - w|^2.$$

Indeed we have

$$|u - w|^2 = (u - w) \cdot (u - w)$$
$$= u \cdot u + w \cdot w - 2u \cdot w$$
$$= |u|^2 + |w|^2 \quad \text{because} \quad u \cdot w = 0. \qquad \square$$

The theorem about the angle in a semicircle, from section 5.2, is another that follows by an easy calculation.

Angle in a semicircle. *If u and −u are the ends of the diameter of a circle with center* **0***, and if v is any point between them on the circle, then the directions from v to u and −u are perpendicular.*

Proof. Consider the inner product of the directions, $v - u$ and $v + u$, to v from u and $-u$, respectively.

$$(v - u) \cdot (v + u) = v \cdot v - u \cdot u = |v|^2 - |u|^2 = 0$$

because $|u|$, $|v|$ are both radii of the circle, hence equal.

Thus $v - u$ and $v + u$ are perpendicular. $\qquad \square$

Finally, here is a theorem we have not yet proved, because it is hard to prove by traditional geometric methods. Yet it is very easy with the inner product.

Concurrence of altitudes. *The perpendiculars from each vertex of a triangle to the opposite side (its* altitudes*) have a common point.*

Proof. Let the triangle have vertices u, v, w and choose the origin at the intersection of the altitudes from u and v (figure 5.30).

Then the direction of u from $\mathbf{0}$ is perpendicular to the direction $w - v$ of the opposite side, so

$$u \cdot (w - v) = 0, \quad \text{that is,} \quad u \cdot v = u \cdot w.$$

Similarly, the altitude from v gives the equation

$$v \cdot (w - u) = 0, \quad \text{that is,} \quad u \cdot v = v \cdot w.$$

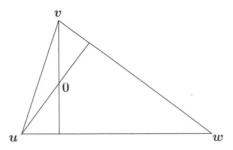

Figure 5.30: Altitudes of a triangle.

Subtracting the second equation from the first gives

$$0 = u \cdot w - v \cdot w = w \cdot (u - v).$$

The latter equation says that the line from w through 0 is perpendicular to the side from u to v. That is, 0 is also on the altitude from w. Thus 0 is the common point of all three altitudes. □

5.9 Constructible Number Fields

Vector spaces have another role to play in geometry: distinguishing between the numbers constructible by straightedge and compass and those that are not. The classical example is $\sqrt[3]{2}$. It arose in the ancient problem known as *duplicating the cube*, which asked for the construction of a cube twice the volume of a given cube. Taking the side of the given cube as 1, this problem amounts to constructing the number $\sqrt[3]{2}$. And since the allowable construction tools are straightedge and compass, this amounts, by section 5.6, to constructing the number $\sqrt[3]{2}$ from 1 by the operations of $+$, $-$, $\cdot$, $\div$, and $\sqrt{}$.

Given a number α constructed by these operations, it is helpful to look at the field $\mathbb{Q}(\alpha)$ obtained by adjoining α to $\mathbb{Q}$. As we have seen in section 4.9, $\mathbb{Q}(\alpha)$ is a vector space over $\mathbb{Q}$ whose dimension equals the *degree* of α. It turns out, by the Dedekind product theorem of section 4.9, that this dimension is severely limited for constructible numbers α: it must be a power of 2. An example will make it clear why this is so.

Consider the constructible number $\alpha = \sqrt{1 + \sqrt{3}}$. The field $\mathbb{Q}(\alpha)$ is conveniently built in two steps from $\mathbb{Q}$. The first step is to adjoin

$\sqrt{3}$ to $\mathbb{Q}$, obtaining $\mathbb{F} = \mathbb{Q}(\sqrt{3})$. Since $\sqrt{3}$ is irrational, $\mathbb{F}$ is of degree 2, hence dimension 2, over $\mathbb{Q}$. The second step is to adjoin to $\mathbb{F}$ the square root of its member $1 + \sqrt{3}$. The number $\alpha = \sqrt{1 + \sqrt{3}}$ is not in $\mathbb{F}$, so $\mathbb{F}(\alpha)$ is an extension of $\mathbb{F}$. The extension is of degree 2, since α satisfies the equation $x^2 - (1 + \sqrt{3}) = 0$ and $1 + \sqrt{3}$ is in $\mathbb{F}$. Altogether, we have $\mathbb{F}$ of dimension 2 over $\mathbb{Q}$ and $\mathbb{F}(\alpha) = \mathbb{Q}(\alpha)$ of dimension 2 over $\mathbb{F}$. So, by the Dedekind product theorem, $\mathbb{Q}(\alpha)$ is of dimension $2 \cdot 2 = 4$ over $\mathbb{Q}$.

If α is any constructible number, we can build $\mathbb{Q}(\alpha)$ similarly, by a series of extensions by square roots. This is a series of extensions of dimension 2 so, by the Dedekind product theorem, *the dimension of* $\mathbb{Q}(\alpha)$ *over* $\mathbb{Q}$, *for any constructible number* α, *is a power of* 2.

Now we can definitively show that $\sqrt[3]{2}$ is not a constructible number by establishing the following.

Degree of $\sqrt[3]{2}$. *The degree of $\sqrt[3]{2}$ is 3.*

Proof. The number $\sqrt[3]{2}$ satisfies the equation $x^3 - 2 = 0$, so it suffices to show that the polynomial $p(x) = x^3 - 2$ is irreducible over $\mathbb{Q}$. If not, then $p(x)$ has rational factors of lower degree, one of which must be linear. Without loss of generality we can assume that it is $x - m/n$, where m and n are integers.

But then $0 = p(m/n) = \frac{m^3}{n^3} - 2$ or, equivalently, $2n^3 = m^3$.

The last equation contradicts unique prime factorization, because the number of factors of 2 is different in the prime factorizations of the two sides. In m^3, the number of factors of 2 is a multiple of 3 (three times the number of 2s in the prime factorization of m), whereas in $2n^3$ the number of factors of 2 in the prime factorization is 1 *plus* a multiple of 3 (the visible factor 2 plus three times the number of 2s in the prime factorization of n). $\qquad\square$

Thus $\sqrt[3]{2}$ is not a constructible number, and hence the problem of duplicating the cube is not solvable by straightedge and compass. The negative solution of this ancient problem was first given by Wantzel (1837), more than 2000 years after the problem was first posed! At the same time, Wantzel also gave a negative solution of another ancient problem: *trisection of the angle.*

The trisection problem asks for a straightedge and compass construction that divides any angle into three equal parts.[4] If there were such a construction, then it would be possible to construct an angle $\pi/9$ by trisecting the angle $\pi/3$ occurring in the equilateral triangle (section 5.4).

But then we could construct the length $\cos\frac{\pi}{9}$ by constructing a right-angled triangle with angle $\pi/9$ and hypotenuse 1. It can be shown that $x = \cos\frac{\pi}{9}$ satisfies the equation

$$8x^3 - 6x - 1 = 0.$$

The cubic polynomial $8x^3 - 6x - 1$ is in fact irreducible, though this is not as easy to show as it was for $x^3 - 2$. Once this is done, however, we know that $\cos\frac{\pi}{9}$ is of degree 3, so it is not constructible, and therefore the trisection problem is not solvable by straightedge and compass.

5.10 Historical Remarks

Euclid's *Elements* is the most influential mathematics book of all time, and it gave mathematics a geometric slant that persisted until the twentieth century. Until quite late in the twentieth century, students were introduced to mathematical proof in the style of the *Elements*, and indeed it is hard to argue with a method that survived over 2000 years. However, we now know that Euclidean geometry has an algebraic description (vector space with an inner product), so we have another "eye" with which to view geometry, which we surely should use. To see how this new viewpoint came about, we review the history of elementary geometry.

Euclid's geometry, illustrated in sections 5.2 to 5.4, has delighted thinkers through the ages with its combination of visual intuition, logic, and surprise. We can visualize what Euclid is talking about (points, lines, areas), we are impressed that so many theorems follow from so few axioms, and we are *convinced* by the proofs—even when the conclusion is unexpected. Perhaps the biggest surprise is the

---

[4] There is a straightedge and compass construction that divides an arbitrary angle into *two* equal parts. It follows easily from the process for bisecting a line segment given in section 5.4.

Pythagorean theorem. Who expected that the side lengths of a triangle would be related via their squares?

The Pythagorean theorem reverberates through Euclid's geometry and all its descendants, changing from a theorem to a definition on the way (in a Euclidean space it holds virtually by definition of the inner product). For the ancient Greeks, the Pythagorean theorem led to $\sqrt{2}$ and hence to the unwelcome discovery of irrational lengths. As mentioned in section 5.3, the Greeks drew the conclusion that lengths in general are not numbers and that they cannot be multiplied like numbers. Instead, the "product" of two lengths is a rectangle, and equality of products must be established by cutting and reassembling areas. This was complicated, though interesting and successful for the theory of area. However, as mentioned in section 5.3, one needs to cut volumes into· *infinitely* many pieces for a satisfactory theory of volume.

It should be emphasized that the subject of the *Elements* is not just geometry, but also number theory (Books VII to IX on primes and divisibility) and an embryonic theory of real numbers (Book V), in which arbitrary lengths are compared by means of their rational approximations. Later advances in geometry, particularly Descartes (1637) and Hilbert (1899), work towards fusing all three subjects of the *Elements* into a unified whole.

The first major conceptual advance in geometry after Euclid came with the introduction of coordinates and algebra, by Fermat and Descartes in the 1620s. These two mathematicians seem to have arrived at the same idea independently, with very similar results. For example, they both discovered that the curves with equations of degree 2 are precisely the conic sections (ellipse, parabolas, and hyperbolas). Thus, they achieved not only a unification of geometry and algebra, but also a unification of Euclid's geometry with the *Conics* of Apollonius (written some decades after the *Elements*). Indeed, they set the stage for *algebraic geometry*, where curves of arbitrarily high degree could be considered. When calculus emerged, shortly thereafter, algebraic curves of degree 3 and higher provided test problems for the new techniques for finding tangents and areas.

However, algebra and calculus did not merely create new geometric objects; they also threw new light on Euclid's geometry. One of Descartes' first steps was to break the taboo on multiplying lengths, which he did with the similar triangle constructions for product, quotient, and square root of lengths described in sections 5.5 and 5.6. Thus, for all algebraic purposes, lengths could now be viewed as numbers. And by giving an algebraic description of numbers constructible by straightedge and compass, Descartes paved the way for the nineteenth-century proofs of *non*constructibility, such as the one given in section 5.9.

Descartes did not intend to construct a new foundation for geometry, with "points" being ordered pairs (x, y) of numbers, "lines" being point sets satisfying linear equations, and so on. He was content to take points and lines as Euclid described them, and mainly wished to solve geometric problems more simply with the help of algebra. (Sometimes, in fact, he tried to solve algebraic equations with the help of geometry.) The question of new foundations for geometry came up only when Euclid's geometry was challenged by *non-Euclidean* geometry in the 1820s. We say more about non-Euclidean geometry in the subsection below. The challenge of non-Euclidean geometry was not noticed at first, because the geometry was conjectural and not taken seriously by most mathematicians. This all changed when Beltrami (1868), building on some ideas of Gauss and Riemann, constructed *models* of non-Euclidean geometry, thereby showing that its axioms were just as consistent as those of Euclid.

Beltrami's discovery was an earthquake that displaced Euclidean geometry from its long-held position at the foundation of mathematics. It strengthened the case for *arithmetization*: the program of founding mathematics on the basis of arithmetic, including the real numbers, instead of geometry. Arithmetization was already under way in calculus and, thanks to Descartes, arithmetic was a ready-made foundation for Euclidean geometry. Beltrami's models completed the triumph of arithmetization in geometry, because they too were founded on the real numbers and calculus.

By and large, geometry remains arithmetized today, with both Euclidean and non-Euclidean "spaces" situated among a great variety of *manifolds* that locally resemble $\mathbb{R}^n$, but with "curvature." Among them, Euclidean geometry retains a somewhat privileged position as the one with zero curvature at all points. In this sense, Euclidean geometry is the simplest geometry, and the non-Euclidean spaces of Beltrami are among the next simplest, with constant (but negative) curvature at all points. In the modern geometry of curved spaces, Euclidean spaces have a special place as *tangent spaces*. A curved manifold has a tangent space at each point, and one often works in the tangent space to take advantage of its simpler structure (particularly, its nature as a vector space).

It is all very well to base geometry on the theory of real numbers, but how well do we understand the real numbers? What is *their* foundation? Hilbert (1899) raised this question, and he had an interesting answer: the real numbers can be based on geometry! More precisely, they can be based on a "completed" version of Euclid's geometry. Hilbert embarked on the project of completing Euclid's axioms in the early 1890s, first with the aim of filling in some missing steps in Euclid's proofs. As the project developed, he noticed that addition and multiplication arise from his axioms in a quite unexpected way, so that the field concept can be given a completely geometric foundation. See section 5.11. Then, by adding an axiom guaranteeing that the line has no gaps, he was able to recover a complete "number line" with the usual properties of $\mathbb{R}$.

*Non-Euclidean Geometry

In the 1820s, Janos Bolyai[5] and Nikolai Lobachevsky independently developed a rival geometry to Euclid's: a *non*-Euclidean geometry that satisfied all of Euclid's axioms except the parallel axiom. The parallel axiom has a different character from the other axioms, which describe the outcome of finite constructions or "experiments" one can imagine carrying out:

1. Given two points, draw the line segment between them.
2. Extend a line segment for any given distance.
3. Draw a circle with given center and radius.
4. Any two right angles are equal (that is, one can be moved to coincide with the other).

The parallel axiom, on the other hand, requires an experiment that involves an indefinite wait:

---

[5] Son of the Bolyai in the Bolyai-Gerwien theorem.

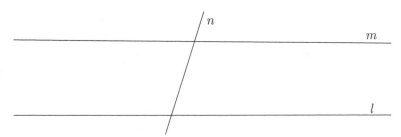

Figure 5.31: Two lines that are supposed to meet.

5. Given two lines *l* and *m*, crossed by another line *n* making interior angles with *l* and *m* together less than two right angles, the lines *l* and *m* will meet, *if produced indefinitely* (figure 5.31).

Since the time of Euclid, mathematicians have been unhappy with the parallel axiom, and have tried to prove it from the other, more constructive, axioms. The most determined attempt was made by Saccheri (1733), who got as far as showing that, if non-diverging lines *l* and *m* did *not* meet, they would have a common perpendicular at infinity. This, Saccheri thought, was "repugnant to the nature of straight lines." But it was not a contradiction, and in fact there is a geometry in which lines behave in precisely this fashion.

Bolyai and Lobachevsky worked out a large and coherent body of theorems that follow from Euclid's axioms 1 to 4 together with the axiom:

5'. There exist two lines *l* and *m*, which do not meet, although they are crossed by another line *n* making interior angles with *l* and *m* together less than two right angles.

Their results were eventually published in Lobachevsky (1829) and Bolyai (1832) (the latter an appendix to a book by Bolyai's father). They found no contradiction arising from this set of axioms, and Beltrami (1868) showed that no contradiction exists, because axioms 1, 2, 3, 4, and 5' are satisfied under a suitable interpretation of the words "point," "line," and "angle." (We sketch one such interpretation below.) Thus Euclid's geometry had a rival, and deciding how to interpret the terms "point," "line," "distance," and "angle" became an issue.

As we have seen, Euclid's axioms had a ready-made interpretation in the coordinate geometry of Descartes: a "point" is an ordered pair (x, y) of real numbers, a "line" consists of the points satisfying an equation $ax + by + c = 0$, and the "distance" between (x_1, y_1) and (x_2, y_2) equals $\sqrt{(x_2 - x_1)^2 + (y_2 - y_2)^2}$.

For Bolyai's and Lobachevsky's axioms, Beltrami found several elegant interpretations, admittedly with a somewhat complicated concept of "distance." The simplest is probably the *half-plane model*, in which:

- "points" are points of the upper half-plane; that is, pairs (x, y) with $y > 0$,

- "lines" are the open semicircles in the upper half-plane with their centers on the x-axis, and the open half-lines $\{(x, y) : x = a, y > 0\}$,

- the "distance" between "points" P and Q is the integral of $\sqrt{dx^2 + dy^2}/y$ over the "line" connecting P and Q.

It turns out that "angle" in this model is just the ordinary angle between curves; that is, the angle between their tangents. This leads to some beautiful pictures of non-Euclidean geometric configurations, such as the one in figure 5.32. It shows a tiling of the half-plane by triangles which are "congruent" in the sense of non-Euclidean distance. In particular, they each have angles $\pi/2$, $\pi/3$, $\pi/7$. Knowing that they are congruent one can get a sense of non-Euclidean distance. One can see that the x-axis is infinitely far away—which explains why it is not included in the model—and perhaps also see that a "line" is the shortest path joining its endpoints, if one estimates distance by counting the number of triangles along a path.

It is also clear that the parallel axiom fails in the model. Take for example the "line" that goes straight up the center of the picture and the point on the far left with seven "lines" passing through it. Several of the latter "lines" do not meet the center line.

Vector Space Geometry

Grassmann (1861), the *Lehrbuch der Arithmetik* mentioned in section 1.9, was not Grassmann's only great contribution to the

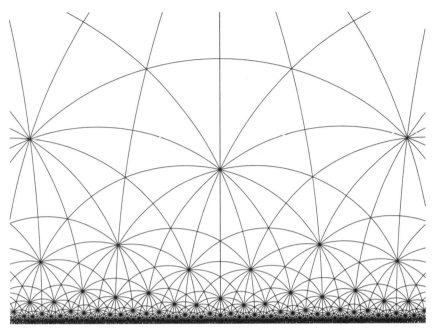

Figure 5.32: Congruent triangles in the half-plane model.

foundations of mathematics. The first was his *Ausdehnungslehre* ("extension theory"), Grassmann (1844), in which he based Euclidean geometry on vector spaces. The *Ausdehnungslehre*, like the unfortunate *Lehrbuch*, was greeted at first with total incomprehension. The only person to review it was Grassmann himself, and its virtually unsold first edition was destroyed by the publisher. The full story of the *Ausdehnungslehre*, its genesis and aftermath, is in the Grassmann biography by Petsche (2009).

Grassmann was let down by an extremely obscure style, and terminology of his own invention, in attempting to explain an utterly new and complex idea: that of a real n-dimensional vector space with an *outer product*.[6] The simpler concept of *inner* product was in Grassmann's view an offshoot of the outer product—one he planned to expound in *Ausdehnungslehre*, volume two. Not surprisingly, the second volume was abandoned after the failure of the first.

---

[6] We will not define the concept of outer product, but it underlies the concept of determinant, then at the center of what was called "determinant theory" and now at a less central position in today's linear algebra.

Thus, Grassmann's contribution to geometry might well have been lost—if not for a marvelous stroke of luck. In 1846, the Jablonowskian Society of Science in Leipzig offered an essay prize on a question only Grassmann was ready to answer: developing a sketchy idea of Leibniz about "symbolic geometry." (The aim of the prize was to commemorate the 200th anniversary of Leibniz's birth.) Grassmann (1847) duly won the prize with a revised version of his 1844 theory of vector spaces— one that put the inner product and its geometric interpretation at the center of the theory. He pointed out that his definition of inner product was motivated by the Pythagorean theorem, but that, once the definition is given, all geometric theorems follow from it by pure algebra.

Despite its greater clarity, Grassmann's essay was not an overnight success. However, his ideas gathered enough momentum to justify a new version of the *Ausdehnungslehre*, Grassmann (1862), and they were gradually adopted by other mathematicians. Peano was among the first to appreciate Grassmann's ideas, and was inspired by him to create the first axiom system for real vector spaces in Peano (1888), section 72. Klein (1909) brought Grassmann's geometry to a wider audience by restricting it to three dimensions. Klein mentioned the inner product, but his version of Grassmann relied mainly on the determinant concept, which gives convenient formulas for areas and volumes.

5.11 Philosophical Remarks

Non-Euclidean Geometry

In this book I have made the judgement that non-Euclidean geometry is more advanced than Euclidean. There is ample historical reason to support this call, since non-Euclidean geometry was discovered more than 2000 years after Euclid. The "points" and "lines" of non-Euclidean geometry can be modeled by Euclidean objects, so they are not advanced in themselves, but the concept of non-Euclidean distance surely is.

One way to see this is to map a portion of the non-Euclidean plane onto a piece of a surface S in $\mathbb{R}^3$ in such a way that distance is preserved. Then ask: how simple is S? Well, the simplest possible S is the

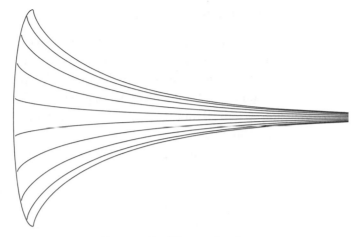

Figure 5.33: The pseudosphere.

trumpet-shaped surface shown in figure 5.33 and known as the *pseudosphere*. It is obtained by rotating the *tractrix* curve, with equation

$$x = \ln \frac{1 + \sqrt{1 - y^2}}{y} - \sqrt{1 - y^2},$$

about the x-axis.

This formula is complicated enough, but the conceptual complication is much greater. It is possible to compare only small pieces of the non-Euclidean plane with small pieces of a surface in $\mathbb{R}^3$, because a complete non-Euclidean plane does not "fit" smoothly in $\mathbb{R}^3$. This was proved by Hilbert (1901). The pseudosphere, for example, represents just a thin wedge of the non-Euclidean plane, the edges of which are two non-Euclidean lines that approach each other at infinity. These two edges are joined together to form the tapering tube shown in figure 5.33.

In contrast, Euclidean geometry is modeled by the simplest possible surface in $\mathbb{R}^3$—the plane![7]

---

[7] It may be thought unfair to the hyperbolic plane to force it into the Euclidean straightjacket of $\mathbb{R}^3$. Might not the Euclidean plane look equally bad if forced to live in non-Euclidean space? Actually, this is not the case. Beltrami showed that the Euclidean plane fits beautifully into non-Euclidean space, where it is a "sphere with center at infinity."

*Numbers and Geometry

Now let us return to the axioms of Hilbert (1899), and what they tell us about the relationship between numbers and geometry. Hilbert's axioms probably capture Euclid's concept of the line (with the finer structure explored in Book V of the *Elements*). So, given the Descartes "model" $\mathbb{R}^2$ of Euclid's axioms, Hilbert has shown Euclidean geometry to be essentially *equivalent* to the algebra of $\mathbb{R}$. However, algebraists and logicians now prefer not to use the full set $\mathbb{R}$ in geometry. They point out that the set of constructible numbers suffices, because Euclid's geometry "sees" only the points arising from straightedge and compass constructions. Thus one can get by with an algebraically defined set of points, which is only a "potential" infinity, in contrast to the "actual" infinity $\mathbb{R}$. Logicians also prefer the theory of constructible numbers because its "consistency strength" is less than that of the theory of $\mathbb{R}$.

That is, it is easier to prove the consistency of the theory of constructible numbers (and hence the consistency of Euclid's axioms) than it is to prove the consistency of the theory of $\mathbb{R}$ (and hence the consistency of Hilbert's axioms).

*Geometry and "Reverse Mathematics"

In recent decades, mathematical logicians have developed a field called *reverse mathematics*, whose motivation was stated by Friedman (1975) as follows:

> When the theorem is proved from the right axioms, the axioms can be proved from the theorem.

As logicians understand it, reverse mathematics is a technical field, concerned mainly with theorems about the real numbers (see section 9.9). However, if we understand reverse mathematics more broadly as the search for the "right axioms," then reverse mathematics began with Euclid.

He saw that the parallel axiom is the right axiom to prove the Pythagorean theorem, and perhaps the reverse—that the Pythagorean theorem proves the parallel axiom (given his other axioms). The same is true of many other theorems of Euclidean geometry, such as the theorem of Thales and the theorem that the angle sum of a triangle is π. All of these theorems are equivalent to the parallel axiom, so it is the "right axiom" to prove them.

To formalize this and other investigations in reverse mathematics we choose a *base theory* containing the most basic and obvious assumptions about some area of mathematics. It is to be expected that the base theory will fail to prove certain interesting but less obvious theorems. We then seek the "right" axiom or axioms to prove these theorems, judging an axiom to be "right" if it implies the theorem, and conversely, using only assumptions from the base theory.

Euclid began with a base theory now known as *neutral geometry*. It contains basic assumptions about points, lines, and congruence of triangles but *not* the parallel axiom. He proved as many theorems as he could before introducing the parallel axiom—only when it was needed to prove theorems about the area of parallelograms and ultimately the Pythagorean theorem. He also needed the parallel axiom to prove the theorem of Thales and that the angle sum of a triangle is π. We now know, conversely, that all of these theorems imply the parallel axiom in neutral geometry, so the latter is the "right" axiom to prove them.

Neutral geometry is also a base theory for *non*-Euclidean geometry, because the latter is obtained by adding to neutral geometry the "non-Euclidean parallel axiom" stating that there is more than one parallel to a given line through a given point.

Grassmann's theory of real vector spaces, as we have seen, can also be taken as a base theory for Euclidean geometry. It is quite different from the base theory of neutral geometry because the Euclidean parallel axiom *holds* in real vector spaces, and so does the theorem of Thales. Nevertheless, this new base theory is not strong enough to prove the Pythagorean theorem, or indeed to say anything about angles. Relative to the theory of real vector spaces, the "right" axiom to prove the Pythagorean theorem is existence of the inner product, because we can reverse the implication by using the Pythagorean theorem to define distance, hence angle and cosine, and then define the inner product by

$$\boldsymbol{u} \cdot \boldsymbol{v} = |\boldsymbol{u}| \cdot |\boldsymbol{v}| \cos \theta.$$

This raises the possibility of adding a different axiom to the theory of real vector spaces and obtaining a different kind of geometry, just as we obtain non-Euclidean geometry from neutral geometry by adding a different parallel axiom. Indeed we can, and simply by asserting the existence of a different kind of inner product. The inner product introduced by Grassmann is what we now call a *positive-definite* inner product, characterized by the property that $u \cdot u = 0$ only if u is the zero vector.

Non-positive-definite inner products also arise quite naturally. Probably the most famous is the one on the vector space $\mathbb{R}^4$ that defines the *Minkowski space* of Minkowski (1908). If we write the typical vector in $\mathbb{R}^4$ as $u = (w, x, y, z)$ then the Minkowski inner product is defined by

$$u_1 \cdot u_2 = -w_1 w_2 + x_1 x_2 + y_1 y_2 + z_1 z_2.$$

In particular, the length $|u|$ of a vector u in Minkowski space is given by

$$|u|^2 = u \cdot u = -w^2 + x^2 + y^2 + z^2,$$

so $|u|$ can certainly be zero when u is not the zero vector.

Minkowski space is famous as the geometric model of Einstein's *special relativity theory*. In this model, known as *flat spacetime*, x, y, and z are the coordinates of ordinary three-dimensional space and $w = ct$, where t is the time coordinate and c is the speed of light. As Minkowski (1908) said:

> The views of space and time which I wish to lay before you have sprung from the soil of experimental physics, and therein lies their strength. They are radical. Henceforth space by itself, and time by itself, are doomed to fade away into mere shadows, and only a kind of union of the two will preserve an independent reality.

Undoubtedly, relativity theory put non-positive-definite inner products on the map, making them as real and important as the ancient concept of distance. But in fact such inner products had already been considered by mathematicians, and one of them is involved in a model of non-Euclidean geometry discovered by Poincaré (1881)—the so-called *hyperboloid model*.

Figure 5.34: Triangle tessellation of the hyperboloid model.

To see where the hyperboloid comes from, consider the three-dimensional Minkowski space of vectors $\boldsymbol{u} = (w, x, y)$ with one time coordinate w and two space coordinates x, y. In this space, where $|\boldsymbol{u}|^2 = -w^2 + x^2 + y^2$, we consider the "sphere of imaginary radius"

$$\{\boldsymbol{u} : |\boldsymbol{u}| = \sqrt{-1}\} = \{(w, x, y) : -w^2 + x^2 + y^2 = -1\}.$$

This "sphere"[8] consists of the points (w, x, y) in $\mathbb{R}^3$ such that

$$w^2 - x^2 - y^2 = 1,$$

so it is actually a hyperboloid; namely, the surface obtained by rotating the hyperbola $w^2 - y^2 = 1$ in the (w, y)-plane about the w axis. If we take "distance" on either sheet of the hyperboloid to be the Minkowski distance, it turns out to be a model of the non-Euclidean plane. As with the other models of the non-Euclidean plane, calculating distance is a little troublesome so I omit the details. Instead, I offer figure 5.34—a black-and-white version of a picture due to Konrad Polthier of the Freie Universität, Berlin—which shows what the triangle tessellation of

---

[8] Long before the development of Minkowski space, or even the development of non-Euclidean geometry, Lambert (1766) speculated about the geometry of a "sphere of imaginary radius." In particular, he guessed that the angle sum of a triangle would be less than π on such a sphere—which is essentially the non-Euclidean parallel axiom.

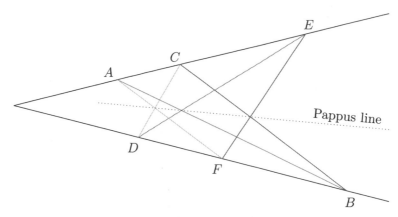

Figure 5.35: The Pappus configuration.

figure 5.32 looks like in the hyperboloid model. (It also relates the hyperboloid model to another model—the *conformal disk* model—which is a kind of intermediary between the hyperboloid model and the half-plane model shown in figure 5.32.)

This elegant relationship between Minkowski space and the non-Euclidean plane has been used for some textbook treatments of non-Euclidean geometry, such as Ryan (1986). Just as the positive-definite inner product is the "right axiom" to develop Euclidean geometry over the base theory of real vector spaces, the Minkowski inner product is the "right axiom" to develop non-Euclidean geometry.

*Projective Geometry

Another wonderfully "right" axiom is the theorem discovered by Pappus a few hundred years after Euclid. Pappus viewed his theorem as part of Euclidean geometry, but it does not really belong there. It is unlike typical Euclidean theorems in making no mention of length or angle, so its home should be a geometry that does not involve these concepts. The statement of the theorem is the following, which refers to the configuration shown in figure 5.35.

Theorem of Pappus. *If A, B, C, D, E, F are points of the plane lying alternately on two lines, then the intersections of the pairs of lines AB and DE, BC and EF, CD and FA, lie on a line.*

The Pappus theorem has a Euclidean *proof* using the concept of length, and also a coordinate proof using linear equations to define lines. But it seems to have no proof using only concepts "appropriate" to its statement: points, lines, and the membership of points in lines. The appropriate setting for the Pappus theorem is the *projective geometry* of the plane, a geometry which tries to capture the behavior of points and lines in a plane without regard to length or angle. In projective geometry, configurations of points and lines are considered the same if one can be projected onto the other. Projection can of course change lengths and angles, but the straightness of lines remains, as does the membership of points in lines.

If one seeks axioms for projective plane geometry, the following come easily to mind:

1. Any two points belong to a unique line.
2. Any two lines have a unique point in common.
3. There are four points, no three of which are in the same line.

The first axiom is also one of Euclid's. The second disagrees with Euclid in the case of parallel lines, but projective geometry demands it, because even parallel lines can be projected so that they meet—on the "horizon." The third axiom is there to ensure that we really have a "plane," and not merely a line. However, these simple axioms are very weak, and it can be shown that they do not suffice to prove the Pappus theorem. They do, however, form a natural base theory to which other axioms about points and lines can be added.

What are the "right axioms" to prove the Pappus theorem? The answer is no less than the Pappus theorem itself, thanks to what the Pappus theorem implies; namely, that the abstract plane of "points" and "lines" can be given *coordinates* which form a *field* as defined in section 4.3. Thus geometry springs fully armed from the Pappus theorem! The Pappus axiom (as we should now call it) is the right axiom to prove coordinatization by a field, because such a coordinatization allows us to prove the Pappus axiom. As remarked above, this follows by using linear equations to define lines. The field properties then enable us to find intersections of lines by solving linear equations, and to verify that points of intersection lie on a line.

The idea of reversing the coordinate approach to geometry began with von Staudt (1847), who used the Pappus axiom to define addition and multiplication of points on a line. Hilbert (1899) extended this idea to prove that the coordinates form a field, but he had to assume another projective axiom, the so-called *theorem of Desargues*, which was discovered around 1640. (Roughly speaking, the Pappus axiom easily implies that addition and multiplication are commutative, while the Desargues axiom easily implies that they are associative.) Quite remarkably, considering how long the Pappus and Desargues theorems had been around, Hessenberg (1905) discovered that the Pappus theorem implies the Desargues theorem. So the single Pappus axiom is in fact *equivalent* to coordinatization of the plane by a field.

This reversal of the Pappus theorem also tells us something remarkable about algebra: the nine field axioms follow from four geometric axioms—the three projective plane axioms plus Pappus!

6

⤳

Calculus

PREVIEW

Calculus has its origins in the work of Euclid and Archimedes, who used infinite sums to evaluate certain areas and volumes. We begin this chapter by studying the simplest kind of infinite sum, the *geometric series*, which reappears later as a kind of seed that generates other infinite series.

Calculus as we know it today—as a means of *calculating* the outcomes of infinite processes—began by calculating tangents to curves. We do this calculation for the curves of the form $y = x^n$, and use the results to solve the inverse problem of finding the areas beneath these curves. Finding tangents and areas are formalized by the calculus operations of *differentiation* and *integration*, and their inverse relationship is formalized by the *fundamental theorem of calculus*.

But while differentiation does not lead outside the class of known functions, integration often does. The logarithm and circular functions, for example, arise from rational functions by integration. From these, in combination with rational and algebraic functions, we obtain a large class called the *elementary functions*. All rational functions may be integrated in terms of elementary functions.

This result suggests that there might be an *elementary calculus* in which only elementary functions are studied. Unfortunately, this restricted calculus is still not completely "elementary." There remain some intuitively obvious facts that are hard to prove, such as the theorem that a function with zero derivative is a constant. Because of this, in calculus one must assume certain plausible results without proof, or else occasionally cross the line into advanced calculus.

In this chapter we have opted to occasionally cross the line, in order to make it clear where the line lies. The sections containing advanced arguments or assumptions have been marked with a star (*).

6.1 Geometric Series

Before any distance can be traversed half the distance must be traversed, and these half distances are infinite in number.

Aristotle, *Physics*, 263a5

The simplest and most natural geometric series arises from the situation described above (rather tersely) by Aristotle: going halfway to a destination. At the halfway point we still have to traverse half the remaining distance (which is one quarter of the whole), then half of that (one eighth), and so on. Thus the whole is the sum of the fractions

$$\frac{1}{2} + \frac{1}{4} + \frac{1}{8} + \frac{1}{16} + \cdots.$$

This sum, in which each term is half of the one before, is an example of an *infinite geometric series*. It is infinite because at every finite stage we are a little short of the destination.

However, it is also clear that we pass any position short of the destination at some finite stage, because the distance remaining becomes arbitrarily small. So the infinite sum cannot be less than the whole distance. In other words, we must have

$$\frac{1}{2} + \frac{1}{4} + \frac{1}{8} + \frac{1}{16} + \cdots = 1.$$

Infinite processes, such as this one, are at the heart of calculus, and we will not try to avoid them. However, the meaning of the infinite sum can be clarified by comparing it with the finite sum

$$S_n = \frac{1}{2} + \frac{1}{4} + \frac{1}{8} + \cdots + \frac{1}{2^n}.$$

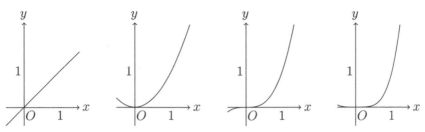

Figure 6.1: Graphs of $y = x$, $y = x^2$, $y = x^3$, and $y = x^4$.

Because of the restriction on r, these formulas are valid only for $|x| < 1$. For these values of x, we call these formulas *power series expansions of the functions* $\frac{1}{1+x}$ *and* $\frac{1}{1+x^2}$. It may seem idiotic to replace the simple finite expression $\frac{1}{1+x}$ by the infinite one $1 - x + x^2 - x^3 + \cdots$, but there are some things we can do more easily with the powers x, x^2, x^3, ... than we can with $\frac{1}{1+x}$. In fact, the aim of this chapter is to develop an understanding of *elementary functions* such as e^x, $\sin x$, and $\cos x$ through an understanding of powers of x.

This was the approach to calculus developed by Newton around 1665, and it remains the best elementary approach today. It is true that calculus is not an entirely elementary subject, and there are delicate questions concerning the meaning of infinite series in particular. However, for the part of calculus we deal with, these questions are generally easy to answer, or else the plausible answer can be justified with some extra work. We will point out such questions as we go along, but will not let them derail our train of thought.

6.2 Tangents and Differentiation

We begin our study of the powers of x by looking at their graphs $y = x^n$. Figure 6.1 shows the first few examples: $y = x$, $y = x^2$, $y = x^3$, and $y = x^4$. For each of these, a fundamental geometric question, whose answer is not obvious except for $y = x$, is: how do we find the tangent to the graph at a given point? This question is answered by first studying the simpler problem of finding the chord between two given points on the curve.

We illustrate the method by finding the tangent to the parabola $y = x^2$ at the point $P = (1, 1)$. To do this we consider the chord

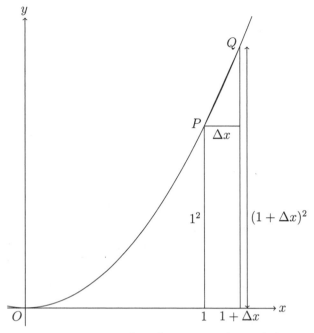

Figure 6.2: Approaching the tangent to the parabola.

between P and a neighboring point[1] $Q = (1 + \Delta x, (1 + \Delta x)^2)$ as shown in figure 6.2.

As Δx approaches 0, the point Q approaches P and the line through P and Q approaches the tangent. In fact, the slope of PQ approaches 2, as is clear when we calculate the slope in terms of Δx:

$$\text{slope of } P\,Q = \frac{\text{change in } y\text{-value}}{\text{change in } x\text{-value}}$$

$$= \frac{(1 + \Delta x)^2 - 1^2}{(1 + \Delta x) - 1}$$

$$= \frac{1^2 + 2\Delta x + (\Delta x)^2 - 1^2}{\Delta x}$$

---

[1] We denote the horizontal distance to the neighboring point by Δx, because it represents a *difference* (initial letter d or Δ in Greek) in the x direction. This symbolism may seem unnecessarily fancy in the present situation, but it fits later situations better, so it is worth getting used to it.

$$= \frac{2\Delta x + (\Delta x)^2}{\Delta x}$$

$$= 2 + \Delta x,$$

which certainly approaches 2 as Δx approaches 0.

A similar calculation finds the slope of the tangent at any point $P = (x, x^2)$ on $y = x^2$, by calculating the slope to the nearby point $Q = (x + \Delta x, (x + \Delta x)^2)$

$$\text{slope of } PQ = \frac{(x + \Delta x)^2 - x^2}{(x + \Delta x) - x}$$

$$= \frac{x^2 + 2x \cdot \Delta x + (\Delta x)^2 - x^2}{\Delta x} = \frac{2x \cdot \Delta x + (\Delta x)^2}{\Delta x}$$

$$= 2x + \Delta x,$$

which approaches $2x$ as Δx approaches 0.

The same method applies to $y = x^3$ and higher powers of x, though with longer calculations. For example, on $y = x^3$ the slope from the typical point $P = (x, x^2)$ to the nearby point $Q = (x + \Delta x, (x + \Delta x)^3)$ is

$$\frac{(x + \Delta x)^3 - x^3}{(x + \Delta x) - x} = \frac{x^3 + 3x^2 \cdot \Delta x + 3x \cdot (\Delta x)^2 + (\Delta x)^3 - x^3}{\Delta x}$$

$$= 3x^2 + 3x \cdot \Delta x + (\Delta x)^2.$$

As Δx approaches 0, so do the terms $3x \cdot \Delta x$ and $(\Delta x)^2$, hence the slope of PQ approaches $3x^2$.

Similar calculations show that

for $y = x^4$ the slope of the tangent at (x, x^4) equals $4x^3$,

for $y = x^5$ the slope of the tangent at (x, x^5) equals $5x^4$,

$$\cdots$$

for $y = x^n$ the slope of the tangent at (x, x^n) equals nx^{n-1}.

To do the calculation for an arbitrary positive integer n the most direct method is to use the binomial theorem of section 1.6, according to

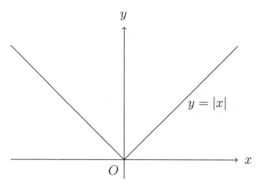

Figure 6.3: Graph of the function $f(x) = |x|$.

which

$$(x + \Delta x)^n = x^n + nx^{n-1} \cdot \Delta x + \frac{n(n-1)}{2} x^{n-2} \cdot (\Delta x)^2 + \cdots + (\Delta x)^n.$$

The function nx^{n-1} that gives the slope of the tangent to $y = x^n$ for any value of x is called the *derivative* of x^n, and so we have the theorem.

Derivative of x^n. *The derivative of x^n, for every positive integer n, is* nx^{n-1}. □

For the moment, these are the only derivatives that we need, though we will make a more general definition for future use.

Definition. A function f is called *differentiable* (for a certain domain of x values) if

$$\frac{f(x + \Delta x) - f(x)}{\Delta x}$$

approaches a definite value $f'(x)$ as Δx approaches 0, for each x in the domain of f. The function $f'(x)$ is called the *derivative* of $f(x)$.

In particular, we call f differentiable at $x = a$ if $f'(a)$ exists. It is quite common for a function to have no derivative at certain points. For example, the function $f(x) = |x|$ has no derivative at $x = 0$, because the slope from O to any point on the positive side of the graph is $+1$, while the slope from O to any point on the negative side is -1. Thus there is no common value for all slopes to approach (figure 6.3).

When $f'(x)$ exists we call it the *limit* of $\frac{f(x+\Delta x)-f(x)}{h}$ as Δx approaches 0, and we write this statement concisely as

$$f'(x) = \lim_{\Delta x \to 0} \frac{f(x+\Delta x) - f(x)}{\Delta x},$$

also introducing the symbol $\to$ for "approaches." It is possible to go further and define the concepts of "approaching" and "limit," but we will not do so here, since we are concerned mainly with cases where it is clear *which* value is being approached. For example, it is clear that $2 + \Delta x$ approaches 2 as Δx approaches 0—as clear as it was in the previous section that $1 - \frac{1}{2^n}$ approaches 1. In elementary calculus, limits are usually as clear as these, so they cannot be made clearer by a deeper explanation of the limit concept. That can wait for advanced calculus, though advanced calculus is not very distant from elementary calculus, as we will see.

As a further example in which the limit is clear, take the function $f(x) = \frac{1}{x}$. In this case

$$\frac{f(x+\Delta x) - f(x)}{\Delta x} = \frac{1}{\Delta x}\left(\frac{1}{x+\Delta x} - \frac{1}{x}\right)$$

$$= \frac{1}{\Delta x} \cdot \frac{x - (x+\Delta x)}{(x+\Delta x)x} = \frac{-1}{(x+\Delta x)x},$$

so

$$f'(x) = \lim_{\Delta x \to 0} \frac{-1}{(x+\Delta x)x} = \frac{-1}{x^2}.$$

This shows, incidentally, that the above formula for the derivative of x^n is also correct for $n = -1$. In fact, it holds for all real values of n, though we will not need this result.

Other Instances of the Derivative Concept

Expressions like $\frac{f(x+\Delta x)-f(x)}{\Delta x}$, and their limits, occur not only in geometry, but in any situation where a quantity is changing and we wish to measure its *rate of change*. Thus, slope is the rate of change of height with respect to horizontal distance and, as another example, *speed* is the rate of change of position with respect to time.

We estimate the speed of an object by measuring its positions $p(t + \Delta t)$ and $p(t)$ at nearby times $t + \Delta t$ and t. This gives the average speed between these times as

$$\frac{\text{distance traveled}}{\text{time taken}} = \frac{p(t + \Delta t) - p(t)}{\Delta t},$$

and the speed at the instant t is the limit of this expression as $\Delta t \to 0$:

$$\text{speed at time } t = p'(t) = \lim_{\Delta t \to 0} \frac{p(t + \Delta t) - p(t)}{\Delta t}.$$

Similarly, *acceleration* is the rate of change of speed with respect to time. Consequently, acceleration is the derivative of speed, which makes it the *second* derivative (the "derivative of the derivative") of position. The second derivative of position may seem to be an esoteric concept, but it is one we can feel! By Newton's second law of motion, acceleration is perceived as *force*, as is the case when a car starts or stops suddenly. Later we will see that functions like e^x, $\sin x$, and $\cos x$ have infinitely many derivatives, and all of them are important.

6.3 Calculating Derivatives

As its name suggests, calculus is a system for calculation. Its first great success is the calculation of derivatives, due to the fact that the derivatives of the simplest functions are obvious, and there are simple rules for calculating the derivative of any reasonable combination of functions whose derivatives are known. Another contribution to the success of calculus is a notation, due to Leibniz, which reflects the origin of the derivative concept in the calculation of limits of fractions. When $y = f(x)$, it is often helpful to write

$$f'(x) = \frac{dy}{dx}.$$

Since $\frac{dy}{dx}$ is the limit of the fraction $\frac{\Delta y}{\Delta x}$, it often behaves like a fraction, and this makes it easier to remember and to derive the rules for differentiation.

It is also convenient to denote the differentiation operation by $\frac{d}{dx}$, since this naturally applies to function expressions on the left (which the $'$ symbol does not). For example, we can write

$$\frac{d}{dx}x^2 = 2x.$$

The simplest functions are the *constant* functions $f(x) = k$, each of which has derivative 0, and the *identity* function $f(x) = x$, which has derivative 1. From these we obtain the derivatives of many other functions by the following rules.

- The derivatives of the sum, difference, product, and quotient of differentiable functions u and v are given by:

$$\frac{d}{dx}(u+v) = \frac{du}{dx} + \frac{dv}{dx}, \quad \frac{d}{dx}(u-v) = \frac{du}{dx} - \frac{dv}{dx},$$

$$\frac{d}{dx}(u \cdot v) = u\frac{dv}{dx} + v\frac{du}{dx}, \quad \frac{d}{dx}\left(\frac{u}{v}\right) = \frac{v\frac{du}{dx} - u\frac{dv}{dx}}{v^2} \quad \text{for } v \neq 0.$$

- If $\frac{dy}{dx}$ is the derivative of a function $y = f(x)$, and $\frac{dy}{dx} \neq 0$, then the derivative $\frac{dx}{dy}$ of the inverse function $x = f^{-1}(y)$ is $1 / \frac{dy}{dx}$.

- If $z = f(y)$ has derivative $\frac{dz}{dy}$ (with respect to y) and $y = g(x)$ has derivative $\frac{dy}{dx}$, then $z = f(g(x))$ has derivative with respect to x given by $\frac{dz}{dx} = \frac{dz}{dy} \cdot \frac{dy}{dx}$ (the *chain rule*).

The last two rules show that the symbol $\frac{dy}{dx}$ is apt, because $\frac{dy}{dx}$ behaves like a fraction. This is no surprise, because $\frac{dy}{dx}$ is the limit[2] of the actual fraction $\frac{\Delta y}{\Delta x}$. Indeed, the proofs of the rules above are basically manipulations of fractions, with some care to avoid zero denominators, followed by passage to the limit. The other important ingredient in the limits is:

Continuity of differentiable functions. *If $y = f(x)$ is differentiable at $x = a$, then $f(x)$ is continuous at $x = a$; that is, $f(x) \to f(a)$ as $x \to a$.*

---

[2] In the words of philosopher George Berkeley, dx and dy are "ghosts of departed quantities."

Proof. The value of the derivative at $x = a$ is

$$\lim_{x \to a} \frac{f(x) - f(a)}{x - a}.$$

Since the denominator $x - a \to 0$, the limit can exist only if the numerator also approaches 0; that is, if $f(x) \to f(a)$. □

Before applying this result, we should highlight the concept of continuity, which has quietly slipped in here.

Definitions. A function f is called *continuous at the point* $x = a$ if $f(x) \to f(a)$ as $x \to a$. We say that f is *continuous*, on some domain, if f is continuous at each point x of that domain.

We now illustrate the application of fractions, limits, and continuity to the *product rule* for differentiation:

$$\frac{d}{dx}(u \cdot v) = u\frac{dv}{dx} + v\frac{du}{dx}.$$

By definition, the left side is the limit, as $\Delta x \to 0$, of the fraction

$$\frac{u(x + \Delta x) \cdot v(x + \Delta x) - u(x) \cdot v(x)}{\Delta x}.$$

To create $\frac{\Delta u}{\Delta x}$ and $\frac{\Delta v}{\Delta x}$ in the fraction we add and subtract the term $u(x + \Delta x) \cdot v(x)$ in the top line, obtaining

$$\frac{u(x + \Delta x) \cdot v(x + \Delta x) + u(x + \Delta x) \cdot v(x) - u(x + \Delta x) \cdot v(x) - u(x) \cdot v(x)}{\Delta x}$$

$$= u(x + \Delta x)\frac{v(x + \Delta x) - v(x)}{\Delta x} + v(x)\frac{u(x + \Delta x) - u(x)}{\Delta x}$$

$$= u(x + \Delta x)\frac{\Delta v}{\Delta x} + v(x)\frac{\Delta u}{\Delta x}.$$

Finally, letting $\Delta x \to 0$, we get $u(x)\frac{dv}{dx} + v(x)\frac{du}{dx}$, because $u(x + \Delta x) \to u(x)$ by the continuity of the differentiable function u.

The functions obtained from constant and identity functions by $+, -, \cdot, \div$ are the large class of rational functions. They include, for example,

$$x^2, \quad 3x^2, \quad 1 + 3x^2, \quad \frac{x}{1 + 3x^2}, \quad x^3 - \frac{x}{1 + 3x^2}, \quad \dots,$$

all of which we can now differentiate using the rules for $+, -, \cdot, \div$ of differentiable functions.

With the rule for inverse functions we can differentiate algebraic functions such as $\sqrt{x}$. We know that $y = f(x) = x^2$ has derivative

$$\frac{dy}{dx} = 2x,$$

so the derivative $\frac{dx}{dy}$ (with respect to y) of the inverse function $x = \sqrt{y}$ is

$$\frac{dx}{dy} = 1 \Big/ \frac{dy}{dx} = \frac{1}{2x} = \frac{1}{2\sqrt{y}} = \frac{1}{2} y^{-1/2}.$$

Renaming the variable as x, this says

$$\frac{d}{dx} x^{1/2} = \frac{1}{2} x^{-1/2},$$

which conforms to the rule $\frac{d}{dx} x^n = nx^{n-1}$ already observed for positive integer values of n.

*A Hard Question about Derivatives

All of the above material, in my opinion, qualifies as elementary calculus. It is basically elementary algebra with occasional routine applications of the limit concept. However, there is a question arising from this material that has quite a different character: *if $f(x)$ has derivative 0, is $f(x)$ necessarily constant?*

The answer seems to be: obviously yes! How can a function change value if its *rate* of change is zero? Yet, perhaps our words deceive us, because it is *not* obvious how to connect the derivative of a function with the totality of its values. This question is perhaps the simplest example of a problem that is typical of advanced calculus: going from

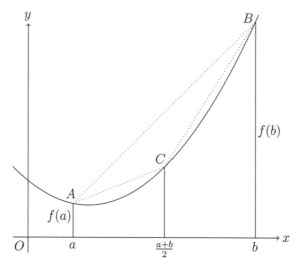

Figure 6.4: Curve with "global slope" from A to B.

a "local" assumption (slope at each point $= 0$) to a "global" conclusion (everywhere constant function).

The assumption $f'(x) = 0$ is "local" in the sense that it tells us only how f behaves near individual points: at each point P the tangent is horizontal, but the chord PQ to another point Q need not be. All we can say is that as $Q \rightarrow P$ the slope of $PQ \rightarrow 0$. It turns out that this assumption is enough to work with, but it takes (in my opinion) an *advanced* argument to draw the global conclusion that $f(x) =$ constant. Here it is.

Zero derivative theorem. *If $f'(x) = 0$ at each point of some interval, then $f(x)$ is constant on that interval.*

Proof. Suppose that the differentiable curve $y = f(x)$ has points A, B of different heights. We condense this "global slope" to a "local slope" (at least as large) of the tangent at a point P between A and B. The point P is found by an infinite process called *repeated bisection*.

We can assume that $A = (a, f(a))$, $B = (b, f(b))$ as shown in figure 6.4, and slope $AB = 1$ (if necessary, multiply f by a suitable constant).

We divide the interval $I_1 = [a, b] = \{x : a \leq x \leq b\}$ in half at $x = \frac{a+b}{2}$, and let $C = \left(\frac{a+b}{2}, f\left(\frac{a+b}{2} \right) \right)$. Then, as is clear from figure 6.4, at

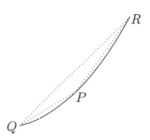

Figure 6.5: Local slope at P.

least one of the chords AC, CB has slope ≥ 1. We let I_2 be the half of I_1 on which the greater slope occurs (if the slope is the same on both halves, take the left half to be definite), and repeat the process on I_2.

In this way we obtain an infinite sequence of nested intervals,

$$[a, b] = I_1 \supseteq I_2 \supseteq I_3 \supseteq \cdots ,$$

each one half the length of the one before. And on each interval the slope of the curve from left end to right end is ≥ 1. Now, *there is exactly one point P common to all these intervals*. And, by construction, P lies between arbitrarily close points Q and R on the curve with slope $QR \geq 1$ (figure 6.5); namely, the points $(x, f(x))$ for x at the ends of a sufficiently small interval I_k.

It follows that P has arbitrarily close neighbors (Q or R) on the curve to which the slope of the chord from P is ≥ 1. Hence the slope of the tangent at P (which exists by the assumption that $f(x)$ is differentiable) is also ≥ 1.

Thus, if f is differentiable and the curve $y = f(x)$ has a nonzero slope, then the curve has a tangent of nonzero slope. Consequently, if the tangent always has zero slope, then the slope between any two points of the curve is also zero. That is, $f(x)$ is constant. $\square$

The advanced features of this argument are the construction of infinitely many nested intervals and the assumption that the intervals have a common point. These are tied to the structure of the real number system $\mathbb{R}$—specifically its so-called *completeness*—which we discuss further in the Historical and Philosophical Remarks.

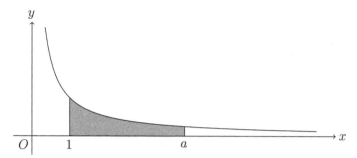

Figure 6.6: Area under the hyperbola $y = \frac{1}{x}$.

6.4 Curved Areas

Another problem about curves, even older than finding their tangents, concerns the *areas* they bound. In section 1.5 we saw the first significant contribution to this problem: Archimedes' determination of the area of a parabolic segment. His solution was based on an ingenious "exhaustion" of the parabolic segment by triangles, and there is no guarantee that the idea will work for the curves $y = x^3$, $y = x^4$, and so on. It was not until the seventeenth century that calculus triumphed over the problem, giving a simple and uniform determination of the areas bounded by all these curves. We present the solution in the next section.

In the meantime, we want to present a less transparent example— the area bounded by the hyperbola $y = \frac{1}{x}$—because it shows that the area concept is deeper than the tangent concept, and that it gives unexpected insights. We consider the area between the curve $y = \frac{1}{x}$, the x-axis, and the vertical lines through $x = 1$ and $x = a$. We will call it the area "under $y = \frac{1}{x}$ between $x = 1$ and $x = a$" (shown shaded in figure 6.6). We define this curved area in terms of known areas by approximating it from above and below by rectangles (figure 6.7).

The difference between the upper and lower approximations is the sum of the small white rectangles through which the curve passes in figure 6.7. This difference can clearly be made arbitrarily small by making the rectangles sufficiently narrow. So either the upper or lower rectangles can be used to approach the area bounded by the curve.

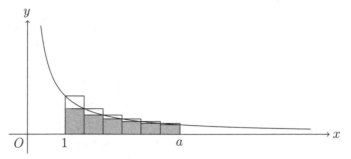

Figure 6.7: Approximating the area by rectangles.

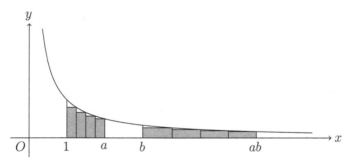

Figure 6.8: Proving the product property of the logarithm.

However, if we calculate, say, the sum of the lower rectangles obtained by dividing the interval from 1 to a into n equal parts, we do not see any obvious value as the limit of this sum—even though it is geometrically clear that the limit exists. The reason is that the area under $y = \frac{1}{x}$ from 1 to a is a function of a we have not yet considered. It is in fact the *natural logarithm* function, $\ln a$, best known for its characteristic property of turning products into sums:

$$\ln ab - \ln a + \ln b.$$

This property of the logarithm function is something we can prove easily by approximating curved areas by rectangles. Consider figure 6.8, which compares the area under $y = \frac{1}{x}$ from 1 to a with the area from b to ab.

For convenience we have chosen $b > a$ and divided each of the intervals, from 1 to a and from b to ab, into four equal parts. Exactly the same argument applies if the intervals are divided into n equal parts.

Bearing in mind that $y = \frac{1}{x}$, we see that each rectangle between b and ab has height $1/b$ times the height of the corresponding rectangle between 1 and a. However, the width of each rectangle between b and ab is b times the width of the corresponding rectangle between 1 and a. Therefore, *the total area of each set of rectangles is the same.*

This remains true if we divide the intervals between 1 and a and between b and ab into n equal parts, for any positive integer n. And we know that if we let n grow indefinitely then each of these equal sums approaches the area under the curve. Therefore,

$$\text{area under } y = \frac{1}{x} \text{ between 1 and } a \text{ equals}$$

$$\text{area under } y = \frac{1}{x} \text{ between } b \text{ and } ab.$$

The left-hand side of this equation is $\ln a$, by definition of the ln function. And the right-hand side is $\ln ab - \ln b$ (area from 1 to ab minus area from 1 to b). This gives the equation

$$\ln a = \ln ab - \ln b,$$

and so

$$\ln ab = \ln a + \ln b,$$

as claimed above.

The Logarithm Function in Nature

If follows from $\ln ab = \ln a + \ln b$ that $\ln(a^n) = n \ln a$, so the ln function "squashes" the exponential growth of the function a^n down to the linear growth of the function $n \ln a$ when a is a constant. Surprisingly, many exponentially growing quantities in nature are perceived by us to grow linearly, and our units of measure for these quantities are essentially their logarithms. There is even a term for this perception of linearity: the Weber-Fechner law of psychophysics.

For example, the natural measure of the *pitch* of a sound is the number of vibrations per second, but our ears measure pitch in *octaves* (or in subdivisions of the octave such as tones or semitones).

But raising pitch by one octave corresponds to *multiplying* frequency by 2, so an increase in pitch by n octaves corresponds to multiplication of frequency by a factor of 2^n.

It is similar with volume, or intensity, of sound. It is natural to measure intensity in units of power, such as watts. But we measure in *decibels* which better correspond to the way we perceive sound intensity. Adding 10 decibels to the intensity of a sound corresponds to *multiplying* its power by 10.

Likewise with brightness of light. The brightness of stars is measured on a scale called *magnitudes*, with magnitude increasing as brightness decreases. For example, the brightest star in the sky, Sirius, has magnitude -1.46, the second brightest, Canopus, has magnitude -0.72, and the brightest star in the constellation of Orion, Rigel, has magnitude 0.12. Brighter than all of these is the planet Venus, which has magnitude -4.6 at its brightest. Towards the other end of the scale, the seven stars normally visible in the Pleiades cluster have magnitudes between 2.86 and 5. Magnitude 6 is about the limit of typical human vision. An increase of 5 in the magnitude scale is equivalent to a 100-fold decrease in light power.

Finally, perhaps the best known example is the measurement of earthquake power by the *Richter scale*. The strongest earthquakes have magnitude around 9 on the Richter scale, and earthquakes of magnitude as low as 2 on the scale can be felt by humans. Yet a one point increase on the Richter scale corresponds to a 30-fold increase in power!

I don't know which is more remarkable: that nature produces phenomena that tend to vary over exponential scales, or that the human senses are able to squash these phenomena down to linear scales.

6.5 The Area under $y = x^n$

The approach to area in the previous section—approximating it by upper and lower rectangles—works particularly well for the parabola $y = x^2$ and the other curves of the form $y = x^n$. We show this only for $y = x^2$, because the idea works, with only minor changes, for the other positive integer powers. To find the area under the curve $y = t^2$ between $t = 0$ and $t = x$, the secret is to let x vary. It then turns out that area is a

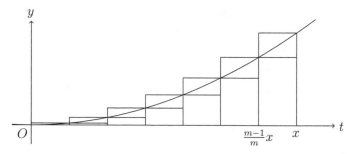

Figure 6.9: Upper and lower approximations to the area under $y = t^2$.

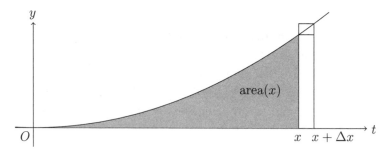

Figure 6.10: Area as a function of x.

function of x we can *differentiate* (this is why we view y as a function of t—to make the *area* a function of x), and this allows us to find exactly what the area function is.

Figure 6.9 shows the upper and lower approximations to the parabola when the interval from 0 to x is divided into m equal parts.

The difference between the upper and lower rectangle is greatest between $\frac{m-1}{m}x$ and x, where the difference in height is

$$x^2 - \left(\frac{m-1}{m}x\right)^2 = x^2\left(1 - \frac{m^2 - 2m + 1}{m^2}\right) = x^2 \frac{2m-1}{m^2}.$$

Since the width of each rectangle is $1/m$, the difference in area between the end rectangles is $\frac{2m-1}{m^3}x^2$. It follows, since there are m rectangles altogether, that the difference between the upper and lower approximations to the area under the curve is at most $\frac{2m-1}{m^2}x^2$, which approaches 0 as m increases. Therefore, there is a well-defined *area* beneath $y = t^2$ between $t = 0$ and $t = x$, which we will call area(x) (figure 6.10).

Also shown in figure 6.10 is an extra strip between x and $x + \Delta x$, because we now want to find the *derivative* of area(x) with respect to x:

$$\frac{d}{dx}\text{area}(x) = \lim_{\Delta x \to 0} \frac{\text{area}(x + \Delta x) - \text{area}(x)}{\Delta x}.$$

The numerator area$(x + \Delta x) - \text{area}(x)$ represents the area beneath $y = t^2$ from $t = x$ to $t = x + \Delta x$. In this interval, the height of the curve lies between x^2 and $(x + \Delta x)^2$, so the comparison between the area under the curve and the lower and upper rectangles reads

$$x^2 \cdot \Delta x \le \text{area}(x + \Delta x) - \text{area}(x) \le (x + \Delta x)^2 \cdot \Delta x.$$

Then, dividing by Δx,

$$x^2 \le \frac{\text{area}(x + \Delta x) - \text{area}(x)}{\Delta x} \le (x + \Delta x)^2.$$

It follows, since both ends of this inequality approach x^2 as $\Delta x \to 0$, that

$$\frac{d}{dx}\text{area}(x) = \lim_{\Delta x \to 0} \frac{\text{area}(x + \Delta x) - \text{area}(x)}{\Delta x} = x^2.$$

Thus area(x) is a function whose derivative is x^2. We already know one such function, namely $\frac{1}{3}x^3$, because

$$\frac{d}{dx}\left(\frac{1}{3}x^3\right) = \frac{1}{3}\frac{d}{dx}x^3 = \frac{1}{3} \cdot 3x^2 = x^2,$$

by the rules for differentiation in section 6.3. The only *other* functions with derivative x^2 are those of the form $\frac{1}{3}x^3 + k$, for constant k. This is because of the zero derivative theorem in section 6.3: the difference between two functions with derivative x^2 has zero derivative and hence is constant.

Since we obviously have area$(x) = 0$ when $x = 0$, the correct area function must be $\frac{1}{3}x^3$. To sum up, we have:

Area under $y = t^2$. *The area under $y = t^2$ between $t = 0$ and $t = x$ is well-defined and equals $\frac{1}{3}x^3$.* ☐

A very similar argument, using the fact that $\frac{d}{dx}x^{n+1} = (n+1)x^n$, gives:

Area under $y = t^n$. *For any positive integer n, the area under $y = t^n$ between $t = 0$ and $t = x$ is well-defined and equals $\frac{1}{n+1}x^{n+1}$.* □

It is possible to avoid using the hard zero derivative theorem in finding these areas, but only at the cost of considerable algebra to find exact formulas for the sums of upper and lower rectangles. This gets harder and harder as n increases, and in any case it turns a blind eye to the really important insight of these proofs: finding area is in some sense *inverse* to finding derivatives. This insight is worth elaborating, because it leads to a fundamental theorem.

6.6 *The Fundamental Theorem of Calculus

The idea of area under a curve is captured in calculus by the concept of *integral*. There are several concepts of integral, but elementary calculus deals only with the simplest one, the *Riemann integral*, and applies it only to continuous functions.

Given a function $y = f(t)$, continuous from $t = a$ to $t = b$, the integral of f from a to b is written

$$\int_a^b f(t)\, dt,$$

and it is defined in the same way that we defined "area under the curve $y = f(t)$" for certain functions f in the previous section. Namely, we divide the interval $[a, b]$ into finitely many parts and approximate the graph of f from above and below by rectangles erected on these parts (figure 6.11).

If it is possible to make the difference between the upper and lower approximations arbitrarily small, then there is a single number that lies between them (by the completeness of $\mathbb{R}$), and this number is the value of $\int_a^b f(t)\, dt$.

It is very plausible that, for a continuous function, the difference between upper and lower approximations can be made arbitrarily small. But proving this is a delicate matter, much like the proof of the zero

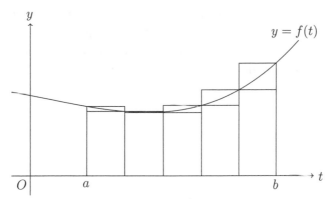

Figure 6.11: Approximating $\int_a^b f(t)\,dt$ by rectangles.

derivative theorem in section 6.3. (In fact, the heart of the proof is an infinite bisection process, like the one used there.) For this reason, proving the existence of the Riemann integral for continuous functions belongs to advanced calculus.

However, if we make the plausible assumption that the integral exists, we can proceed to differentiate this integral, just as we differentiated the special area functions in the previous section. This gives:

Fundamental theorem of calculus. *If f is continuous on an interval $[a, b]$, and*

$$F(x) = \int_a^x f(t)\,dt,$$

then $F'(x) = f(x)$. $\qquad\qquad\qquad\qquad\qquad\qquad$ □

The fundamental theorem can be used to identify functions $F(x)$ defined by integrals in the case where we know a function $G(x)$ whose derivative is $f(x)$ (just as we did for the area functions in the previous section). In this case, the zero derivative theorem tells us that $F(x)$ differs from $G(x)$ only by a constant.

The fundamental theorem is also useful in the case where $F(x)$ is *not* among the functions whose derivatives we already know (for example, the function $\ln x = \int_1^x dt/t$ is not among the algebraic functions). In this case, the fundamental theorem can be viewed as a new differentiation rule, telling us how to differentiate functions defined by integrals. When combined with the other differentiation rules, this greatly extends the class of functions we are able to differentiate.

The following subsection shows how the basic facts about logarithm and exponential functions unfold when we apply differentiation rules to $\ln x = \int_1^x dt/t$ and its inverse function.

Logarithm and Exponential Functions

Given $u = \ln x = \int_1^x \frac{dt}{t}$, the fundamental theorem of calculus says that

$$\frac{du}{dx} = \frac{1}{x}.$$

The inverse function of $u = \ln x$ is called exp (the *exponential function*), so $x = \exp(u)$ and

$$\frac{dx}{du} = 1 \Big/ \frac{du}{dx} = x$$

by the rule for the derivative of an inverse function (section 6.3).

That is, $\frac{d}{du} \exp(u) = \exp(u)$, so exp has the remarkable property of *being equal to its own derivative*. This property has important "real-life" implications, as we point out below. But first we need to get a better grip on the exp function; in particular, we need to understand why it is called "exponential."

We know from section 6.4 that ln has the property $\ln ab = \ln a + \ln b$. What does this say about exp? Well,

$$\text{if } \ln a = A \text{ then } \exp(A) = a,$$
$$\text{if } \ln b = B \text{ then } \exp(B) = b,$$
$$\text{if } \ln ab = C \text{ then } \exp(C) = ab = \exp(A)\exp(B),$$
$$\text{and } C = A + B,$$
$$\text{so } \exp(A + B) = \exp(A)\exp(B).$$

This is the property we usually write as the *law of exponents*,

$$e^{A+B} = e^A e^B.$$

In fact, we can take $\exp(u)$ to be the definition of e^u.

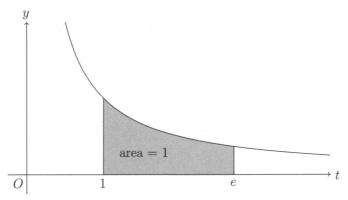

Figure 6.12: Geometric interpretation of e.

Thus $\exp(u)$ is a certain number e raised to the power u. But what is e? Well, $e = e^1 = \exp(1)$, so if we take $x = \exp(1) = e$ in the equation

$$\ln x = \int_1^x \frac{dt}{t},$$

we get

$$1 = \int_1^e \frac{dt}{t}$$

since ln and exp are inverse to each other and hence $\ln(\exp(1)) = 1$. In other words, *e is the number such that the area under $y = \frac{1}{t}$ between 1 and e equals 1* (figure 6.12). From this interpretation we can estimate that e is about 2.718, so e^u is a function that indeed grows "exponentially"; it grows faster than 2^u, for example.

Exponential growth (or decay) rates occur in nature wherever the growth rate of a quantity is proportional to its size. For example, if there are no constraints on living space or food supply, population grows at a rate proportional to its size. If $p(t)$ denotes population at time t, the growth rate $\frac{d}{dt} p(t)$ is a (positive) constant b times $p(t)$. This leads to a solution

$$p(t) = ae^{bt} \quad \text{for some constant } a,$$

because $\frac{d}{dt} e^t = e^t$.

This is a tragic equation, because the geometry of the universe prevents population growing at a rate faster than a constant times t^3

(that is, proportional to the volume of a sphere expanding at constant speed), assuming that the geometry is approximately Euclidean. (However, exponential growth is sustainable in non-Euclidean geometry. See the explosive growth of triangles in figure 5.32.)

6.7 Power Series for the Logarithm

Our definition of the natural logarithm,

$$\ln x = \int_1^x \frac{dt}{t},$$

goes back to the early days of calculus. It is in the book *Logarithmotechnia* by Mercator (1668), which makes a clever application of geometric series to express the logarithm function as an infinite series in powers of x:

$$\ln(1+x) = x - \frac{x^2}{2} + \frac{x^3}{3} - \frac{x^4}{4} + \cdots \qquad \text{for } |x| < 1.$$

The train of thought leading to this formula is the following. Replacing x by $x + 1$ in the definition gives $\ln(x+1) = \int_1^{x+1} \frac{dt}{t}$, then replacing t by $t + 1$ we get

$$\ln(x+1) = \int_0^x \frac{dt}{t+1}$$
$$= \int_0^x (1 - t + t^2 - t^3 + \cdots)\, dx,$$

which we know from section 6.1 is valid for $|t| < 1$. Presumably,

$$\int_0^x (1 - t + t^2 - t^3 + \cdots)\, dt = \int_0^x 1\, dt - \int_0^x t\, dt$$
$$+ \int_0^x t^2\, dt - \int_0^x t^3\, dt + \cdots.$$

So, since each term $\int_0^x t^n \, dt = \frac{x^{n+1}}{n+1}$ by section 6.5, we finally have

$$\ln(1+x) = x - \frac{x^2}{2} + \frac{x^3}{3} - \frac{x^4}{4} + \cdots.$$

But we are presuming that the integral of an infinite sum is the infinite sum of the corresponding integrals, and we know nothing about integrals of infinite sums. In elementary calculus it is usual to gloss over such problems, but in this case the problem has an elementary solution. We can write $\frac{1}{1+t}$ as the *finite* sum

$$\frac{1}{1+t} = 1 - t + t^2 - t^3 + \cdots \pm t^n \mp \frac{t^{n+1}}{1+t},$$

as can be checked by summing the infinite geometric series of terms that come after $\pm t^n$, namely,

$$t^{n+1} - t^{n+2} + t^{n+3} - \cdots = t^{n+1}(1 - t + t^2 - \cdots) = \frac{t^{n+1}}{1+t}.$$

And it follows easily from the definition of integral that the integral of a sum of two functions (and hence of any finite number) is the sum of their integrals.

Therefore,

$$\int_0^x \frac{dt}{1+t} = \int_0^x \left(1 - t + t^2 - t^3 + \cdots \pm t^n \mp \frac{t^{n+1}}{1+t}\right) dt$$

$$= x - \frac{x^2}{2} + \frac{x^3}{3} - \cdots \pm \frac{x^{n+1}}{x+1} \mp \int_0^x \frac{t^{n+1}}{1+t} \, dt. \qquad (*)$$

We do not know the exact value of $\int_0^x \frac{t^{n+1}}{1+t} \, dt$, but we know enough to prove that it approaches 0 as n increases. Certainly, if $x \geq 0$ then $1 + t \geq 1$ and so

$$\int_0^x \frac{t^{n+1}}{1+t} \, dt \leq \int_0^x t^{n+1} \, dt = \frac{x^{n+2}}{n+2} \to 0 \quad \text{as } n \text{ increases.}$$

But also, if $x < 0$ we must have $x > -1$ for the integral to exist, in which case t varies from 0 down to some value $-1 + \delta$ with $0 < \delta < 1$. In this case $1 + t \geq \delta$ so

$$\int_0^x \frac{t^{n+1}}{1+t}\, dt \leq \frac{1}{\delta} \int_0^x t^{n+1}\, dt = \frac{1}{\delta} \frac{x^{n+2}}{n+2} \to 0 \quad \text{as } n \text{ increases.}$$

This allows us to conclude from (*) that

$$x - \frac{x^2}{2} + \frac{x^3}{3} - \cdots \pm \frac{x^{n+1}}{n+1} \to \int_0^x \frac{dt}{1+t} = \ln(1+x) \quad \text{as } n \text{ increases,}$$

which is precisely what it means to say

$$\ln(1+x) = x - \frac{x^2}{2} + \frac{x^3}{3} - \frac{x^4}{4} + \cdots . \qquad (**)$$

The careful proof that $\int_0^x \frac{t^{n+1}}{1+t}\, dt \to 0$ is not just virtuous, it also brings an extra reward: *it is also valid for $x = 1$* (because $x^{n+1} = 1$ in this case). Thus the formula (**) is valid for $x = 1$, which gives the remarkable formula

$$\ln 2 = 1 - \frac{1}{2} + \frac{1}{3} - \frac{1}{4} + \cdots .$$

Error in Stopping at the nth Term

A nice feature of the series (**) for $\ln(1+x)$ is that it gives an easy way to compute logarithms to a required degree of accuracy. This is because we know that *when $x \leq 1$ is positive, the error in truncating the series at the nth term is less in absolute value than the $(n+1)$st term.*

To see why, first note that

$$x > \frac{x^2}{2} > \frac{x^3}{3} > \frac{x^4}{4} > \cdots$$

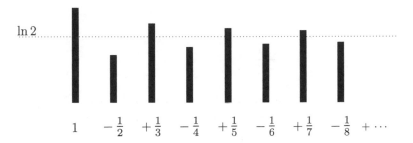

Figure 6.13: Sums of an oscillating series.

when $x \leq 1$ is positive. It follows that the sums

$$x,$$

$$x - \frac{x^2}{2},$$

$$x - \frac{x^2}{2} + \frac{x^3}{3},$$

$$x - \frac{x^2}{2} + \frac{x^3}{3} - \frac{x^4}{4},$$

$$\vdots$$

oscillate up and down in value—up when the last term has a plus sign, and down when the last term has a minus sign. And, since the size of terms steadily decreases, the size of the oscillations steadily decreases too. This implies that the sum of the series always lies between the last high value and the last low value. In other words, the true value of the logarithm always differs from the sum of n terms by an amount less than the $(n+1)$st term. See figure 6.13, which illustrates the oscillation when $x = 1$.

In particular, when $x \leq 1$ is positive, the error in approximating $\ln(1 + x)$ by x is less than $\frac{x^2}{2}$. We will use this estimate of the error later, in section 10.7.

Another Power Series for the Logarithm

We now have a nice simple power series for $\ln(1 + x)$, valid for $|x| < 1$. But the series is definitely *not* valid for $x > 1$. For example, with $x = 2$

we get the series

$$2 - \frac{2^2}{2} + \frac{2^3}{3} - \frac{2^4}{4} + \cdots .$$

This infinite sum is not meaningful, because the nth term $\pm 2^n/n$ grows beyond all bounds, so the sum of the first n terms oscillates wildly between large negative and large positive values.

To obtain a series for the logarithm of any positive number, we combine the series

$$\ln(1 + x) = x - \frac{x^2}{2} + \frac{x^3}{3} - \frac{x^4}{4} + \cdots$$

with series obtained by replacing x by $-x$, namely,

$$\ln(1 - x) = -x - \frac{x^2}{2} - \frac{x^3}{3} - \frac{x^4}{4} - \cdots .$$

which is also valid for $|x| < 1$. Subtracting the second series from the first gives

$$\ln \frac{1+x}{1-x} = \ln(1 + x) - \ln(1 - x) = 2 \left(x + \frac{x^3}{3} + \frac{x^5}{5} + \frac{x^7}{7} + \cdots \right) .$$

The series for $\ln \frac{1+x}{1-x}$ is again valid only for $|x| < 1$, but we can get *any* positive number as a value of $\frac{1+x}{1-x}$ for some x between -1 and 1. For example,

if $2 = \frac{1+x}{1-x}$ we find $2 - 2x = 1 + x$,   so $1 = 3x$ and $x = \frac{1}{3}$.

It follows that

$$\ln 2 = 2 \left(\frac{1}{3} + \frac{1}{3}\frac{1}{3^3} + \frac{1}{5}\frac{1}{3^5} + \frac{1}{7}\frac{1}{3^7} + \cdots \right) .$$

This series is useful for computing $\ln 2$. Taking just the four terms shown, we get the value $0.69313\ldots$, which differs from the correct value $\ln 2 = 0.69314 \ldots$ only after the fourth decimal place.

*Power Series for the Exponential Function

Mercator's series for the natural logarithm was rediscovered by Newton (1671), who pushed the idea further to discover the power series for the inverse function (the exponential function, though it did not have a name at that time). Newton set

$$y = x - \frac{x^2}{2} + \frac{x^3}{3} - \frac{x^4}{4} + \cdots,$$

then, in an astounding feat of computation, *solved* this equation for x, obtaining

$$x = y + \frac{y^2}{2} + \frac{y^3}{6} + \frac{y^4}{24} + \frac{y^5}{120} + \cdots.$$

He then correctly guessed that the nth term is $\frac{y^n}{n!}$, so

$$x = \frac{y}{1!} + \frac{y^2}{2!} + \frac{y^3}{3!} + \frac{y^4}{4!} + \frac{y^5}{5!} + \cdots,$$

which is the function we call $e^y - 1$.

There are simpler ways to find the power series for the exponential function, though these methods (as indeed does Newton's method) require some advanced calculus for their justification. Here we are content to give a very simple method, without complete justification.

We *assume* that e^x is expressible as a power series,

$$e^x = a_0 + a_1 x + a_2 x^2 + a_3 x^3 + \cdots,$$

and that it is valid to differentiate this series term by term. Then we can find the coefficients $a_0, a_1, a_2, a_3, \ldots$ in succession by repeatedly differentiating the equation above, using $\frac{d}{dx} e^x = e^x$, and setting $x = 0$. Before differentiating at all, setting $x = 0$ gives

$$1 = a_0 + 0 + 0 + 0 + \cdots,$$

so $a_0 = 1$. The first differentiation gives

$$e^x = a_1 + 2 \cdot a_2 x + 3 \cdot a_3 x^2 + 4 \cdot a_4 x^3 + \cdots,$$

then setting $x = 0$ gives

$$1 = a_1.$$

The second differentiation gives

$$e^x = 2 \cdot a_2 + 3 \cdot 2 \cdot a_3 x + 4 \cdot 3 \cdot a_4 x^2 + \cdots,$$

then setting $x = 0$ gives

$$1 = 2 \cdot a_2, \quad \text{so} \quad a_2 = \frac{1}{2}.$$

The third differentiation gives

$$e^x = 3 \cdot 2 \cdot a_3 + 4 \cdot 3 \cdot 2a_4 x + \cdots,$$

then setting $x = 0$ gives

$$1 = 3 \cdot 2a_3, \quad \text{so} \quad a_3 = \frac{1}{3 \cdot 2}.$$

It should now be clear that we will get $a_4 = \frac{1}{4 \cdot 3 \cdot 2}$, $a_5 = \frac{1}{5 \cdot 4 \cdot 3 \cdot 2}$, ... by continuing this process. Hence

$$e^x = 1 + \frac{x}{1!} + \frac{x^2}{2!} + \frac{x^3}{3!} + \frac{x^4}{4!} + \cdots.$$

In particular, $x = 1$ now gives

$$e = 1 + \frac{1}{1!} + \frac{1}{2!} + \frac{1}{3!} + \frac{1}{4!} + \cdots.$$

Remark. Having discovered the power series, we can turn around and make it the *definition* of the exponential function:

$$\exp(x) = 1 + \frac{x}{1!} + \frac{x^2}{2!} + \frac{x^3}{3!} + \frac{x^4}{4!} + \cdots.$$

It is easy to prove that this infinite sum exists for all values of x, by showing that its terms from some point onward are less than the terms in the geometric series $\frac{1}{2} + \frac{1}{2^2} + \frac{1}{2^3} + \cdots$ (see the example in the next subsection). But it is harder to prove it is valid to differentiate term by term and hence to prove that $\frac{d}{dx}\exp(x) = \exp(x)$. This involves the advanced version of the continuity concept called *uniform* continuity, which is normally reserved for advanced calculus courses.

The Irrationality of e

The series $e = 1 + \frac{1}{1!} + \frac{1}{2!} + \frac{1}{3!} + \frac{1}{4!} + \cdots$ is good for computation, because the terms decrease rapidly in size. The rapid decrease is also of theoretical value, because it gives a proof that e is irrational (due to Joseph Fourier around 1815).

Irrationality of e. *The number* e *is not equal to* m/n *for any positive integers* m *and* n.

Proof. Suppose on the contrary that

$$\frac{m}{n} = 1 + \frac{1}{1!} + \frac{1}{2!} + \frac{1}{3!} + \cdots + \frac{1}{n!} + \cdots .$$

If we multiply both sides of this equation by $n!$ we get

$$m \cdot (n-1)! = n! + \frac{n!}{1!} + \frac{n!}{2!} + \frac{n!}{3!} + \cdots + \frac{n!}{n!}$$

$$+ \frac{1}{n+1} + \frac{1}{(n+1)(n+2)} + \frac{1}{(n+1)(n+2)(n+3)} + \cdots$$

$$= \text{integer} + \frac{1}{n+1} + \frac{1}{(n+1)(n+2)}$$

$$+ \frac{1}{(n+1)(n+2)(n+3)} + \cdots .$$

The left-hand side of the equation is also an integer, so we conclude that

$$\frac{1}{n+1} + \frac{1}{(n+1)(n+2)} + \frac{1}{(n+1)(n+2)(n+3)} + \cdots \quad \text{is an integer.}$$

But clearly $n \geq 1$ (as the denominator of a fraction), so

$$\frac{1}{n+1} + \frac{1}{(n+1)(n+2)} + \frac{1}{(n+1)(n+2)(n+3)} + \cdots$$

$$< \frac{1}{2} + \frac{1}{2^2} + \frac{1}{2^3} + \cdots = 1.$$

So $\frac{1}{n+1} + \frac{1}{(n+1)(n+2)} + \frac{1}{(n+1)(n+2)(n+3)} + \cdots$ is *not* an integer, and we have a contradiction. Thus it is false to assume that $e = m/n$. □

6.8 *The Inverse Tangent Function and π

Calculus is needed for geometry at quite an elementary level, to understand the circular functions sine, cosine, and tangent, and to evaluate the number π. To illustrate how calculus contributes to geometry, we are going to derive the simplest possible expression for π, namely,

$$\frac{\pi}{4} = 1 - \frac{1}{3} + \frac{1}{5} - \frac{1}{7} + \cdots .$$

This expression is a consequence of the power series for the inverse tangent function,

$$\arctan x = x - \frac{x^3}{3} + \frac{x^5}{5} - \frac{x^7}{7} + \cdots ,$$

which we derive first.

We view $y = \tan \theta$ as the vertical side of a right-angled triangle with horizontal side 1 and angle θ as shown in figure 6.14. Thus $\tan \theta = y$ is measured along the vertical tangent to the unit circle. Its inverse function $\theta = \arctan y$ is so called because θ is also the arc length on the unit circle spanned by the angle θ.

We now calculate $\frac{d}{dy} \arctan y = \frac{d\theta}{dy}$ by estimating the ratio $\frac{\Delta\theta}{\Delta y}$ when $\Delta\theta$ is a small addition to the angle θ, causing a small addition Δy to the height y on the vertical tangent (figure 6.15).

By the Pythagorean theorem, $OB = \sqrt{1 + y^2}$, so the arc length $A'B$ subtended by the angle $\Delta\theta$ on the circle of radius $\sqrt{1 + y^2}$ is

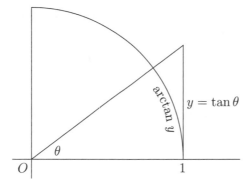

Figure 6.14: Geometry of tan and arctan.

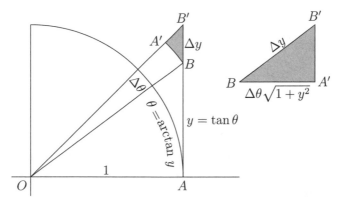

Figure 6.15: Estimating the derivative of arctan y.

$\Delta\theta\sqrt{1+y^2}$. As $\Delta\theta \to 0$ this arc approaches a straight line segment and $BA'B'$ approaches a right-angled triangle similar to OAB. Consequently, the ratio of the sides Δy and $\Delta\theta\sqrt{1+y^2}$ of $BA'B'$ (corresponding to hypotenuse and base) approaches $\sqrt{1+y^2}$, so

$$\frac{\Delta\theta}{\Delta y} \to \frac{1}{1+y^2} \quad \text{as} \quad \Delta\theta \to 0.$$

In other words,

$$\frac{d}{dy}\arctan y = \frac{d\theta}{dy} = \frac{1}{1+y^2}.$$

It follows, integrating both sides, using arctan $0 = 0$ and the zero derivative theorem from section 6.3, that

$$\arctan x = \int_0^x \frac{dy}{1+y^2}.$$

Now we know from the theory of geometric series in section 6.1 that

$$\frac{1}{1+y^2} = 1 - y^2 + y^4 - y^6 + \cdots \quad \text{for} \quad |y| < 1.$$

Using this formula in the integral we get

$$\arctan x = \int_0^x \left(1 - y^2 + y^4 - y^6 + \cdots\right) dy \quad \text{for} \quad |x| < 1$$

$$= \int_0^x 1 \, dy - \int_0^x y^2 \, dy + \int_0^x y^4 \, dy - \int_0^x y^6 \, dy + \cdots$$

$$= x - \frac{x^3}{3} + \frac{x^5}{5} - \frac{x^7}{7} + \cdots.$$

To justify the first step—from the integral of an infinite sum to the infinite sum of integrals—we proceed as we did in section 6.7, using the *finite* series

$$\frac{1}{1+y^2} = 1 - y^2 + y^4 - \cdots \pm y^{2n} \mp \frac{y^{2n+2}}{1+y^2}$$

and showing that $\int_0^x \frac{y^{2n+2}}{1+y^2} \, dy \to 0$ as n increases.

This more rigorous argument has another advantage: *the finite series is also valid for $y = 1$.* So we can conclude that

$$\arctan x = x - \frac{x^3}{3} + \frac{x^5}{5} - \frac{x^7}{7} + \cdots$$

is also valid for $x = 1$, which gives the wonderful formula

$$\frac{\pi}{4} = \arctan 1 = 1 - \frac{1}{3} + \frac{1}{5} - \frac{1}{7} + \cdots.$$

6.9 Elementary Functions

We began this chapter by looking at functions we can differentiate: starting with constant and identity functions and quickly reaching all rational functions, by rules for differentiating $+, -, \cdot, \div$ of functions whose derivatives are known. The next two rules, for inverse functions and composite functions, allow us to differentiate many algebraic functions, such as $\sqrt{x}, \sqrt{1+x^2}, \sqrt[3]{1+x^4}, \ldots$, and so on. Finally, the fundamental theorem of calculus,

$$\frac{d}{dx} \int_a^x f(t)\, dt = f(x),$$

allows us to differentiate functions defined by integrals, such as

$$\ln x = \int_1^x \frac{dt}{t} \quad \text{and} \quad \arctan x = \int_0^x \frac{dt}{1+t^2}.$$

The functions obtainable from the latter two functions and rational functions by inversion, composition, and the rational operations $+, -, \cdot, \div$ are called the *elementary functions*. It may seem rather arbitrary to stop at the ln and arctan integrals—since many other functions are definable as integrals of algebraic functions—but there are good reasons to consider this the right place for elementary calculus to stop.

The main reason is that, assuming the fundamental theorem of algebra, the *integrals of all rational functions reduce to integrals of the* ln *or* arctan *type*. To see why, recall that any rational function of t can be written as a quotient of polynomials, $p(t)/q(t)$. It follows from the fundamental theorem of algebra that $q(t)$ is a product of real linear or quadratic factors

$$at + b \quad \text{or} \quad ct^2 + dt + e.$$

Then we can apply the algebraic technique of *partial fractions* (for an example, see section 7.3) to write $p(x)/q(x)$ as a sum of terms

$$\frac{A}{at + b} \quad \text{and} \quad \frac{Bt + C}{ct^2 + dt + e}.$$

Some further easy transformations then reduce the integral of $p(t)/q(t)$ to integrals of $\frac{1}{t}$ and $\frac{1}{1+t^2}$; that is, to integrals of the ln or arctan type. Thus, *the integral of any rational function is an elementary function.*

This result shows that the elementary functions are not as arbitrary as they look. To reinforce this claim, we look at the elementary functions obtainable from the arctan function. Obviously, they include its inverse function, $\tan\theta$, but they also include the other circular functions, such as $\cos\theta$ and $\sin\theta$. We devote the next subsection to deriving the connection between the sine, cosine, and tangent functions, since it also has a beautiful connection with number theory and geometry.

Rational Points on the Circle

Ever since the discovery of the Pythagorean theorem there has been interest in *Pythagorean triples*: positive integer triples (a, b, c) such that $a^2 + b^2 = c^2$. The simplest Pythagorean triple is (3,4,5), and others are (5,12,13), (7,24,25), and (8,15,17). Specific Pythagorean triples have been known since ancient times in Europe, the Middle East, India, and China. Around 300 BCE, Euclid (*Elements*, Book X, lemma following Proposition 28) found a formula that gives them all:

$$a = (p^2 - q^2)r, \quad b = 2pqr, \quad c = (p^2 + q^2)r,$$

as p, q, r run through all positive integers.

A few centuries later, Diophantus transformed the problem of finding these triples into one about rational numbers, which can be viewed as finding *rational points on the unit circle*. This version of the problem arises because if $a^2 + b^2 = c^2$ then

$$\left(\frac{a}{c}\right)^2 + \left(\frac{b}{c}\right)^2 = 1,$$

so (a, b, c) corresponds to a rational point $\left(\frac{a}{c}, \frac{b}{c}\right)$ on the circle $x^2 + y^2 = 1$. This view of Pythagorean triples is revealing, because the search for them can now be guided by geometry and algebra. See figure 6.16.

If P has rational coordinates, then the slope t from Q to P is a rational number (because the "rise" and "run" are rational). Conversely, if the line PQ has rational slope t we can find P as the common solution

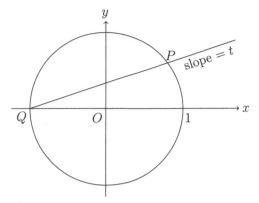

Figure 6.16: A rational point P on the unit circle.

of the linear equation

$$y = t(x+1) \qquad \text{(the equation of } PQ)$$

and the quadratic equation

$$x^2 + y^2 = 1 \qquad \text{(the equation of the unit circle).}$$

Since the coefficients of these two equations are rational, their common solutions are given by a quadratic equation in x with rational coefficients. We know that one of the solutions is $x = -1$ (for Q), so the second solution (giving the x-coordinate of P) must also be rational. We now work out exactly what the second solution is.

Substituting $y = t(x+1)$ in $x^2 + y^2 = 1$ gives the quadratic equation

$$x^2 + t^2(x+1)^2 = 1, \quad \text{that is,} \quad (1+t^2)x^2 + 2t^2x + t^2 - 1 = 0.$$

The quadratic formula gives the solutions

$$x = \frac{-2t^2 \pm \sqrt{(2t^2)^2 - 4(1+t^2)(t^2-1)}}{2(1+t^2)}$$

$$= \frac{-2t^2 \pm \sqrt{4t^4 - 4(t^4 - 1)}}{2(1+t^2)}$$

$$= \frac{-2t^2 \pm 2}{2(1+t^2)}$$

$$= -1, \quad \frac{1-t^2}{1+t^2}.$$

This confirms the solution $x = -1$ for Q, and gives $x = \frac{1-t^2}{1+t^2}$ as the x-coordinate of P. It follows from the equation of PQ that the y-coordinate of P is

$$y = t(x+1) = t\left(\frac{1-t^2+1+t^2}{1+t^2}\right) = \frac{2t}{1+t^2}.$$

Thus, *the rational points on $x^2 + y^2 = 1$ are $(-1, 0)$ and the points $\left(\frac{1-t^2}{1+t^2}, \frac{2t}{1+t^2}\right)$ as t runs through the rational numbers.*

For example, $t = \frac{1}{2}$ gives the point $\left(\frac{3}{5}, \frac{4}{5}\right)$ corresponding to the Pythagorean triple $(3,4,5)$.

If we drop the restriction that t be rational, this makes no difference to the calculation of the coordinates of P,

$$x = \frac{1-t^2}{1+t^2}, \quad y = \frac{2t}{1+t^2}.$$

But now these formulas tell something interesting about the sine, cosine, and tangent functions, due to their association with the circle.

Sine and cosine formulas. *If $t = \tan\frac{\theta}{2}$, then*

$$\cos\theta = \frac{1-t^2}{1+t^2}, \quad \sin\theta = \frac{2t}{1+t^2}.$$

Proof. These formulas can be read off figure 6.17, once it is established that the angles are as shown. If we let θ be the angle between the line OP and the x-axis, then it is certainly true that $P = (\cos\theta, \sin\theta)$, by definition of the sine and cosine functions.

Also, it can be checked that $\theta/2$ is the angle of the line QP, using the facts that OPQ is an isosceles triangle, hence with equal angles at P and Q, and that the angle sum of a triangle is π.

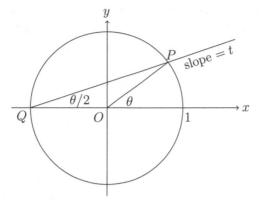

Figure 6.17: Angles in the circle.

Thus the slope t of PQ is $\tan\frac{\theta}{2}$, and hence

$$\left(\frac{1-t^2}{1+t^2}, \frac{2t}{1+t^2}\right) = (\cos\theta, \sin\theta),$$

which gives $\cos\theta = \frac{1-t^2}{1+t^2}$ and $\sin\theta = \frac{2t}{1+t^2}$. □

6.10 Historical Remarks

> Often I have considered the fact that most of the
> difficulties which block the progress of students
> trying to learn analysis stem from this: that although
> they understand little of ordinary algebra, still they
> attempt this more difficult art. From this it follows
> that they remain on the fringes, but in addition they
> entertain strange ideas about the concept of the
> infinite, which they must try to use.

Euler (1748a), preface

Foundations of Calculus

Most of calculus as we know it, at the undergraduate level, was produced in a frenzy of activity in the late seventeenth and early eighteenth century. Both before and after that period there was concern about the infinite processes that underlie the calculus—the foundations of calculus that we now call *analysis*. Infinite processes were already a concern in ancient Greece, and the Greeks devised the method of exhaustion to avoid infinity as far as possible. (Certainly, it avoids *actual* infinity.) As we saw in section 6.1, Archimedes used exhaustion to determine the area of the parabolic segment. It allowed him to conclude, using an "arbitrarily large" but not infinite number of triangles, that the area of the parabolic segment is 4/3 the area of the largest triangle.

The seventeenth-century mathematicians were well-schooled in the method of exhaustion, from Euclid and Archimedes, so they assumed they could fall back on it if necessary. But as Huygens (1659), p. 337, remarked:

> Mathematicians will never have enough time to read all the discoveries in Geometry (a quantity which is increasing from day to day and seems likely in this scientific age to develop to enormous proportions) if they continue to be presented in a rigorous form according to the manner of the ancients.

The idea that calculus could *in principle* be presented "in the manner of the ancients" might have been true in 1659, when mathematicians were still investigating curves individually. But after Newton and Leibniz made calculus truly a method of *calculation*, applicable to seemingly arbitrary formulas, the question of rigor became more acute.

When calculus became this general it was natural to ask: What is a function? What is a continuous function? Is it the same as a differentiable function? Newton, who thought of functions in terms of continuous motion, seemed to assume that continuous functions are both integrable and differentiable. As we now define continuity, continuous functions are integrable (on closed intervals) but some of them are *nowhere* differentiable. These results could not be known until the nineteenth century, when the concept of continuous function was first defined. The foundations of the definition were laid by Bolzano (1817) and Cauchy (1821), but the properties of continuous functions were not rigorously provable until Dedekind gave a definition of the continuum $\mathbb{R}$ of real numbers in 1858 (published in 1872). This started the push to arithmetize geometry and analysis, mentioned in section 5.10, after the discovery of non-Euclidean geometry.

With a definition of $\mathbb{R}$ finally available (Cantor, Meray, and Weierstrass published equivalent definitions of $\mathbb{R}$ about the same time as Dedekind), the calculus was put on a sound foundation by Weierstrass in the 1870s. It can still be asked (and perhaps can never be answered) whether the nineteenth-century foundations of calculus are the best possible, but most of the alternatives proposed so far give the same results, and are not noticeably simpler.

Preparing Students for Calculus

As we have seen, the foundations of calculus involve advanced ideas, particularly about infinity. This was realized long before the full extent of the problem became known, and mathematicians tried to devise appropriate and interesting material to *prepare* students for calculus, without exposing them to all the challenges of infinity. The most brilliant and fascinating book along these lines was the *Introductio in analysin infinitorum* (Introduction to the analysis of the infinite) of Euler (1748a).

In a nutshell, Euler believed that the best preparation for calculus was the study of infinite series. He had good reasons for believing this. As we have seen, infinite geometric series were implicitly used by Euclid and Archimedes, so infinite series were in use long before calculus. Another pointer from history was the contribution of Oresme (1350). He made the important discovery that the so-called *harmonic series*,

$$\frac{1}{2} + \frac{1}{3} + \frac{1}{4} + \frac{1}{5} + \cdots ,$$

does not have a finite sum, even though its terms tend to zero. His proof was by collecting the terms into the following groups,

$$\frac{1}{2} + \left(\frac{1}{3} + \frac{1}{4}\right) + \left(\frac{1}{5} + \frac{1}{6} + \frac{1}{7} + \frac{1}{8}\right)$$

$$+ \left(\frac{1}{9} + \frac{1}{10} + \frac{1}{11} + \frac{1}{12} + \frac{1}{13} + \frac{1}{14} + \frac{1}{15} + \frac{1}{16}\right) + \cdots ,$$

each of which has twice as many terms as the group before. It is easy to see that

$$\frac{1}{3} + \frac{1}{4} > \frac{2}{4} = \frac{1}{2},$$

$$\frac{1}{5} + \frac{1}{6} + \frac{1}{7} + \frac{1}{8} > \frac{4}{8} = \frac{1}{2},$$

$$\frac{1}{9} + \frac{1}{10} + \frac{1}{11} + \frac{1}{12} + \frac{1}{13} + \frac{1}{14} + \frac{1}{15} + \frac{1}{16} > \frac{8}{16} = \frac{1}{2},$$

and so on. Hence each of the infinitely many groups has sum $\geq \frac{1}{2}$, and so the sum of the series grows beyond all bounds. This discovery of Oresme was the first sign that sums of infinite series have some subtleties of their own.

Unknown to Euler, and to anyone else in Europe until the nineteenth century, was a collection of results giving a really spectacular demonstration that infinite series can be studied before calculus. These results, from the Indian mathematician Madhava, who lived around 1350–1425, include

$$\sin x = x - \frac{x^3}{3!} + \frac{x^5}{5!} - \frac{x^7}{7!} + \cdots,$$

$$\cos x = 1 - \frac{x^2}{2!} + \frac{x^4}{4!} - \frac{x^6}{6!} + \cdots,$$

$$\tan^{-1} x = x - \frac{x^3}{3} + \frac{x^5}{5} - \frac{x^7}{7} + \cdots,$$

the last of which gives a famous formula for π by substituting $x = 1$:

$$\frac{\pi}{4} = 1 - \frac{1}{3} + \frac{1}{5} - \frac{1}{7} + \cdots.$$

For more on these results, whose discovery by Indian mathematicians was little known until recently, see Plofker (2009).

Finally, just before the dawn of calculus, Wallis (1655) discovered that

$$\frac{\pi}{2} = \frac{2 \cdot 2}{1 \cdot 3} \cdot \frac{4 \cdot 4}{3 \cdot 5} \cdot \frac{6 \cdot 6}{5 \cdot 7} \cdot \frac{8 \cdot 8}{7 \cdot 9} \cdots,$$

and his colleague Lord Brouncker discovered the mysteriously similar result

$$\frac{\pi}{4} = \cfrac{1}{1 + \cfrac{1^2}{2 + \cfrac{3^2}{2 + \cfrac{5^2}{2 + \cfrac{7^2}{2 + \cfrac{9^2}{2 + \ddots}}}}}}$$

One of Euler's achievements was to explain how the three formulas for π are related. The relation between the series and the continued fraction is explained in section 10.9.

Thus Euler had good reason to think that infinite series, and a few similar infinite processes, are a rich field that can be explored before embarking upon calculus. All of the above results actually appear in his book, together with many of Euler's own. Indeed, Euler is now considered to be the most brilliant and creative virtuoso of infinite series who ever lived, so it is no surprise that his *Introductio* goes well beyond the needs of a prospective calculus student. Nevertheless, we can all marvel at what he did.

For example, by comparing the power series for sine, cosine, and the exponential function, he discovered the formula

$$e^{i\theta} = \cos\theta + i\sin\theta,$$

with its miraculous special case

$$e^{i\pi} = -1.$$

And by multiplying geometric series together he discovered that, for any $s > 1$,

$$\left(\frac{1}{1-2^{-s}}\right)\left(\frac{1}{1-3^{-s}}\right)\left(\frac{1}{1-5^{-s}}\right)\cdots\left(\frac{1}{1-p^{-s}}\right)\cdots$$
$$= \frac{1}{1^s} + \frac{1}{2^s} + \frac{1}{3^s} + \frac{1}{4^s} + \frac{1}{5^s} + \cdots,$$

where p runs through all the prime numbers. The exponent $s > 1$ is there to ensure that the series on the right-hand side (the so-called *zeta* function of s) has a finite sum. For $s = 1$ the sum is infinite by Oresme's result, but this fact too is something we can exploit. If there are only finitely many primes, then the left side is *finite*, so we have a contradiction. Therefore, we have a new proof that there are infinitely many primes!

This result is just the first of a cornucopia of results about primes that fall out of Euler's formula.

The Geometric Series in Hardy's Pure Mathematics

> I have indeed in an examination asked a dozen candidates, including several future Senior Wranglers, to sum the series $1 + x + x^2 + \cdots$, and have not received a single answer that was not practically worthless—and this from men quite capable of solving difficult problems connected with the curvature and torsion of twisted curves.

Hardy (1908), p. vi

The humble geometric series, which we have used as a foundation for much of this chapter, itself depends on a fundamental fact about limits: that $x^n \to 0$ as $n \to \infty$ when $|x| < 1$. For beginners, this fact seems obvious enough, and it was assumed in landmark works on the foundations of calculus, such as Cauchy (1821) and Jordan (1887). However, Hardy (1908), in his famous *Course of Pure Mathematics*, thought it worthwhile to probe more deeply, because his hope for his students was that

> accurate thought in connexion with these matters will become an integral part of their ordinary mathematical habit of mind. It is this conviction that has led me to devote so much space to the most elementary ideas of all connected with limits, to be purposely diffuse about fundamental points, to illustrate them by so elaborate a system of examples, and to write a chapter of fifty pages without advancing beyond the ordinary geometrical series.

Hardy (1908), p. vii

So Hardy embeds his discussion of the geometric series in a long chapter about basic properties of limits. These include some properties of increasing sequences that depend on the completeness of $\mathbb{R}$ (see next section). However, the fact that $x^n \to 0$ when $|x| < 1$ is proved in elementary fashion.

Hardy offers two proofs. The simpler proof, in my opinion, is to consider $0 < x < 1$ and write

$$x = \frac{1}{1+h}, \quad \text{with} \quad h > 0.$$

Now $(1+h)^n \geq 1 + nh$ by the binomial theorem. (This can also be proved directly, by induction on n.) It follows, since $nh \to \infty$ as $n \to \infty$, that

$$(1+h)^n \to \infty \quad \text{and therefore} \quad x^n = \frac{1}{(1+h)^n} \to 0.$$

6.11 Philosophical Remarks

The line between elementary and advanced calculus is typically crossed in two different ways:

1. By considering functions beyond the elementary functions of section 6.9.

   These usually arise from more complicated integrals, so we can exclude them by allowing only integrals of rational functions. As explained in section 6.9, this is quite a natural place to draw the line between elementary and advanced.
2. By deeper study of fundamental concepts, such as the real numbers, continuous functions, and differentiability; that is, by studying *analysis*.

   So, there is a line between calculus and analysis; but the line is difficult to draw. As we saw in section 6.3, the elementary theorem that a constant function has zero derivative has a difficult converse—seemingly on the advanced side of the line. In cases like this (and there are many of them in analysis) we should at least try to glimpse what lies on the other side.

Here then are a few remarks on some key issues of beginning analysis.

*Completeness of $\mathbb{R}$

To get anywhere in analysis we need to know that the system $\mathbb{R}$ of real numbers is a good model of the line. In particular, $\mathbb{R}$ must have "no gaps," so that there is a real number for every point on the line. This property of $\mathbb{R}$ is called *completeness*. The concept can be made precise in two equivalent ways, the second of which involves the nested intervals we have already used to prove the zero derivative theorem in section 6.3. This idea goes back to Bolzano, and we will see it again in the so-called *Bolzano-Weierstrass theorem* in section 7.9.

Least upper bound property. Any bounded set S has a least upper bound; that is, a number $l \geq$ each member of S, but such that if $k < l$ then $k <$ some member of S.

Nested interval property. If I_n is the closed interval

$$[a_n, b_n] = \{x \in \mathbb{R} : a_n \leq x \leq b_n\}$$

and if $I_1 \supseteq I_2 \supseteq I_3 \supseteq \cdots$ then there is an x common to all of $I_1, I_2, I_3, \ldots$.

This property is needed to prove many theorems about the existence of seemingly obvious points, such as the *intermediate value theorem*, which says that a continuous function that takes both negative and positive values also takes the value zero at some point.

A definition of $\mathbb{R}$ that ensures completeness was first given by Dedekind in 1858. The definition is quite simple, in terms of sets of rational numbers, but also quite profound, because it is defining an "actual infinity" of numbers. Because of the involvement of infinity, we postpone this definition until chapter 9 on logic. It is doubtful that a careful study of the real numbers can really be elementary, because the careful study of infinity is not elementary. But at least we can get closer to it from the viewpoint of logic.

*Continuity

The concept of continuity is related to the completeness of $\mathbb{R}$, but in a very subtle way. Our intuition is that a function f is continuous if its graph $y = f(x)$ has "no gaps," like a wiggly version of $\mathbb{R}$. But to make this intuition precise we must, unfortunately, express the "absence of gaps" quite indirectly. The usual way in analysis is to first say what it means for f to be *continuous at the point* $x = a$. We defined this in section 6.3 by saying that $f(x) \to f(a)$ as $x \to a$. In other words, we are saying that $f(x)$ can be made as close as we please to $f(a)$ (within distance ε, say) by choosing x sufficiently close to a (within distance δ, say).

Then we said that f is continuous if it is continuous at all points (in its domain, which need not be all of $\mathbb{R}$). The property of "absence of gaps in the graph" is a *consequence* of this definition, partly expressed by the intermediate value theorem.

The definition of continuous function arose from the work of Bolzano (1817) (attempting to prove the intermediate value theorem) and Cauchy (1821). But it was not possible to prove that continuous functions had the expected properties until 1858, when Dedekind's definition of real numbers made the completeness of $\mathbb{R}$ provable. One can hardly expect to prove that a continuous graph has no gaps until one has proved that $\mathbb{R}$ itself has no gaps!

Given the close relationship between continuity and the completeness of $\mathbb{R}$, one would not expect continuity to be an elementary concept. This is supported by two other observations about continuity:

1. To prove that certain definitions of continuity[3] are equivalent requires the *axiom of choice*—an axiom of advanced set theory. (For more on the axiom of choice, see section 9.10.)
2. The constructivist school of mathematicians believe that discontinuous functions are not well-defined.

While I consider the constructivist view to be extreme, as do most

---

[3] One very natural definition, used in the excellent analysis book of Abbott (2001), can be stated here. A function f is continuous at $x = a$ if $f(a_n) \to f(a)$ as $n \to \infty$, for any sequence $a_1, a_2, a_3, \ldots$ such that $a_n \to a$ as $n \to \infty$. To prove that this definition ("sequential continuity") is equivalent to the one given in section 6.3 requires the so-called *countable axiom of choice*.

mathematicians, I believe that constructivists play the role of "canary in the coal mine" for mathematics. If constructivists have doubts about a concept or a theorem, it is a sign of a deep idea. Concepts that worry constructivists probably belong to advanced mathematics.

Another sign that continuity is an advanced concept appears in the famous book of G. H. Hardy, *A Course of Pure Mathematics*. On p. 185 of the 8th edition, Hardy (1941) motivates the above definition of continuity by claiming:

> To be able to define continuity *for all values of* x we must first define continuity *for any particular value of* x.

This sounds very plausible, but it is not true. Hausdorff (1914), p. 361, introduced a very general approach to continuity via *open sets*. For example, any open set in $\mathbb{R}$ is a union of open intervals, where an open interval is a set of the form $(a, b) = \{x \in \mathbb{R} : a < x < b\}$. Hausdorff defines f to be continuous *at all points of its domain* if each open set in the range of f is the image of an open set in the domain of f.[4] Hardy was one of the leading analysts of his time so, if he could be this mistaken about continuity, it must be an advanced concept!

---

[4] This is now a standard definition of continuity; if not in analysis, then certainly in topology.

7

⤚⤳

Combinatorics

PREVIEW

Combinatorics is often described by terms such as "finite," "discrete," or "counting." As such, it is hard to distinguish from arithmetic. Both look like the theory of finite sets, though this is more obvious for combinatorics, which often uses low-level set concepts such as membership and containment. The key concepts of arithmetic—addition, multiplication, and their algebraic structure—seem to lie at a higher level, since they are not so easily expressed in set-theoretic terms (though this *can* be done, as we will see in chapter 9).

Thus combinatorics emerges as a field, more elementary than arithmetic, with the potential to clarify other parts of mathematics by identifying their combinatorial content. This is mainly what we do in this chapter: first exhibiting combinatorial content in arithmetic, then (at greater length) in geometry.

It was the discovery of combinatorial content in geometry—with the *Euler polyhedron formula* of 1752—that led to the now vast field of topology, in which combinatorics ultimately joined forces with ideas from analysis, such as limits and continuity. The part of topology involving the most elementary concepts—*graph theory*—is now the largest subfield of combinatorics.

Analysis itself has interesting combinatorial content. As befits a subject that studies infinite processes, the combinatorics in analysis can itself be infinite, but it is enlightening nevertheless. The simplest theorem of infinite graph theory, the *Kőnig infinity lemma*, expresses the combinatorial content of the *Bolzano-Weierstrass theorem*. The latter theorem, in combination with *Sperner's lemma* from finite graph theory, proves the *Brouwer fixed point theorem*. This famous theorem of topology was thought to be quite difficult before its combinatorial content was uncovered.

7.1 The Infinitude of Primes

A new proof that there are infinitely many primes, with a distinctly combinatorial flavor, was given by Thue (1897).

Infinitude of primes. *There are infinitely many prime numbers.*

Proof. We assume the result mentioned in section 2.3, that every positive integer $n > 1$ has prime factorization. Now suppose (aiming for a contradiction) that there are only k prime numbers: 2, 3, ..., p. Then it follows from prime factorization that every positive integer $n > 1$ can be written in the form

$$n = 2^{a_1} 3^{a_2} \cdots p^{a_k} \quad \text{for integers } a_1, a_2, \ldots, a_k \geq 0.$$

Obviously, if $n < 2^m$ then $a_1, a_2, \ldots, a_k < m$. But the number of sequences of k positive integers $a_i < m$ is just m^k, because each term in the sequence takes m different values: 0, 1, ..., $m - 1$.

  Since k is fixed, $m^k < 2^m$ for m sufficiently large; in fact $m^k < 2^{m-1}$. This means that, for m sufficiently large, there are not enough sequences $a_1, a_2, \ldots, a_k$ to represent all the numbers $n < 2^m$ in the form $2^{a_1} 3^{a_2} \cdots p^{a_k}$, so we have a contradiction. $\qquad \square$

  This proof illustrates the "counting" feature often used to describe combinatorics. In this case we count how many positive integers can be written as products of a fixed number of primes with bounded exponents, and conclude that the number of products does not grow fast enough to keep up with the number of positive integers. More precisely, for exponents less than m we cannot keep up with the numbers with less than m binary digits.

7.2 Binomial Coefficients and Fermat's Little Theorem

Fermat seems to have found the theorem that we write as $a^{p-1} \equiv 1$ (mod p) from considerations involving binomial coefficients. For more on the history of the theorem, see Weil (1984). Fermat actually proved the equivalent result $a^p \equiv a$ (mod p), which is true for *all* integers a. It can be obtained from the binomial theorem as follows.

Fermat's version of his little theorem. *If p is prime and a is any natural number, then*

$$a^p \equiv a \ (mod \ p).$$

Proof. The cases $a = 0$ and $a = 1$ are obvious, so consider $a = 2$. Because of the binomial theorem,

$$2^p = (1+1)^p = 1 + \binom{p}{1} + \binom{p}{2} + \cdots + \binom{p}{p-1} + 1. \quad (*)$$

Also, by the combinatorial interpretation of the binomial coefficients, as in section 1.6,

$$\binom{p}{k} = \frac{p(p-1)\cdots(p-k+1)}{k!}.$$

Since $\binom{p}{k}$ is an integer, all the factors in $k!$ must divide factors in the numerator. But, since $k < p$ and p is prime, the factors in the denominator do *not* divide p. Therefore, p divides $\binom{p}{k}$ for $k = 1, 2, \ldots, p-1$, and so it follows from (*) that

$$2^p \equiv 1 + 1 = 2 \ (mod \ p).$$

Now, supposing inductively that we have proved $a^p \equiv a$ for $a = n$, it follows for $a = n+1$ because

$$(n+1)^p = n^p + \binom{p}{1} n^{p-1} + \cdots$$

$$+ \binom{p}{p-1} n + 1 \quad \text{by the binomial theorem again,}$$

$$\equiv n^p + 1 \;(\text{mod } p), \quad \text{because } p \text{ divides } \binom{p}{k},$$

$$\equiv n + 1 \;(\text{mod } p) \quad \text{by the induction hypothesis.}$$

Hence it follows by induction that $a^p \equiv a \;(\text{mod } p)$ for all natural numbers a (and hence also for negative integers, since each of them is congruent to a natural number, mod p). $\qquad \square$

The benefits that combinatorics brings to this proof are not only the formula

$$\binom{n}{k} = \frac{n(n-1) \cdots (n-k+1)}{k!},$$

but also the interpretation (as the number of ways of choosing k things from n) which makes it clear that $\frac{n(n-1)\cdots(n-k+1)}{k!}$ is an integer. A proof of the latter fact using conventional arguments about divisibility is quite difficult. In fact, Gauss (1801), article 127, took the trouble to give such a pure number theory proof, because "until now, as far as we know, no one has demonstrated it directly." Today, it seems obvious that combinatorics owns the "best" proof of this theorem, not number theory. It is an early example of the way in which proofs can be simplified by importing ideas from outside those in the statement of a theorem.

7.3 Generating Functions

Combinatorics often produces sequences of numbers in new ways, so it is a challenge to describe these sequences in conventional arithmetic or algebraic terms. A powerful technique for bringing combinatorics and algebra together is the so-called method of *generating functions*. We illustrate this method with two examples: the sequence of binomial coefficients and the sequence of Fibonacci numbers.

The Binomial Coefficients

For each n we have already defined the sequence of binomial coefficients, $\binom{n}{k}$, each of which equals the number of ways of choosing k things from a set of n things. We have also seen, in section 1.6, that these numbers are the coefficients in the expansion of $(a+b)^n$ (hence the name "binomial"). Now, for the sake of uniformity, we slightly change the setup so as to use one variable, x, rather than the two variables a and b. Namely, we package the sequence of numbers

$$\binom{n}{0}, \quad \binom{n}{1}, \quad \binom{n}{2}, \quad \ldots, \quad \binom{n}{n}$$

in the polynomial

$$\binom{n}{0} + \binom{n}{1}x + \binom{n}{2}x^2 + \cdots + \binom{n}{n}x^n.$$

As we know from the binomial theorem, this polynomial has a more concise description, namely

$$(1+x)^n.$$

This concise description leads to easy proofs of certain properties of the binomial coefficients. For example, we can prove the famous property of Pascal's triangle,

$$\binom{n}{k} = \binom{n-1}{k-1} + \binom{n-1}{k},$$

simply by writing

$$(1+x)^n = 1 \cdot (1+x)^{n-1} + x \cdot (1+x)^{n-1}.$$

This is essentially what we did in section 1.6, and it clearly illustrates the benefits of packaging a sequence of numbers by an algebraically simple generating function. Our next example shows that a simple generating function may exist even for an infinite sequence of numbers.

The Fibonacci Sequence

In the *Liber abaci* (Book of calculation), Fibonacci (1202) introduced the sequence of numbers

$$0, 1, 1, 2, 3, 5, 8, 13, 21, 34, 55, 89, 144, 233, 377, 610, 987, 1597, 2584, \ldots$$

in which each term, from the third onwards, is the sum of the previous two. In the time-honored tradition of bogus "applications," he invented a situation (the "rabbit problem") in which these numbers arise. But really, the sequence is an exercise in adding Arabic numerals, in keeping with the purpose of the *Liber abaci*.

At any rate, the Fibonacci sequence continued to fascinate mathematicians long after they learned how to add. This simple process of iterated addition produces a sequence which is surprisingly complex and enigmatic. In particular, a formula for the nth term is so elusive that it was not found for over 500 years. And, surprisingly, the formula involves the irrational number $\sqrt{5}$.

The formula was found, apparently independently, by Daniel Bernoulli (1728) and de Moivre (1730). Moreover, their proofs used the same idea, which amounts to finding and manipulating the generating function for the Fibonacci numbers:

$$F(x) = 0 \cdot 1 + 1 \cdot x + 1 \cdot x^2 + 2x^3 + 3x^4 + 5x^5 + 8x^6 + 13x^7 + \cdots .$$

To study $F(x)$ we inductively introduce the following symbols for the Fibonacci numbers:

$$F_0 = 0, \quad F_1 = 1, \quad F_n = F_{n-1} + F_{n-2} \text{ for } n \geq 2.$$

Then the Fibonacci generating function can be rewritten as

$$F(x) = F_0 + F_1 x + F_2 x^2 + F_3 x^3 + \cdots .$$

First we compare $F(x)$ with $xF(x)$ and $x^2 F(x)$, which are:

$$xF(x) = \qquad F_0 x + F_1 x^2 + \cdots + F_{n-1} x^n + \cdots$$
$$x^2 F(x) = \qquad F_0 x^2 + \cdots + F_{n-2} x^n + \cdots .$$

Then, subtracting the series for $xF(x)$ and $x^2 F(x)$ from $F(x)$ gives

$$(1 - x - x^2)F(x) = F_0 + (F_1 - F_0)x + \cdots$$
$$+ (F_n - F_{n-1} - F_{n-2})x^n + \cdots$$
$$= x \qquad \text{because } F_0 = 0, \; F_1 = 1,$$
$$\text{and } F_n - F_{n-1} - F_{n-2} = 0.$$

Therefore

$$F(x) = \frac{x}{1 - x - x^2}.$$

Formula for F_n. *The coefficient of x^n in $F(x)$ is given by*

$$F_n = \frac{1}{\sqrt{5}}\left[\left(\frac{1+\sqrt{5}}{2}\right)^n - \left(\frac{1-\sqrt{5}}{2}\right)^n\right].$$

Proof. To extract information from $F(x) = \frac{x}{1-x-x^2}$ we split its denominator into linear factors, using the roots of $1 - x - x^2 = 0$. By the quadratic formula, these are

$$x = -\frac{1-\sqrt{5}}{2}, \; -\frac{1+\sqrt{5}}{2},$$

which, since $(1 - \sqrt{5})(1 + \sqrt{5}) = -4$, can be seen to equal

$$x = \frac{2}{1+\sqrt{5}}, \; \frac{2}{1-\sqrt{5}}.$$

So we have the factorization

$$1 - x - x^2 = \left(1 - \frac{1+\sqrt{5}}{2}x\right)\left(1 - \frac{1-\sqrt{5}}{2}x\right).$$

Next we set

$$\frac{x}{1-x-x^2} = \frac{x}{\left(1-\frac{1+\sqrt{5}}{2}x\right)\left(1-\frac{1-\sqrt{5}}{2}x\right)} = \frac{A}{1-\frac{1+\sqrt{5}}{2}x} + \frac{B}{1-\frac{1-\sqrt{5}}{2}x}.$$

Comparing coefficients in the numerator after taking a common denominator on the right, we find $A = -B = \frac{1}{\sqrt{5}}$. This gives the *partial fraction* decomposition

$$F(x) = \frac{x}{1-x-x^2} = \frac{1}{\sqrt{5}}\left(\frac{1}{1-\frac{1+\sqrt{5}}{2}x} - \frac{1}{1-\frac{1-\sqrt{5}}{2}x}\right).$$

Finally, we expand the fractions $\frac{1}{1-\frac{1+\sqrt{5}}{2}x}$ and $\frac{1}{1-\frac{1-\sqrt{5}}{2}x}$ as geometric series, using

$$\frac{1}{1-a} = 1 + a + a^2 + \cdots.$$

We find that the coefficient of x^n, which is F_n by definition of $F(x)$, is

$$F_n = \frac{1}{\sqrt{5}}\left[\left(\frac{1+\sqrt{5}}{2}\right)^n - \left(\frac{1-\sqrt{5}}{2}\right)^n\right]. \qquad \square$$

7.4 Graph Theory

Graphs are perhaps the simplest finite *geometric* objects. Certainly, graph theory is the most visual part of combinatorics, and the part most accessible to geometrically minded mathematicians. Figure 7.1 shows an example of a graph in the way they are normally presented.

The points marked v_1, v_2, v_3, v_4 are some of the *vertices* of the graph, the lines marked e_1, e_2, e_3 are some of its *edges*, and the sequence $v_1, e_1, v_2, e_2, v_3, e_3, v_4$ is an example of a *path*. This path is called *simple* because it contains no repeated vertex. The formal definition of a graph is as follows.

Definition. A (finite) *graph* G consists of a finite set of objects v_i called its *vertices*, and a set of pairs $e_k = \{v_i, v_j\}$ with $v_i \neq v_j$, called its *edges*.

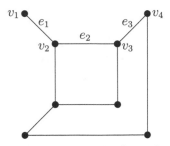

Figure 7.1: Example of a graph.

Thus, in principle, the vertices of a graph can be any kind of mathematical objects, such as natural numbers. In practice, we think of vertices as points in the plane or space. Also an edge is completely determined by a pair $\{v_i, v_j\}$ of vertices (its "endpoints"), though in practice we view the edge as a curve joining the points v_i and v_j. It follows that there is at most one edge with given endpoints, and there is no edge whose endpoints are the same.

It also follows that many different pictures can represent the same graph. The positions of the vertices (as long as distinct vertices are in different positions) and the lengths and shapes of the edges are irrelevant. This means that the "geometry" we are doing is actually *topology*. One of the important concepts of graph theory in particular, and topology in general, is the concept of path. We have already illustrated this concept in figure 7.1, and we formalize it (and related concepts) as follows.

Definitions. A *path* in a graph G is a sequence of vertices and edges of G of the form

$$v_1 e_1 v_2 e_2 v_3 \cdots v_n e_n v_{n+1},$$

where each $e_i = \{v_i, v_{i+1}\}$. The path is called *closed* if $v_1 = v_{n+1}$, and *simple* if no vertex occurs in it twice, except in the case $v_1 = v_{n+1}$. Finally, G is called *connected* if any pair of its vertices occur in a path.

In practice we are concerned mainly with connected graphs, because any graph is a union of connected graphs (called its "connected components") which can be considered separately.

Finally, we wish to define the concept of *valency* of a vertex, so-called because it is similar to the concept of that name in chemistry.

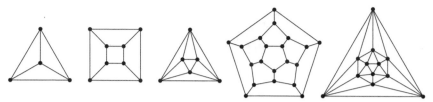

Figure 7.2: The graphs of the regular polyhedra.

Definition. The *valency* of a vertex v in a graph G is the number of edges of G to which v belongs.

For example, in figure 7.1,

$$\text{valency}(v_1) = 1,$$
$$\text{valency}(v_2) = 3,$$
$$\text{valency}(v_3) = 3.$$

The valency is often called the *degree* in combinatorics literature, but I think that "valency" is clearly preferable, because it is not used elsewhere in mathematics, whereas "degree" is used to excess. The concept of valency appears in the first theorem of graph theory:

Total valency. *In any graph, the sum of the valencies of the vertices is even.*

Proof. The sum of the valencies is the sum of contributions from the edges. Each edge contributes 2 to the sum, so the sum of the valencies is even. □

We conclude this section with pictures of some particularly beautiful graphs, the *regular polyhedral* graphs, which come from projecting regular polyhedra onto the plane (figure 7.2). In section 7.6 we will say more about the regular polyhedra, and give a graph-theoretic proof that there are only five of them.

7.5 Trees

A *tree* is a connected graph containing no simple closed path. Some examples of trees are shown in figure 7.3.

So, yes, these graphs do look like trees—more or less.

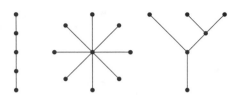

Figure 7.3: Some trees.

Since a tree contains no simple closed path, it is possible to draw any tree in the plane without edges crossing. Intuitively, one imagines building a picture of the tree one edge at a time. Each edge leads to a new vertex (otherwise a simple closed path would be created), so it can be drawn without crossing the edges previously drawn. This intuition can be justified with the help of the following result, which leads to a process for building each tree one edge at a time.

Valency in trees. *Each tree with more than one vertex has a vertex of valency 1.*

Proof. Pick any vertex v_1 of the tree T. If v_1 has valency 1 we are done. If not, follow an edge e_1 out of v_1 to a vertex v_2 one edge away. If v_2 is not of valency 1, follow an edge $e_2 \neq e_1$ out of v_2, and continue in this way. This creates a simple path $v_1 e_1 v_2 e_2 \cdots$ in T. Since T is finite and without closed paths, $v_1 e_1 v_2 e_2 \cdots$ must terminate, necessarily at a vertex of valency 1.

(Incidentally, if v_1 does have valency 1, the above process leads to a *second* vertex of valency 1. And if v_1 has valency > 1, by taking an edge $e_1' \neq e_1$ out of v_1 we again arrive at a second vertex of valency 1. So in fact a tree has at least two vertices of valency 1.) □

Once we have found a vertex v of valency 1 we can remove v, and the edge e to which it belongs, without disconnecting the graph. Thus the graph that remains is still a tree, and we can repeat the process until the tree is reduced to a single vertex. By doing this process in reverse, we can build the tree one edge at a time, and hence make a drawing of it in the plane without edges crossing. For this reason, trees are said to be *planar* graphs. We study planar graphs more generally in the next section, but trees are the foundation for this study, because of the following theorem.

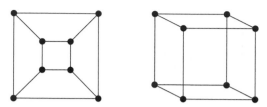

Figure 7.4: Plane and nonplane views of the cube graph.

Characteristic number of a tree. *If a tree has V vertices and E edges, then*

$$V - E = 1.$$

Proof. This is true of the smallest tree, which has one vertex and no edges. It remains true when we add one vertex v and one edge e. So, since any tree can be built by adding one vertex and one edge at a time, $V - E = 1$ for any tree. □

The invariant number $1 = V - E$ is sometimes called the *Euler characteristic* of the tree. The idea extends to the Euler characteristic of graphs in the plane, which is related to a characteristic number for polyhedra, discovered by Euler in 1752. How all these ideas fit together will be explained in the next section.

7.6 Planar Graphs

In this section we clarify what it means to "draw" a graph in the plane without edges crossing. We wish to distinguish, for example, between the two pictures shown in figure 7.4.

Both can be recognized as views of the same graph, whose vertices and edges correspond to the vertices and edges of a cube. But the first is an *embedding* of these vertices and edges in the plane, in the sense that distinct points go to distinct points. We also call it a *plane graph*. The second picture is not a plane graph, because in two cases (where edges cross) two distinct points on edges of the cube go to the same point in the plane.

The cube graph embeds quite naturally in the plane because we can project a cube onto the plane without edges crossing, by suitably choos-

Figure 7.5: The five regular polyhedra (images from Wikimedia Commons).

ing the point of projection. This also explains why the plane cube graph has straight edges. It is in fact true that, if a graph can be embedded in the plane, there is an embedding with straight edges, though we will not make a detour to prove this fact. Instead, we will allow the embedded edges to be *polygonal*; that is, simple paths consisting of finitely many line segments. (We may as well allow this, since a simple closed path in the graph is going to be polygonal in any case.)

Definition. A *plane graph* G is one whose vertices are points of $\mathbb{R}^2$ and whose edges are polygonal paths, meeting only where they have common endpoints.

To capture the idea of a plane graph being a "picture" of another graph we need the concept of *graph isomorphism*.

Definition. Graphs G and G' are *isomorphic* if there is a one-to-one correspondence $v_i \leftrightarrow v_i'$ between the vertices v_i of G and the vertices v_i' of G' such that $\{v_i, v_j\}$ is an edge of G if and only if $\{v_i', v_j'\}$ is an edge of G'.

Since we are now considering concrete graphs, whose edges may be line segments or polygonal paths, the "edge $\{v_i, v_j\}$" should be interpreted as the line segment or polygonal path with endpoints v_i and v_j. Under this interpretation, the two graphs shown in figure 7.4 are isomorphic, and there are several suitable correspondences between their vertices.

Now, finally, we can define *planarity*.

Definition. A graph G is *planar* if G is isomorphic to a plane graph G'.

Many graphs that are naturally realized by points and line segments in $\mathbb{R}^3$ are in fact planar. The most famous examples are the graphs of the regular polyhedra (figure 7.5), all of which can be projected one-to-one

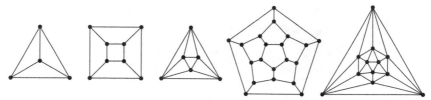

Figure 7.6: The plane graphs of the regular polyhedra.

onto the plane. One of them is the cube, whose plane graph has already been shown in figure 7.4. Indeed, plane graphs for all of them have been shown in figure 7.2.

To help you compare, figure 7.6 shows these plane graphs again.

In the *Elements* (Book XIII, Proposition 18), Euclid showed that these five polyhedra are the only regular ones, using the fact that the angle sum of an n-sided polygon is $(n-2)\pi$. (This follows easily from the theorem of section 5.2 that the angle sum of a triangle is π.) It follows that, if the polygon is regular, each of its angles is $\frac{n-2}{n}\pi$. Further, if the polyhedron is regular, so that the number of polygons meeting at a vertex is a constant number m, one finds that the angle sum at a vertex is $< 2\pi$ only in the following cases:

$$n=3, \quad m=3, 4, 5, \quad \text{(tetrahedron, octahedron, icosahedron)}$$

$$n=4, \quad m=3, \quad \text{(cube)}$$

$$n=5, \quad m=3. \quad \text{(dodecahedron)}$$

Thus the five known regular polyhedra are the only ones possible.

Interestingly, one does not need concepts of Euclidean geometry to show that the above values of m and n are the only ones possible. They follow from pure graph theory, as one of the consequences of a fundamental theorem about plane graphs.

7.7 The Euler Polyhedron Formula

Euler (1752) discovered that each polyhedron has a characteristic number $2 = V - E + F$, where V, E, and F are the numbers of vertices, edges, and faces, respectively. For example, for the cube we

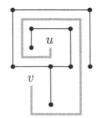

Figure 7.7: Connecting points outside the tree.

have $V - E + F = 8 - 12 + 6 = 2$, and for the tetrahedron $V - E + F = 4 - 6 + 4 = 2$. The theorem holds more generally for connected plane graphs, if we use the appropriate concepts of "edge" and "face," and we will prove it in that more general context. A special case of the theorem for plane graphs is the result about trees proved in section 7.5, that $V - E = 1$, because $F = 1$ for trees, as we will see. In fact, we prove that $V - E + F = 2$ in general by reducing to the special case of trees.

Euler plane graph formula. *If G is a connected plane graph with V vertices, E edges, and F faces, then*

$$V - E + F = 2.$$

The term "face" of course specializes to the face of a polyhedron, or to the polygon in the plane which results from projecting the polyhedron. In the case of an arbitrary connected plane graph G, the "faces" are the connected components of $\mathbb{R}^2 - G$, where points u and v are in the same connected component if they may be connected by a path in $\mathbb{R}^2$ that does not meet G. In particular, if G is a tree then G has *one* face, because any two points not in G may be connected by a path that does not meet the tree. See, for example, figure 7.7.

Another important example is where G is a simple polygon. In this case there are *two* faces, corresponding to the "inside" and "outside" of G. To expedite the proof of the formula, we assume these two results about faces, but we comment further on them in the subsection below.

Proof of the Euler plane graph formula. Suppose we are given a connected plane graph G with V vertices, E edges, and F faces. If G is a tree, then we are done, because $V - E = 1$ by section 7.5, and $F = 1$ by the assumption above.

If G is not a tree, then G contains a simple closed path p. We can remove any edge e from p, and the graph remains connected because vertices connected using the edge e can still be connected by replacing e by what remains of the path p.

Notice also that when e is removed:

- V remains the same, because we remove only e, not its endpoints.

- E is reduced by 1.

- F is reduced by 1, because the two faces on opposite sides of e (which are distinct by the second assumption above) become one when e is removed.

Thus removal of an edge from a closed path does not change $V - E + F$, nor does it disconnect the graph, so we can repeat the process.

But the process does reduce the number of simple closed paths by 1, so after finitely many removals we have a tree. At this stage we know that $V - E + F = 2$, so $V - E + F = 2$ for the original graph G as well. □

The formula $V - E + F = 2$ applies, in particular, to any convex polyhedron (since such a polyhedron can be projected onto $\mathbb{R}^2$ from a point just outside the middle of any face), and more generally to any polyhedron whose graph is planar. Intuitively speaking, these are the polyhedra without "holes." For these, we call it the *Euler polyhedron formula* after Euler (1752).

We now use the formula to show that the five regular polyhedra in figure 7.5 are the only polyhedra that are regular in the combinatorial sense. That is, they are the polyhedra in which each vertex is the meeting point of m faces and each face has n edges, for constant m and n.

Enumeration of regular polyhedra. *If a polyhedron has m faces meeting at each vertex, and each face has n edges, then the possible values of (m, n) are:*

$$(3, 3), \quad (3, 4), \quad (4, 3), \quad (3, 5), \quad (5, 3).$$

Proof. Suppose the polyhedron has V vertices, E edges, and F faces. Since each face has n edges, we have

$$E = nF/2 \quad \text{because each edge is shared by exactly two faces.}$$

Similarly

$$V = nF/m \quad \text{because each vertex is shared by exactly } m \text{ faces.}$$

Substituting these expressions for V and E in $V - E + F = 2$, we get

$$2 = \frac{nF}{m} - \frac{nF}{2} + F = F\left(\frac{n}{m} - \frac{n}{2} + 1\right) = F\frac{2n - mn + 2m}{2m},$$

and therefore

$$F = \frac{4m}{2m + 2n - mn}.$$

Since F must be positive, we must have

$$2m + 2n - mn > 0 \quad \text{or, equivalently,} \quad mn - 2m - 2n < 0. \quad (*)$$

Now we notice that $mn - 2m - 2n + 4 = (m - 2)(n - 2)$. So, adding 4 to both sides of $(*)$ gives

$$(m - 2)(n - 2) < 4.$$

Also, $n \geq 3$ because a polygon has at least three sides, and $m \geq 3$ because at least three faces meet at each vertex of a polyhedron. Thus it remains to solve

$$(m - 2)(n - 2) < 4 \quad \text{for } m \geq 3 \text{ and } n \geq 3. \quad (**)$$

Routine checking shows that the solutions of $(**)$ are

$$(m, n) = (3, 3), (3, 4), (4, 3), (3, 5), (5, 3). \qquad \square$$

Conversely, plane graphs actually exist for all these pairs of values (m, n), as figure 7.6 shows. This gives a combinatorial analogue of Euclid's Proposition 18, Book XIII, in the *Elements*, mentioned in the previous section. Interestingly, the actual construction of the polyhedra/plane graphs is the more difficult part of both theorems. However, the plane graphs are easier than the polyhedra.

*Face Numbers for Trees and Polygons

There is a story about the English mathematician G. H. Hardy that goes as follows (it is probably a myth, but still it strikes a chord[1]):

> In the course of a lecture, Hardy started to say "It is obvious that ...," then checked himself. After a long pause, he left the lecture theatre and paced up and down the corridor. Returning after about 15 minutes, he resumed his lecture by saying "Yes, it *is* obvious."

We are in a similar bind concerning the face numbers of trees and polygons. They look obvious, and with a closer look they can be proved by constructions involving a finite number of simple steps. Yet they are not *immediately* obvious. One has to think for a long time before their obviousness becomes, well, obvious. To my mind, this makes them "advanced" rather than elementary.

Face number of a tree. *A plane graph of a tree has one face.*

Proof. Let T be a plane graph which is a tree. So the vertices of T are points of $\mathbb{R}^2$ and the edges of T are simple polygonal paths in $\mathbb{R}^2$, which meet only at common endpoints. By considering all corners of polygonal paths as vertices, we can reimagine T as a plane tree T^* whose edges are line segments.

To prove that T^* has only one face, it suffices to show that any two points u, v in $\mathbb{R}^2$ but not in T^* can be joined by a polygonal path not meeting T^*. To do this we construct the ε-*neighborhood* $N_\varepsilon(T^*)$ of T^*, the set of points within distance at most ε from T^*, where ε is chosen small enough that the boundary of $N_\varepsilon(T^*)$ does not intersect itself or T^*, and so that u, v are not in $N_\varepsilon(T^*)$. Figure 7.8 shows an example.

The boundary of $N_\varepsilon(T^*)$ is a single closed curve, as can be seen by building T^* one edge at a time, as described in section 7.5, and building $N_\varepsilon(T^*)$ along with it. At first, when T^* is a single vertex, $N_\varepsilon(T^*)$ is

---

[1] An eyewitness account of a very similar incident in a lecture by Emil Artin is given in the book Ostermann and Wanner (2012), p. 7.

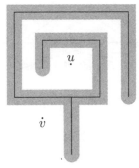

Figure 7.8: Tree T^* and its ε-neighborhood.

bounded by a circle of radius ε. As each edge is added, a "switchback" is added to the previous boundary curve, so the boundary continues to be a single curve.

Finally, to connect u and v we can extend lines from u and v until they hit the boundary curve of $N_\varepsilon(T^*)$, at u' and v' say, and then connect u' to v' by the piece of the boundary curve between them. (Strictly speaking, this path may not be polygonal, since it can contain some small circular arcs. But these arcs can be replaced by polygonal paths that avoid T^*.) □

A sign that this proof is advanced is the business of choosing ε "sufficiently small," which is typical of arguments in analysis and topology. The same idea appears in the proof that a polygon graph has two faces, but an extra ingredient is needed—the idea of "deforming" a path in a series of "small" steps.

Face number of a polygon.[2] *A plane polygon graph has two faces.*

Proof. Given any polygon P, let e be one of its edges and let $P - e$ be the graph obtained by removing e (but not its endpoints) from P. Then $P - e$ is a tree, so it follows from the previous proof that any two points u, v not on P (and hence not on $P - e$) can be joined by a polygonal path p not meeting $P - e$.

---

[2] This theorem is also known as the *Jordan curve theorem* for polygons. It is a special case of the more difficult theorem of Jordan (1887) that any simple closed curve in the plane divides the plane into two regions.

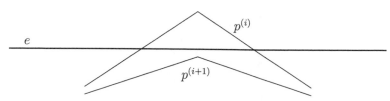

Figure 7.9: Removing two crossings.

Of course, the path p from u to v may cross the edge e. But, by making a series of small changes in p, deforming it to $p^{(1)}$, $p^{(2)}$, ..., $p^{(k)} = q$ we can remove crossings two at a time while keeping the deformed path away from $P - e$. The number of crossings finally obtained is therefore 0 or 1. Figure 7.9 shows how two crossings disappear in a small change from $p^{(i)}$ to $p^{(i+1)}$ near e.

A path that crosses e just once cannot be deformed to a path not crossing e, because it can be arranged that each change is by *two* crossings (namely, by ensuring that no edge of $p^{(i)}$ has the same direction as e and preventing corners of $p^{(i)}$ from touching $P - e$).

Now let a and b be points not on P that are connected by a path crossing e exactly once. We can obviously find such points by choosing a and b sufficiently close to e. Then it follows, for any point w not on P that exactly one of the following possibilities holds.

1. w can be joined to a by a path not crossing e,
2. w can be joined to b by a path not crossing e.

At least one possibility holds, because paths from w to a and b, each crossing e once, give a path from a to b (via w) crossing e twice, which is impossible. And we cannot have both, because that gives a path from a to b (via w) not crossing e at all.

This says that P divides the plane into regions: one containing a and the other containing b. □

The extra, advanced, ingredient in this proof is the idea of dividing a process into "sufficiently small" steps, so as to minimize the amount of change that can occur in a step (in this case, a change of ± 2 in the number of crossings). This, too, is typical of arguments in topology.

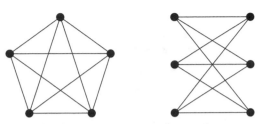

Figure 7.10: Two nonplanar graphs.

7.8 Nonplanar Graphs

The surprising power of the Euler plane graph formula can be demonstrated by using it to prove that certain graphs are *nonplanar*. Two famous examples are shown in figure 7.10.

The one on the left is called K_5, the *complete graph on five vertices*. Along with its five vertices, it has the ten edges connecting all pairs of distinct vertices. The one on the right is called $K_{3,3}$, the *complete bipartite graph on two sets of three vertices*. It contains an edge from each vertex in the first set of three to each vertex in the second set of three. $K_{3,3}$ is sometimes called the "utilities graph" because of its role in the following puzzle: given three houses and three utilities (gas, water, electricity), is it possible to connect each house to each utility without the lines crossing?

Any attempt to draw K_5 or $K_{3,3}$ in the plane without edges crossing will fail—but it is hard to know when all possible placements of the edges have been exhausted. The Euler plane graph formula removes all doubt.

Nonplanarity of K_5 and $K_{3,3}$. *Neither K_5 nor $K_{3,3}$ is a planar graph.*

Proof. First suppose (looking for a contradiction) that there is a plane version of K_5. For K_5 we know that $V = 5$ and $E = 10$, so the number F of faces in the plane version of K_5 satisfies

$$5 - 10 + F = 2 \quad \text{by the Euler plane graph formula.}$$

Thus $F = 7$. But each face must have at least three edges, so $F = 7$ implies

$$E \geq 3F/2 = 3 \cdot 7/2 > 10.$$

This contradicts $E = 10$, so K_5 is not a planar graph.

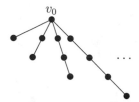

Figure 7.11: An infinite tree without infinite simple paths.

Again suppose (looking for a contradiction) that there is a plane version of $K_{3,3}$. For $K_{3,3}$ we know that $V = 6$ and $E = 9$, so the number F of faces in the plane version of $K_{3,3}$ satisfies

$$6 - 9 + F = 2 \quad \text{by the Euler plane graph formula.}$$

Thus $F = 5$. Now $K_{3,3}$ *contains no triangles*, because if vertices A, B are both connected to C then A, B are in the same set of three, so there is no edge AB. Therefore, any face of the plane version of $K_{3,3}$ must have at least four edges, so $F = 5$ implies

$$E \geq 4F/2 = 4 \cdot 5/2 > 9.$$

This contradicts $E = 9$, so $K_{3,3}$ is not a planar graph. □

7.9 *The Kőnig Infinity Lemma

The first book on graph theory, *Theorie der endlichen und unendlichen Graphen (Theory of Finite and Infinite Graphs)* by Kőnig (1936), already recognized the importance of infinite graphs. In it, Kőnig proved a fundamental theorem about infinite graphs, which at the same time isolated the combinatorial content of many theorems in analysis.

Kőnig infinity lemma. *A tree with infinitely many vertices, each of which has finite valency, contains an infinite simple path.*

Before giving the proof we remark that the definition of infinite tree is exactly the same as the definition of finite tree, except that the set of vertices is infinite. We also note that the finite valency condition in the lemma is necessary. Without it, we have the counterexample shown in figure 7.11.

In this tree there are infinitely many vertices, and arbitrarily long simple paths, but every simple path is finite. This happens because the valency of the top vertex v_0 is infinite.

Proof of the Kőnig infinity lemma. Let T be a tree with infinitely many vertices v_0, v_1, v_2, . . ., each of which has finite valency. Since T is connected, the edges out of v_0 lead to all other vertices, so at least one of these finitely many edges *itself* leads to infinitely many vertices of T. Choose such an edge $\{v_0, v_i\}$ (say, with the minimum possible i) as the first edge of a simple path p.

At v_i we repeat the argument. The finitely many edges out of v_i, other than $\{v_0, v_i\}$, together lead to infinitely many vertices of T, so one of them itself leads to infinitely many. Choose such an edge $\{v_i, v_j\}$ (with minimal possible j if you want to be specific) as the second edge of the path p.

We can repeat the argument at v_j, and we can continue indefinitely. Thus we obtain an infinite sequence of distinct vertices, v_0, v_i, v_j, . . ., each attached to the one before by an edge, so p is an infinite simple path in T. □

Underlying the above proof is a combinatorial principle even simpler than the lemma itself: the *infinite pigeonhole argument*. This principle states that if infinitely many objects are placed in finitely many boxes (pigeonholes), then one of the boxes contains infinitely many objects. The infinite pigeonhole principle seems very close to elementary mathematics—but it is not quite there, in my opinion. Some important theorems of analysis are proved along the same lines, and they are generally regarded as advanced. We give an example in the following subsection.

What the Kőnig lemma and infinite pigeonhole principle do show, I think, is that infinite combinatorics marks part of the boundary between elementary and advanced mathematics.

*The Bolzano-Weierstrass Theorem

A special case of the Kőnig infinity lemma, in which T is a subtree of the *infinite binary tree* shown in figure 7.12, is the combinatorial essence of many proofs in analysis.

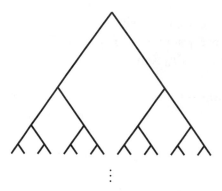

Figure 7.12: The infinite binary tree.

Typically one seeks a *limit point* of a set of real numbers, rather than an infinite path in a tree, but the underlying idea is the same.

Definition. If S is an infinite set of real numbers then l is a *limit point* of S if there are members of S arbitrarily close to l.

The simplest theorem about limits, and the closest in essence to the Kőnig infinity lemma, is the following.

Bolzano-Weierstrass theorem. *If S is an infinite set of points in the unit interval $[0, 1] = \{x : 0 \leq x \leq 1\}$, then S has a limit point.*

Proof. Since S is infinite, at least one half of $[0,1]$ contains infinitely many members of S. Let I_1 be the leftmost half of $[0,1]$ (including endpoints, so that I_1 is *closed*, as defined in section 6.11) that contains infinitely many points of S, and repeat the argument in I_1.

Since I_1 contains infinitely many points of S, at least one half of I_1 contains infinitely many points of S. Let I_2 be the leftmost half of I_1 (including endpoints) that contains infinitely many points of S, and repeat the argument in I_2.

In this way we obtain an infinite sequence of nested closed intervals

$$[0, 1] = I_0 \supset I_1 \supset I_2 \supset \cdots ,$$

each of which contains infinitely many points of S. Also, each interval is half the length of the one before, so, by the completeness of $\mathbb{R}$, there

I_0

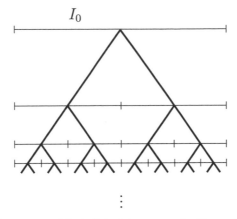

Figure 7.13: The infinite binary tree of subintervals.

is exactly one point l common to them all. The point l is a limit point of S, because the endpoints of the intervals approach l arbitrarily closely, and each of these intervals contains points of S. □

This proof implicitly involves an infinite binary "tree of subintervals" of $I_0 = [0, 1]$. I_0 is the top vertex; its left and right halves are the two vertices below I_0, and so on (see figure 7.13).

In the proof we consider the subtree T of subintervals that contain infinitely many points of S. The pigeonhole principle implies that the subtree is infinite. So, by the Kőnig infinity lemma, T contains an infinite simple path. The *leftmost* such path is the nested sequence of intervals $I_0 \supset I_i \supset I_j \supset \cdots$ that leads to the limit point l.

The Bolzano-Weierstrass theorem is easily generalized to two or more dimensions. In the plane, for example, one can prove that an infinite set of points in the square or a triangle has a limit point. The proof proceeds similarly, by repeatedly dividing the region into a finite number of parts (say, quarters) but making the parts arbitrarily small, and repeatedly applying the infinite pigeonhole principle. In the next section we will show how the Bolzano-Weierstrass theorem for a triangle can be combined with some elementary (finite) graph theory to prove a famous theorem of topology.

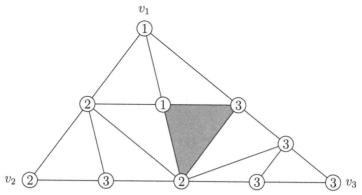

Figure 7.14: Example of a triangle subdivision.

7.10 Sperner's Lemma

Sperner's lemma is a strangely simple result about graphs with labeled vertices. Sperner (1928) devised the lemma to give a new proof of a theorem of Brouwer about the invariance of dimension under continuous maps. It can also be used to prove another famous theorem of Brouwer (1910) about continuous maps, his *fixed point theorem*. Here we use the lemma to direct a limit process, through an infinite sequence of graphs, so as to prove the Brouwer fixed point theorem for the plane. (There are similar proofs for three or more dimensions.)

The planar version of the lemma concerns graphs obtained by subdividing a triangle $v_1 v_2 v_3$ into subtriangles, and labeling their vertices 1, 2, or 3 according to certain rules. Figure 7.14 shows one such subdivision, with labeling that obeys the following rules.

1. The vertices v_1, v_2, v_3 are labeled 1, 2, 3, respectively.
2. Vertices on edge $v_1 v_2$ are labeled 1 or 2.
3. Vertices on edge $v_2 v_3$ are labeled 2 or 3.
4. Vertices on edge $v_3 v_1$ are labeled 3 or 1.

Sperner's lemma. *If triangle $v_1 v_2 v_3$ is divided into subtriangles whose vertices are labeled according to the rules above, then at least one subtriangle has vertices with all three labels.*

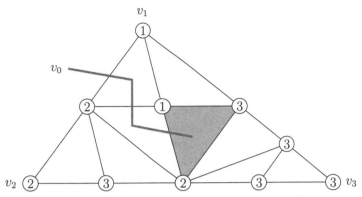

Figure 7.15: Triangle subdivision plus edges of G.

Proof. Given the subdivision of $v_1 v_2 v_3$, we construct a graph G with the following vertices and edges.

- A vertex inside each triangle, and also a vertex v_0 in the region outside triangle $v_1 v_2 v_3$.

- An edge joining any two of the vertices u, v just described, provided that u and v lie on opposite sides of an edge e (in the original subdivision) whose vertices are labeled 1 and 2. In this case the edge uv crosses the edge e.

For the subdivided triangle in figure 7.14, the edges of G are the thick gray line segments in figure 7.15 (and the vertices of G are the ends of these segments).

In particular, the edges in G from the vertex v_0 outside $v_1 v_2 v_3$ can cross only subedges of the side $v_1 v_2$, because only these have ends labeled 1 and 2. Also, there are an *odd number* of such subedges, because on any of them the label changes from one end to the other, and there must be an odd number of changes—otherwise v_1 and v_2 would be labeled the same. Consequently, *the valency of v_0 is odd*.

For any other vertex u of G the valency is one of the following:

- 0 if u lies in a subtriangle whose vertices lack one of the labels 1, 2,

- 1 if u lies in a subtriangle whose vertices have all the labels 1, 2, 3,

- 2 if u lies in a subtriangle whose vertices have the labels 1, 2 only (because two edges have both labels in that case).

Now recall that the sum of the valencies is even, by the theorem in section 7.4, so G *must have an even number of vertices of odd valency*. Apart from the vertex v_0, the only vertices of odd valency are those of valency 1—namely, those inside subtriangles with labels 1, 2, 3. So the number of such triangles is odd, and therefore *nonzero*. □

*The Brouwer Fixed Point Theorem

In this subsection we apply Sperner's lemma to a theorem about continuous functions on the plane. To make the application as easy as possible we assume that the function maps an equilateral triangle into itself, but the argument can be transferred to many regions in the plane, such as a circular disk.

Brouwer fixed point theorem. *If f is a continous map of the equilateral triangle into itself, then there is a point p of the triangle such that $f(p) = p$.*

Proof. For convenience we take the equilateral triangle T in $\mathbb{R}^3$ with vertices $v_1 = (1, 0, 0)$, $v_2 = (0, 1, 0)$, and $v_3 = (0, 0, 1)$. The beauty of this triangle is that its points are precisely those $x = (x_1, x_2, x_3)$ such that

$$0 \le x_1, x_2, x_3 \le 1 \quad \text{and} \quad x_1 + x_2 + x_3 = 1.$$

It follows that if $f(x) = (f(x)_1, f(x)_2, f(x)_3)$ and *if f has no fixed point*, then f always decreases at least one coordinate; that is, either

$$f(x)_1 < x_1, \quad \text{or} \quad f(x)_2 < x_2, \quad \text{or} \quad f(x)_3 < x_3.$$

We label each point x in T with the *least i* such that $f(x)_i < x_i$.
Then the following conditions hold:

1. The vertices v_1, v_2, v_3 are labeled 1, 2, 3, respectively. For example, $x_1 = 1$ at v_1, and we must have $f(v_1)_1 < 1$, since $f(v_1) \ne v_1$ by hypothesis.

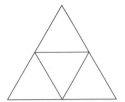 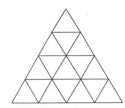 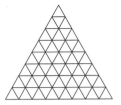

Figure 7.16: Increasingly refined subdivisions of the triangle T.

2. Vertices on $v_1 v_2$ are labeled 1 or 2. Points on $v_1 v_2$ do not get label 3, because they are the points for which $x_3 = 0$, which cannot decrease.

3. Vertices on $v_2 v_3$ are labeled 2 or 3, similarly.

4. Vertices on $v_3 v_1$ are labeled 3 or 1, similarly.

Thus the labeling of points satisfies the conditions of Sperner's lemma. Now suppose we divide T into subtriangles in the (infinitely many) ways suggested by figure 7.16. By Sperner's lemma, each subdivision contains a subtriangle with vertices labeled 1, 2, 3. These triangles become arbitrarily small and their vertices make up an infinite set of points in the triangle T. So, by the Bolzano-Weierstrass theorem of section 7.9, this set has a limit point p.

This means that any neighborhood of the point $p = (p_1, p_2, p_3)$ contains a triangle with labels 1, 2, 3. Now if the label of p is 1, for example, then

$$f(p)_1 < p_1 \quad \text{and either} \quad f(p)_2 > p_2 \quad \text{or} \quad f(p)_3 > p_3.$$

Since f is continuous, we have, for *any* point $q = (q_1, q_2, q_3)$ sufficiently close to p,

$$f(q)_1 < q_1 \quad \text{and either} \quad f(q)_2 > q_2 \quad \text{or} \quad f(q)_3 > q_3.$$

In that case, q also has label 1, contradicting the fact that points arbitrarily close to p have labels 1, 2, and 3. (And similarly if p has label 2 or 3.)

This contradiction shows that our initial assumption, that f has no fixed point, was false. □

7.11 Historical Remarks

Pascal's Triangle

Pascal's triangle has a place in the history of combinatorics like that held by the Pythagorean theorem in the history of geometry. It is very old, was discovered independently in several cultures, and has since become one of the foundations of the subject.

It seems to have been observed in India, perhaps as early as 200 BCE, in the writings of Pingala on literary composition. The binomial coefficients come up in counting the number of combinations of heavy and light syllables. Later, the idea was taken up by Indian mathematicians, and it was transmitted to the Muslim world by al-Biruni in the eleventh century.

In China, the binomial coefficients were discovered in their algebraic setting as the coefficients in the expansion of $(a + b)^n$. Medieval Chinese mathematicians used such expansions in a sophisticated method for numerical solution of polynomial equations—a method that became known in the West, much later, as *Horner's method* after its rediscovery by Horner in 1819. The picture of Pascal's triangle by Zhu Shijie (1303), shown in section 1.6, comes from this flourishing period of Chinese mathematics.

In Italy, Pascal's triangle is sometimes known as "Tartaglia's triangle," due to its discovery by Niccolò Tartaglia (also one of the discoverers of a solution to cubic equations). Tartaglia tells us that he made the discovery "on the first day of Lent, 1523, in Verona." He eventually published his triangle, which includes the coefficients up to the 12th row, in Tartaglia (1556). By this time, the triangle had been rediscovered by Michael Stifel (1544) in Germany.

Blaise Pascal himself wrote his treatise on the arithmetic triangle in 1654, having probably learned of the triangle from Father Marin Mersenne. So Pascal was by no means the discoverer of the triangle. Nevertheless, *The Arithmetic Triangle* broke new ground with his masterly treatment of the subject, using the first really modern proofs by induction, and by making the first application of binomial coefficients to probability theory.

In the next chapter we further discuss the role of the binomial coefficients in probability, where they are also of fundamental importance.

Graph Theory

Graph theory, although it began with Euler, was a fringe mathematical topic until the twentieth century. With the exception of the Euler polyhedron formula, its results had little influence on other branches of mathematics. Indeed, the field of topology, which was largely inspired by the Euler formula, rapidly overtook graph theory in the early twentieth century, reaching a height from which certain topologists could look down on graph theory as "the slums of topology." This early history of graph theory is recounted in the engaging book *Graph Theory, 1736–1936* of Biggs et al. (1976).

However, about the time when the book came out, graph theory was entering a new era of respectability. The landmark event of 1976 was the proof of the *four-color theorem*, solving a problem that had been open since 1852. Coloring maps with a minimum of colors seemed at first to be one of those popular puzzles that graph theory might dispose of quite easily. In fact Kempe (1879) offered a proof that four colors suffice, which was accepted for more than a decade. Mission accomplished. Then Heawood (1890) pointed out a flaw in Kempe's proof, which he could repair only as far as proving that five colors suffice.

Their illusions shattered, graph theorists and topologists began a long and difficult search for a rigorous proof that four colors suffice.

The search ended in 1976, though not without controversy. Appel and Haken (1976) gave a proof, but it involved an unexpectedly long investigation of separate cases, and required over 1000 hours of computer time. Mathematicians were dismayed: both by the lack of insight in the proof and its reliance on computation. It was felt that such a proof was unreliable, due to the possibility of programming errors. Also it was disappointing that the proof gave little understanding *why* the theorem is true.

The possibility of programming errors has now been virtually eliminated, by a "computer-verifiable" proof of the theorem written by Georges Gonthier in 2005. It remains unknown whether an insightful proof is possible, but this perhaps increases the mystique of the four-color theorem. Is it possible that graph theory contains a theorem that is beyond human comprehension?

Also in the 1970s it was found that many algorithmic graph theory problems are hard for another reason: they are NP-complete. Three of the best known are:

Hamiltonian path. Given a finite graph G, decide whether G contains a "path" (connected sequence of edges) that includes each vertex exactly once.

Traveling salesman. Given a finite graph G, with integer "length" values for its edges, and an integer L, decide whether there is a path of length $< L$ that includes all the vertices.

Vertex three-coloring. Given a finite graph G, decide whether there is an assignment of three colors to its vertices such that the end vertices of each edge have different colors.

Another key problem—in a sense *the* fundamental problem of graph theory—is the problem of deciding whether two graphs are the "same"; that is, *isomorphic* in the sense of section 7.6.

Graph isomorphism. Given two finite graphs G and G', decide whether G is isomorphic to G'.

This problem is not known to be in P; nor is it known to be NP-complete.

With these problems, graph theory has become inseparable from the theory of computation, and hence a fundamental part of mathematics today.

7.12 Philosophical Remarks

Combinatorics and Arithmetic

The addition property that defines Pascal's triangle makes it easy to compute whole rows of binomial coefficients by repeated addition. The first investigators of Pascal's triangle in Europe delighted in such computations, with Tartaglia computing all the $\binom{n}{k}$ up to $n = 12$ and Mersenne up to $n = 25$.

To compute an individual coefficient one has the formula derived in section 1.6,

$$\binom{n}{k} = \frac{n(n-1)(n-2)\cdots(n-k+1)}{k!}, \qquad (*)$$

which also calls on multiplication and division. In fact, since the binomial coefficient is an integer, immediately from its definition, the formula (*) expresses a theorem of arithmetic:

$$k! \text{ divides } n(n-1)(n-2)\cdots(n-k+1). \qquad (**)$$

When expressed in this way, as a theorem about addition, multiplication, and division, the integrality of $\binom{n}{k}$ is not at all obvious.

Indeed, Gauss (1801), article 127, made heavy weather of proving (**) by pure arithmetic reasoning. Dirichlet (1863), §15, also had a hard time giving a purely arithmetic proof. Thus a hard theorem of arithmetic can be an easy theorem of combinatorics. To put it another way, arithmetic can become more elementary when viewed from a combinatorial standpoint. In section 10.1 we give a more advanced example of the way combinatorics can simplify arithmetic.

This suggests that both arithmetic and combinatorics might benefit from a unified viewpoint—one that allows reasoning about finite sets as well as about numbers. In fact, finite set theory itself is capable of reasoning about numbers, as we will explain in chapter 9.

Discrete or Continuous?

Is the physical world discrete or continuous? Combinatorialists like to cite the history of modern physics—from the discovery of atoms onward—as evidence that the world is fundamentally discrete and hence that mathematics should be fundamentally combinatorial. Be that as it may, a believer in continuity can still appreciate how combinatorics gives insight into continuous structures. The proof of the Brouwer fixed point theorem by Sperner's lemma is a splendid example of cooperation between discrete and continuous mathematics.

Examples like this are common in topology, which is officially the theory of continuous functions, but historically an offshoot of combinatorics. The Euler polyhedron formula was first proved for essentially discrete objects, "polyhedra," determined by a finite number of points called "vertices." The "edges" of a polyhedron can be viewed simply as pairs of vertices, and its "faces" as finite sequences of edges (the "face boundary paths"). However, the formula remains true for any surface, in continuous one-to-one correspondence with the sphere, when "edges" are arbitrary continuous arcs on the surface, meeting only at endpoints, and "faces" are the pieces obtained by removing the edges from the surface. Given the potential complexity of curves, the Euler formula $V - E + F = 2$ could hardly have been foreseen without guidance from its discrete forerunner.

Indeed, the whole vast field of algebraic topology today owes its existence to clues provided by its precursor "combinatorial topology," of which the Euler polyhedron formula is a shining example. Poincaré (1895) launched the subject of algebraic topology in an astonishing long paper in which he boldly drew conclusions about continuous structures from theorems about discrete ones. Poincaré tried to prove that certain continuous manifolds could be given a combinatorial structure. But it took another 20 years of work, by other mathematicians such as Brouwer, before all of Poincaré's conclusions could be justified.

*Infinite Graph Theory

The Kőnig infinity lemma, as we have seen in sections 7.9 and 7.10, underlies some important constructions in analysis and topology. More generally, as Kőnig (1927) himself described it, it is a method of reasoning from the finite to the infinite. He gave the example: *if every finite planar graph is four-colorable, then so is every infinite planar graph*. This was related to the problem of four-coloring maps, which was still open at the time. The map coloring problem is translated into a graph vertex coloring problem by placing a vertex in each region of the given map M, and connecting two vertices by an edge if the corresponding regions in M have a common boundary. The resulting graph G is *planar* because it can be drawn in the plane without edges crossing, and the vertices of G can be four-colored if and only if the regions of M can.

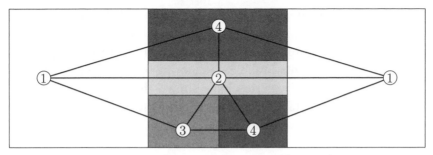

Figure 7.17:   A map and the corresponding graph.

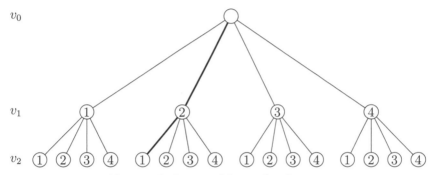

Figure 7.18:   The tree of colorings of the graph with vertices v_1, v_2, ...

An example of a finite map and the corresponding finite graph is shown in figure 7.17. Vertex numbers name the colors of the corresponding map regions.

We pass from finite to infinite as follows. Given an infinite planar graph G with vertices v_1, v_2, v_3, . . ., one builds an infinite tree T with a top vertex v_0 and remaining vertices below it on *levels* labeled v_1, v_2, v_3, ... The aim is to display all possible four-colorings of G by coloring the vertices of T with colors 1, 2, 3, 4. The first three levels of T are shown in figure 7.18. Each vertex has four vertices attached below it, so it has finite valency. The path with thick edges represents a coloring in which v_1 has color 2 and v_2 has color 1.

We now *prune the tree* T by terminating each path at the first level n for which the corresponding coloring of v_1, v_2, . . . , v_n is not valid; that is, when two adjacent vertices of G receive the same color. After pruning, *the tree remains infinite.* In fact, for each n there is a path that goes below level n, because there is a valid four-coloring of the finite graph with vertices v_1, v_2, . . . , v_n, by assumption. Then it follows by the Kőnig infinity lemma that there is an infinite path in the pruned tree. This path represents a valid four-coloring of the infinite graph G.

In recent decades, logicians have shown that the Kőnig infinity lemma is a fundamental principle of mathematical reasoning. We say more precisely why this is so in section 9.9.

8

⌒

Probability

PREVIEW

Like combinatorics, probability is a big field with many methods and many varieties of subject matter. However, we can show how probability evolves from elementary to advanced by pursuing a single question: describing the outcome of n coin tosses. We begin with an experimental setup, the *Galton board*, which collects and displays the results of n coin tosses for modest values of n. It is a physical realization of the *binomial* probability distribution, so called because the probability of getting k heads in n tosses is proportional to $\binom{n}{k}$.

Next, we solve one of the simplest problems involving repeated coin tosses, the *gambler's ruin* problem, which involves a method also of interest in combinatorics: *recurrence relations*.

This is followed by a *random walk* view of coin tossing, in which the number (number of heads — number of tails) "walks" along a line—changing by 1 for each head and by -1 for each tail. We show by algebra that, on average, the number has absolute value $\leq \sqrt{n}$ for a sequence of n tosses. Since $\sqrt{n}$ is much smaller than n for large n, this result suggests a "law of large numbers," saying that heads probably occur "about half" of the time. Using the concepts of *mean*, *variance*, and *standard deviation* we prove a precise version of this law in section 8.4.

Finally, we return to the binomial distribution introduced in section 8.1. In section 8.5 we discuss, without proof, how the "shape" of the binomial coefficients $\binom{n}{k}$, as a function of k, tends to the curve $y = e^{-x^2}$ as n increases. Thus probability theory is like analysis in its reliance on the limit concept, especially limits as $n \to \infty$. It is generally a clear sign of the onset of advanced probability theory when analysis enters the picture.

8.1 Probability and Combinatorics

The "counting" aspect of combinatorics is very useful in finite probability theory, where the probability of an event is found by counting the number of favorable cases and comparing that number to the total number of cases. For example, the probability of throwing a total of 12 with two dice is 1/36, since there is one favorable case (6 with each die) out of 36 possibilities. On the other hand, the probability of throwing a total of 8 is 5/36 because there are five favorable cases: 2+6, 3+5, 4+4, 5+3, 6+2, where the first number in each sum is the number on the first die, and the second number is the number on the second die.

More sophisticated counting often involves the binomial coefficients. We have seen one example already—the "problem of points" or division of stakes—in section 1.7. Another example, which gives a striking visual impression of probability, is the so-called *Galton board* (figure 8.1 shows a simplified model).

I first saw this device at the Museum of Science in Boston in the late 1960s. The original example, designed by Sir Francis Galton himself in 1873, is at University College in London. The device is a board with pegs in it like trees in an orchard. When the board is mounted vertically so that the pegs form a lattice pattern, balls dropped on the top peg will travel downward, bouncing randomly left or right as they hit the pegs. Assuming that bounces left and right are equally probable, balls will fall into containers below the bottom row of pegs with probability proportional to the number of possible paths to each container.

The number of paths to the kth peg in the nth row is none other than $\binom{n}{k}$. This can be seen inductively: there is one path to the single peg in row 1 and, for any other peg p, the number of paths to p is the sum of the numbers of paths to the pegs immediately above p to left or right (since a ball can hit peg p only by bouncing off one of these pegs). But these are precisely the rules for generating "Pascal's triangle" of binomial coefficients, so the numbers of paths are identical with the

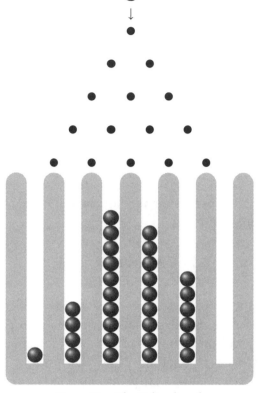

Figure 8.1:   The Galton board.

binomial coefficients. Figure 8.2 shows the first few rows of pegs, each marked by the number of paths that lead to it.

It follows that, if n containers are placed below the gaps in row $n-1$ (including the "gap" at the left and the "gap" at the right), then the number of paths to container k is $\binom{n}{k}$. The probability of a ball falling into container k is therefore proportional to $\binom{n}{k}$.

This distribution of probabilities—with the probability of being in container k proportional to $\binom{n}{k}$—is called the *binomial distribution*. It is found, to a surprisingly high degree of accuracy, in many situations where the outcome depends on a large number of random factors. For example, the heights of adult women, and the scores of students on the SAT exam, both follow a distribution that is close to binomial. Thus the

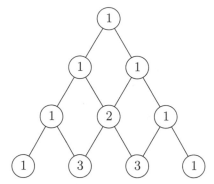

Figure 8.2: The number of paths to each peg.

binomial coefficients are just as fundamental in probability theory as they are in combinatorics.

8.2 Gambler's Ruin

True to its roots, probability theory has many problems motivated by gambling. Here is one that is also of combinatorial interest: the *gambler's ruin* problem. A simple case of the problem goes as follows. Suppose a gambler bets $1 at a time on the toss of a coin, until he has either $0 or $100. What is the probability that he goes broke, after starting with n?

We let $P(n)$ denote the probability of going broke from a start of n. Thus we have the following known values of P:

$$P(0) = 1 \quad \text{and} \quad P(100) = 0.$$

We can also say how $P(k)$ is related to $P(k-1)$ and $P(k+1)$, namely,

$$P(k) = \frac{P(k+1) + P(k-1)}{2}, \qquad (*)$$

because k is equally likely to be followed by $k+1$ or $k-1$.

Now $(*)$ is an example of a *linear recurrence relation*, for which there is a standard method of solution. (The reader may take this method on faith, since the results obtained may be independently verified. If you like, it is a method for *guessing* a solution, and the solution can then be proved correct.)

1. Substitute $P(n) = x^n$ in the recurrence relation, and find which values of x satisfy the resulting polynomial equation.
For the relation (*) we get

$$x^k = \frac{x^{k+1} + x^{k-1}}{2},$$

so

$$2x^k = x^{k+1} + x^{k-1}.$$

Therefore, dividing by x^{k-1},

$$x^2 - 2x + 1 = 0,$$

which has the (double) root $x = 1$.

2. In the case of a double root, a second solution is $P(n) = nx^n$.
Thus we have two solutions of (*), $P(n) = 1^n = 1$ and $P(n) = n1^n = n$.

3. Because the recurrence relation is linear, a constant multiple of a solution is a solution, and so is the sum of two solutions.
In this case, a solution of (*) is

$$P(n) = a + bn, \quad \text{where } a \text{ and } b \text{ are constants.}$$

4. Find the constants by substituting known values of $P(n)$.
In this case, $P(0) = 1$ gives $a = 1$ and then $P(100) = 0$ gives $b = -1/100$.

Thus we get $P(n) = 1 - \frac{n}{100}$, and it can be checked that this solution indeed satisfies the known values and the recurrence relation (*). □

Remark. The method for solving linear recurrence relations gives another way to find the formula for F_n, the nth Fibonacci number, which we found in section 7.3 with the help of generating functions.

We already know two values and a recurrence relation for F_n, namely,

$$F_0 = 0, \quad F_1 = 1, \quad F_{k+2} = F_{k+1} + F_k.$$

Using the method above, we first seek solutions of the recurrence relation of the form $F_n = x^n$. Substituting this in the relation gives

$$x^{k+2} = x^{k+1} + x^k,$$

hence

$$x^2 - x - 1 = 0,$$

which has solutions

$$x = \frac{1 \pm \sqrt{5}}{2}.$$

Thus the general solution of $F_{k+2} = F_{k+1} + F_k$ is

$$F_n = a \left(\frac{1 + \sqrt{5}}{2} \right)^n + b \left(\frac{1 - \sqrt{5}}{2} \right)^n.$$

Using $F_0 = 0$ and $F_1 = 1$ to find the values of a and b gives the formula we found before, namely,

$$F_n = \frac{1}{\sqrt{5}} \left[\left(\frac{1 + \sqrt{5}}{2} \right)^n - \left(\frac{1 - \sqrt{5}}{2} \right)^n \right].$$

8.3 Random Walk

A common process in mathematics and physics is *random walk*: a series of steps which are random in length or direction. In this section we study the simplest case, where the steps are all of length 1 but in a random direction along a line; that is, randomly positive or negative. This case models the behavior of the quantity (number of heads − number of tails) in a series of tosses of a fair coin, since the difference is equally likely to change by $+1$ or -1.

A basic question is: after n steps of a random walk, what is the expected distance from the origin? For the walk along a line—*one-dimensional random walk*—it is easy to give the answer to a related question.

Expected squared length of a random walk. *In a one-dimensional random walk with steps of unit length, the average squared distance from O after n steps is n.*

Proof. Denote the n steps of the random walk by $s_1, s_2, \ldots, s_n$. So each $s_i = \pm 1$ and the final point of the walk is $s_1 + s_2 + \cdots + s_n$. To find the distance $|s_1 + s_2 + \cdots + s_n|$ of the final point from O we consider its square

$$(s_1 + s_2 + \cdots + s_n)^2 = s_1^2 + s_2^2 + \cdots + s_n^2 + \text{ all terms } s_i s_j, \text{ where } i \neq j.$$

The *expected value* of this squared distance is the average of $(s_1 + s_2 + \cdots + s_n)^2$ over all sequences of values $s_i = \pm 1$, because all sequences are equally likely.

To find the average we sum the values of

$$(s_1 + s_2 + \cdots + s_n)^2 = s_1^2 + s_2^2 + \cdots + s_n^2 + \text{ all terms } 2s_i s_j$$

over all 2^n sequences of values $(s_1, s_2, \ldots, s_n)$. Since $s_i = \pm 1$ and $s_j = \pm 1$, the term $s_i s_j$ equals 1 when s_i, s_j have the same sign and it equals -1 when they do not. These two possibilities occur equally often among the sequences of values of $(s_1, s_2, \ldots, s_n)$. Therefore, the values of $s_i s_j$ cancel out, and it remains to find the average value of $s_1^2 + s_2^2 + \cdots + s_n^2$. This of course is n, since $s_1^2 + s_2^2 + \cdots + s_n^2 = n$ regardless of the signs of the s_i.

Thus the expected value of $(s_1 + s_2 + \cdots + s_n)^2$ is n. $\square$

Now, unfortunately, the average length $|s_1 + s_2 + \cdots + s_n|$ is *not* the square root, $\sqrt{n}$, of the average squared length. For example, the average length of a two-step random walk is 1, not $\sqrt{2}$, because there are two walks of length 2 and two walks of length 0. However, we can prove that the average length is *bounded* by $\sqrt{n}$, because of the following inequality.

Averages and squares. *If $x_1, x_2, \ldots, x_n \geq 0$ then the average of the x_i^2 is greater than or equal to the square of the average of the x_i.*

Proof. We wish to prove that

$$\frac{x_1^2 + x_2^2 + \cdots + x_n^2}{n} \geq \left(\frac{x_1 + x_2 + \cdots + x_n}{n}\right)^2$$

or, equivalently, that

$$n(x_1^2 + x_2^2 + \cdots + x_n^2) - (x_1 + x_2 + \cdots + x_n)^2 \geq 0.$$

Well,

$$n(x_1^2 + x_2^2 + \cdots + x_n^2) - (x_1 + x_2 + \cdots + x_n)^2$$
$$= n(x_1^2 + x_2^2 + \cdots + x_n^2) - (x_1^2 + x_2^2 + \cdots + x_n^2 + \text{all terms } x_i x_j \text{ for } i \neq j)$$
$$= (n-1)(x_1^2 + x_2^2 + \cdots + x_n^2) - (\text{all terms } x_i x_j \text{ for } i \neq j)$$
$$= (n-1)(x_1^2 + x_2^2 + \cdots + x_n^2) - (\text{all terms } 2x_i x_j \text{ for } i < j)$$
$$= (\text{all terms } (x_i^2 + x_j^2) \text{ for } i < j) - (\text{all terms } 2x_i x_j \text{ for } i < j)$$

since the terms $x_i^2 + x_j^2$ for $i < j$ include each subscript $n - 1$ times;

namely, paired with the $n - 1$ subscripts unequal to itself

$$= \text{all terms } (x_i^2 - 2x_i x_j + x_j^2) \text{ for } i < j$$
$$= \text{all terms } (x_i - x_j)^2 \text{ for } i < j$$
$$\geq 0, \quad \text{as required.} \qquad \square$$

Coming back to sequences of coin tosses: this theorem shows that, in a sequence of n coin tosses, the expected difference between the number of heads and the number of tails (disregarding sign) is at most $\sqrt{n}$. For large values of n, the value of $\sqrt{n}$ is small in comparison with n, so there is a precise sense in which the expected number of heads is "about half" of the total number of tosses. This is a weak form of the so-called *law of large numbers*, which we will make stronger in the next two sections.

8.4 Mean, Variance, and Standard Deviation

The calculations in the previous section bring to light some important concepts of probability theory. The one-dimensional random walks of

n steps $s_1, s_2, \ldots, s_n$, where each $s_i = \pm 1$ are 2^n in number, so they give 2^n displacements

$$s_1 + s_2 + \cdots + s_n \quad \text{as the } s_i \text{ take all values } +1 \text{ or } -1.$$

The average of these, over all the 2^n possible walks, is of course 0. This average is called the *mean* displacement.

The mean displacement tells us nothing about the expected length of a walk, which is the average of the values

$$|s_1 + s_2 + \cdots + s_n|.$$

We did not find a formula for this average, but we were able to show that the average of the values

$$(s_1 + s_2 + \cdots + s_n)^2$$

is n. This average, which gives a measure of how widely the displacements deviate from the mean, is called the *variance*. Indeed, the square root of the variance ($\sqrt{n}$ in this case) is called the *standard deviation*. The inequality in the previous section, relating averages and squares, shows that

average value of $|s_1 + s_2 + \cdots + s_n| \leq$ standard deviation, $\sqrt{n}$.

These concepts generalize to any sequence $x_1, x_2, \ldots, x_k$ of real numbers as follows.

Definitions. The *mean*, μ, of $x_1, x_2, \ldots, x_k$ is given by

$$\mu = \frac{x_1 + x_2 + \cdots + x_k}{k}.$$

The *variance*, σ^2, is given by

$$\sigma^2 = \frac{(x_1 - \mu)^2 + (x_2 - \mu)^2 + \cdots + (x_k - \mu)^2}{k}.$$

And the *standard deviation*, σ, is given by

$$\sigma = \sqrt{\frac{(x_1 - \mu)^2 + (x_2 - \mu)^2 + \cdots + (x_k - \mu)^2}{k}}.$$

A very simple inequality bounds the probability of a value x_i differing from the mean by more than one standard deviation. To state the inequality we will use the notation $P(x_i)$ to denote the probability of x_i occurring, though we are interested only in the case where $P(x_i) = 1/k$ (that is, we have outcomes x_i of k equiprobable events, such as the displacements for all $k = 2^n$ random walks w_i of n steps). In this case

$$\sigma^2 = \frac{(x_1 - \mu)^2 + (x_2 - \mu)^2 + \cdots + (x_k - \mu)^2}{k}$$

$$= (x_1 - \mu)^2 P(x_1) + (x_2 - \mu)^2 P(x_2) + \cdots + (x_k - \mu)^2 P(x_k).$$

Chebyshev's inequality. *If x is a member of the sequence $x_1, x_2, \ldots, x_k$ with mean μ and variance σ^2, and if $t > \sigma$, then the probability that $|x - \mu| \geq t$ is such that*

$$\text{prob} \, (|x - \mu| \geq t) \leq \frac{\sigma^2}{t^2}.$$

Proof. Given that the probability of x taking the value x_i is $P(x_i)$,

$$\text{prob}(|x - \mu| \geq t) = \text{sum of the terms } P(x_i) \text{ for } |x_i - \mu| \geq t$$

$$\leq \text{sum of the terms } \frac{(x_i - \mu)^2}{t^2} P(x_i) \text{ for } |x_i - \mu| \geq t$$

$$\text{since } \frac{(x_i - \mu)^2}{t^2} \geq 1 \text{ when } |x_i - \mu| \geq t$$

$$\leq \frac{\sigma^2}{t^2} \quad \text{since } \sigma^2 = (x_1 - \mu)^2 P(x_1) + \cdots + (x_k - \mu)^2 P(x_k).$$

$\square$

Notice that the variance enters the proof through squaring the condition $|x_i - t| \geq t$ so as to remove the absolute value sign.

Returning now to the example where x_i is the displacement in the ith random walk of n steps, we have $\mu = 0$ and $\sigma = \sqrt{n}$. According to Chebyshev's inequality, the probability of the length of a random walk being $\geq t$, where $t > \sigma$, is $\leq \sigma^2/t^2$.

Thus for walks of $n = 100$ steps, where $\sigma = 10$,

probability of a walk having length $\geq 20 = 2\sigma$ is $\leq \sigma^2/(2\sigma)^2 = 1/4$,
probability of a walk having length $\geq 30 = 3\sigma$ is $\leq \sigma^2/(3\sigma)^2 = 1/9$,
probability of a walk having length $\geq 40 = 3\sigma$ is $\leq \sigma^2/(4\sigma)^2 = 1/16$.

And for walks of $n = 10000$ steps, where $\sigma = 100$,

probability of a walk having length $\geq 200 = 2\sigma$ is $\leq \sigma^2/(2\sigma)^2 = 1/4$,
probability of a walk having length $\geq 300 = 3\sigma$ is $\leq \sigma^2/(3\sigma)^2 = 1/9$,
probability of a walk having length $\geq 400 = 3\sigma$ is $\leq \sigma^2/(4\sigma)^2 = 1/16$.

This suggests a more precise "law of large numbers," in which we can formalize the idea that "most" random walks have "small" length. We pursue this idea in the next subsection.

The Law of Large Numbers for Random Walks and Coin Tosses

Since $\sigma = \sqrt{n}$ for a random walk of n steps, the probability of a walk having length $\geq m\sigma = m\sqrt{n}$ is $\leq 1/m^2$ by Chebyshev's inequality. We can make this probability less than any given positive ε by suitable choice of m; in fact, for $m > 1/\sqrt{\varepsilon}$. And then we can make the length $m\sqrt{n}$ an arbitrarily small fraction δ of the number of steps, n, in the walk by suitable choice of n; in fact, for $n > m^2/\delta^2$.

Now if the random walk comes from a sequence of coin tosses, with a step of $+1$ for each head and -1 for each tail, making the length of the walk a fraction δ of the total number of steps is equivalent to making the number of heads differ from the number of tails by less than δn. It follows that the fraction of heads differs from $1/2$ by less than δ. To sum up, we have:

Law of large numbers for coin tosses. *For any $\varepsilon > 0$ and $\delta > 0$ there is a number N such that, for any sequence of $n > N$ coin tosses, the probability is less than ε that the fraction of heads differs from $1/2$ by more than δ.* $\qquad\square$

This is called a "weak" law of large numbers, because there are many stronger results in the same vein. However, this example gives the germ of the idea: if an event has probability p then one can prove, in a precise

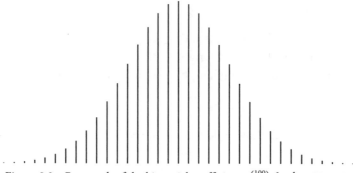

Figure 8.3: Bar graph of the binomial coefficients $\binom{100}{k}$ for $k = 33$ to 67.

sense, that the fraction of successful trials in a "large" sequence of trials will probably be "close" to p.

8.5 *The Bell Curve

The law of large numbers shows the importance of *limit processes* in probability. We expect, and we can prove, that the net result of a large number of trials (such as the fraction of heads in a sequence of coin tosses) tends to a limit in a certain sense. A more spectacular example occurs where a whole *distribution* of probabilities (say, the probabilities of getting k heads in n tosses) tends towards a continuous distribution as the number of trials tends to infinity. We can see this happening if we look at the binomial coefficients $\binom{n}{k}$ for a large value of n, such as $n = 100$ (figure 8.3).

This is a mathematical model of the Galton board described in section 8.1. It seems clear that the graph of binomial coefficients is tending towards a continuous curve shaped like a bell. This curve in fact has the same shape, when the axes are suitably scaled, as the curve $y = e^{-x^2}$ shown in figure 8.4. The probability distribution represented by this curve is called the *normal* distribution.

The two are compared in figure 8.5, showing just how well the curve fits the binomial distribution when $n = 100$.

The wonderful convergence of the graph of binomial coefficients to the graph of $y = e^{-x^2}$ was discovered by de Moivre (1738).

A more precise statement of the result reads as follows.

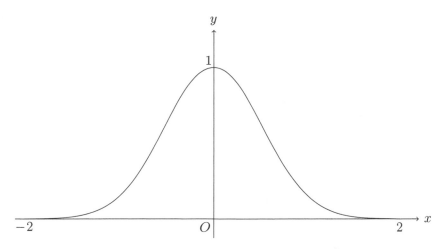

Figure 8.4: The graph of the bell curve $y = e^{-x^2}$.

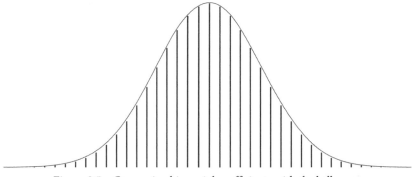

Figure 8.5: Comparing binomial coefficients with the bell curve.

Convergence of the binomial distribution. *As $n \to \infty$ the graph of the binomial coefficients $\binom{n}{k}$, when scaled so that the area under the curve is 1, tends to the graph of*

$$y = \frac{1}{\sqrt{\pi}} e^{-x^2}. \qquad \qquad \square$$

The proof is too ingenious to be considered elementary, so we will not describe it in detail. However, it is related to two facts of independent interest that are each quite close to elementary. We will show in section 10.7 that the second fact (Wallis's product) is in fact sufficient to establish the convergence to $\frac{1}{\sqrt{\pi}}e^{-x^2}$.

1. The *asymptotic formula*

$$n! \sim A\sqrt{n}\left(\frac{n}{e}\right)^n, \quad \text{which says that} \quad \frac{n!}{\sqrt{n}\left(\frac{n}{e}\right)^n} \to$$

constant A as $n \to \infty$.

This formula, discovered by de Moivre, is used to approximate the binomial coefficients $\binom{n}{k} = \frac{n!}{k!(n-k)!}$ as $n \to \infty$.

2. *Wallis's product*

$$\frac{4}{\pi} = \frac{3 \cdot 3 \cdot 5 \cdot 5 \cdot 7 \cdot 7 \cdots}{2 \cdot 4 \cdot 4 \cdot 6 \cdot 6 \cdot 8 \cdots},$$

discovered by Wallis (1655). This formula was used by Stirling (1730) to find the constant A in de Moivre's formula for $n!$. The resulting formula,

$$n! \sim \sqrt{2\pi n}\left(\frac{n}{e}\right)^n,$$

is known as Stirling's formula.

This remarkable formula shows that geometry does not own the number π! It equally belongs to probability theory (and also to combinatorics and number theory).

8.6 Historical Remarks

Probability theory was first associated with gambling, centuries ago, though it is not nearly as old as gambling itself. Gamblers developed intuition about the probability of various outcomes in dice or card games—often quite accurately—but there was no *theory* of probability until around 1500. No doubt this had something to do with superstitions that are common among gamblers, such as the belief that throwing tails becomes more likely after a run of heads. Superstitions like this obviously play havoc with any scientific theory of probability.

Figure 8.6:   Portrait of Luca Pacioli, around 1496.

But even scientists could have doubted that there is such a thing as a "law" of "chance": the two concepts sound incompatible.

Some of the first probability calculations were made in the sixteenth century by the Italian algebraist Girolamo Cardano. He wrote a book on games of chance, *Liber de ludo aleae* (Book of games of chance) around 1550, though it was not published until 1663. In it he used some elementary combinatorics to enumerate outcomes, assumed to be equiprobable, and thereby became the first to calculate a theoretical probability correctly. However, he also made some errors, and was unable to solve the "problem of points" (division of stakes in an unfinished game), which had first been raised by Pacioli (1494). Though not a renowned mathematician, Pacioli made important contributions to subjects from accounting to magic. He is the subject of a splendid mathematical portrait, attributed to Jacopo de' Barbari (figure 8.6).

As mentioned in section 1.7, Pascal (1654) solved the problem of points with the help of the binomial coefficients. He discussed the problem in correspondence with Fermat, who evidently knew the solution independently, having used binomial coefficients to solve problems in number theory (as we saw in section 7.2). Another who apparently solved the problem on his own was the Dutch scientist Christiaan Huygens, who included a systematic treatment of it in his book *De ratiociniis in aleae ludo* (Calculation in games of chance) of 1657. In this book, Huygens correctly solved many finite probability problems, including a formidable instance of the gambler's ruin problem (his fifth problem). With the publication of Huygens's book, elementary finite probability theory at last was on a sound foundation.

Huygens's book was the starting point for Jakob Bernoulli, who reissued it, with a commentary, as the first part of his book *Ars conjectandi* (The art of conjecturing). Bernoulli died in 1705 before his book was published, and it finally appeared in print as Bernoulli (1713), edited by his nephew Nicolaus Bernoulli. The commentary considerably extended the ideas of Huygens, going as far as describing the binomial distribution and proving some results on binomial coefficients that pave the way for the law of large numbers. Jakob Bernoulli proved the law in the fourth part of the *Ars conjectandi*, calling it his "golden theorem." He realized how important it was, because it allows us to *infer* the probability of an event, to arbitrary accuracy and with an arbitrary degree of certainty, by performing a sufficiently large number of trials (now called *Bernoulli trials* in his honor).

An example (which was actually first studied in Bernoulli's time) is the probability of a newborn child being a boy. Records of births in the city of London between 1629 and 1710 served as Bernoulli trials for the event of the newborn being a boy, and they showed a birth ratio of very nearly 18 boys to 17 girls. Thus it could be concluded that the probability of a newborn being a boy is very likely *greater* than 1/2. The law of large numbers was the beginning of statistics as we know it today, but what was missing from Jakob Bernoulli's law was a good estimate of *how many* trials are needed to ensure a close approach to the true probability. He found estimates but they were too large to be useful. So the most important part of his discovery was the *existence* of a limit (the true probability, inferred from the number of successful trials) rather than a way to calculate it. The next step, towards a usable law of large numbers, was made by de Moivre.

Abraham de Moivre first encountered probability theory while at school in France between 1682 and 1684, when he was about 16. He studied Huygens's book on his own while studying logic in Saumur, then learned more mathematics when he moved to Paris in 1684. His studies were interrupted when French protestants lost many of their rights with the revocation of the Edict of Nantes in 1685, and he moved to London as a Huguenot refugee in 1687.

In London, de Moivre became one of the leading mathematicians— a friend of Newton and a member of the Royal Society—but he was unable to obtain a university position, and had to earn his living as a private tutor. He spent much of his time in coffee houses, where he did some of his tutoring, played chess for money, and answered questions from gamblers. His great book on probability, *The Doctrine of Chances*, was first published in 1718, and in expanded editions in 1738 and 1756 (the latter after his death in 1754). The 1756 edition contains his theorem that the limit of the graph of the binomial coefficients, suitably scaled, is the so-called "bell curve" $y = e^{-x^2}$. He had first published this theorem in de Moivre (1733), but the version of 1756 includes further refinements. In chapter 10, when we sample some results beyond the boundaries of elementary mathematics, we will say more about the theorem of de Moivre and its relation to the theorem of Wallis mentioned in section 8.5.

In passing from the binomial coefficients to the function e^{-x^2}, de Moivre made a decisive move towards *analytic* probability theory—a theory that depends on the concepts and methods of calculus. Instead of *adding* binomial coefficients (as one does in solving the problem of points), one finds the *area* under sections of the bell curve, which amounts to finding the integral $\int_a^b e^{-x^2} \, dx$ between given values of a and b. This is a difficult problem, because e^{-x^2} is not the derivative of any elementary function. Whether it was due to the difficulty of the integral, or its origin in what was then a fringe area of mathematics, I do not know; but de Moivre's marvelous discovery remained in obscurity until it was extended and given a new proof by Gauss (1809) and Laplace (1812). The eminence of Gauss and Laplace helped to bring probability theory, and the normal distribution in particular, to the attention of other mathematicians. But it had the rather unjust consequence that the normal distribution is now called "Gaussian," instead of being named after de Moivre.

From the viewpoint of today, de Moivre's limit theorem can be seen as the first version of the so-called *central limit theorem*, whose evolution continued well into the twentieth century. Indeed the theorem was first called "central" by Pólya (1920), because of its central role in probability and statistics. There is a whole book on the history of the central limit theorem, Fischer (2011).

8.7 Philosophical Remarks

The meaning of the probability concept has been much debated by philosophers, but in this book we are content to declare certain simple events to be equally probable, and to calculate the probability of more complicated events as the ratio of favorable cases to the total number of cases. For example, we take it to be equally probable for a coin toss to result in a head (H) or a tail (T) (by symmetry, or by definition of a "fair coin"). It follows, when the coin is tossed twice, that the four outcomes HH, HT, TH, and TT are equally probable, and hence that exactly one head occurs in two cases, so the probability of getting exactly one head is $2/4 = 1/2$.

Thus, for our purposes, probability reduces to counting the number of favorable cases and the total number of cases, which is a problem of combinatorics.

A more interesting problem, to the philosophical mathematician, is that of defining "largeness" and "closeness." A law of large numbers, for example, is supposed to say that the proportion of heads in a "large" sequence of coin tosses is "close" to 1/2 for "most" sequences. Making these ideas precise is really the same as defining the concept of *limit* in calculus, as we do when we define what it means to say "$1/n$ is close to zero when n is large."

Thus finite probability and calculus share an interest in the concept of limit, and probability may even be a better reason to master the limit concept, for some students. At any rate, the occurrence of the limit concept in two quite different contexts reinforces the case for including it in elementary mathematics.

Just beyond the bounds of elementary probability, we can glimpse advanced notions on the horizon when we consider *infinite* sequences of coin tosses. Consider the event of throwing infinitely many heads in succession. This event is *possible*, but in some sense "infinitely unlikely." To express this precisely we say that the event of throwing infinitely many heads in succession has *probability zero*. Thus, in infinite probability theory, probability zero does *not* mean impossible, just infinitely unlikely. Another example of an event of probability zero is the following. Suppose we have an infinitely sharp dart, which will hit just a single point of the plane when we throw it. If we throw the dart at random it is possible that it will hit the origin, but the *probability* of this happening is zero.

While we have the dart in our hand, here is another question: what is the probability of hitting a point (p, q) where p and q are rational? It turns out that this event also has probability zero, for a reason we will discuss in chapter 9. More generally, suppose we throw the dart at an arbitrary set S of points in the unit square $\{(x, y) : 0 \leq x \leq 1, 0 \leq y \leq 1\}$. The unit square has area 1, so the probability of hitting a point in S ought to equal the area of S. If S is the set of points (p, q) in the unit square with p, q rational it is indeed true that S has area zero, which is why the probability of hitting a point of this particular S is zero.

In general, to speak of the probability of hitting a member of a set of points we need a general concept of area, or *measure*, for sets of points. (The measure concept is also needed for advanced calculus, in order to integrate complicated functions, because the integral of a function equals the area of the region under its graph.) We have to ask whether it is even meaningful to speak of the probability of hitting S, because the "area of S" may not be meaningful. The question whether every subset of the unit square has a meaningful area leads to deep questions about set theory and infinity, which are *very* advanced mathematics.

9

⤳

Logic

PREVIEW

Proof, and hence logic, is essential to mathematics, but the logic used in mathematics has several distinctive features. The simplest logic, *propositional logic*, studies the effect of the words AND, OR, and NOT on the truth values of sentences. This logic turns out to have a classical mathematical description: it is simply mod 2 arithmetic, with 0 and 1 representing "false" and "true."

But propositional logic is not expressive enough for mathematics. It has to be enhanced by variables, quantifiers ("for all x" and "there is an x"), and symbols for properties and relations. We briefly discuss how mathematics can be expressed in the resulting logic—*predicate logic*—before moving on to some important axiom systems for mathematics.

The first of these is *Peano arithmetic*, PA, which grew out of the discovery that arithmetic can be based almost entirely on induction, the signature method of mathematical proof that goes back at least as far as Euclid. If we assume, then, that induction is part of elementary mathematics, we can take PA as a good approximation to "elementary mathematics."

To see what is *not* elementary, we study extensions of PA. The first of these is called ZF set theory. By reinventing PA as a certain theory of finite sets, ZF can be viewed as PA plus an axiom stating the existence of an infinite set. Thus one, somewhat crude, description of advanced mathematics is what is added to PA by adding infinity.

A more refined view, called *reverse mathematics*, identifies distinct low levels of advanced mathematics. One of the systems used in reverse mathematics, ACA_0, marks the level of two theorems often seen near the border of elementary mathematics: the completeness of $\mathbb{R}$ and the Bolzano-Weierstrass theorem.

9.1 Propositional Logic

The simplest part of logic, called *propositional logic*, is concerned with finding the truth value of compound propositions from the truth values of their constituent parts, called *atomic propositions*. If we denote the atomic propositions by $p, q, r, \ldots$, then examples of compound propositions are p AND q, p OR q, and NOT p. Initially, we will be concerned only with compounds formed using AND, OR, and NOT. Thus we might wish to know

Is (NOT p) OR (q AND r) true when p is true, q is false, r is true?

Such questions can be answered mechanically with the help of *truth tables*, which give the values of p AND q, p OR q, and NOT p for all possible values of p and q. If we denote "true" by 1 and "false" by 0, then the required truth tables are:

p	q	p AND q
0	0	0
0	1	0
1	0	0
1	1	1

p	q	p OR q
0	0	0
0	1	1
1	0	1
1	1	1

p	NOT p
0	1
1	0

These are also tables of functions in mod 2 arithmetic, namely pq, $pq + p + q$, and $p + 1$, respectively. It is obvious that the function pq has the same values as p AND q, and that $p + 1$ has the same values as NOT p, and it is an easy computation to check that $pq + p + q$ has the same values as p OR q. Thus, finding the truth value of any proposition compounded from AND, OR, and NOT reduces to a computation in mod 2 arithmetic. In particular, finding the value of

$$(\text{NOT } p) \text{ OR } (q \text{ AND } r) \quad \text{when} \quad p = 1, q = 0, r = 1$$

amounts to computing the value of

$$(p+1) \text{ OR } qr = (p+1)qr + (p+1) + qr \quad \text{for} \quad p = 1, q = 0, r = 1.$$

Substituting these values, we get the value

$$(1+1)0 \cdot 1 + (1+1) + 0 \cdot 1 = 0 + 0 + 0 = 0,$$

so in fact (NOT p) OR (q AND r) is false when p is true, q is false, and r is true. This example illustrates a simple case of a far-reaching principle: *logic can be arithmetized.*

However, we do not immediately replace all reasoning by calculations in mod 2 arithmetic. The words AND, OR, and NOT are often more enlightening than addition and multiplication. For example, it is obvious that *any* function whose arguments and values are either 0 or 1 can be expressed in terms of AND, OR, and NOT. An example will show why. Consider the function $F(p, q, r)$ with the following table of values.

p	q	r	$F(p, q, r)$
0	0	0	0
0	0	1	1
0	1	0	1
0	1	1	0
1	0	0	0
1	0	1	0
1	1	0	1
1	1	1	0

The table says that $F(p, q, r)$ is true just in the cases given by lines 2, 3, and 7; that is, when it is true that

(NOT p) AND (NOT q) AND r

OR

(NOT p) AND q AND (NOT r)

OR

p AND q AND (NOT r),

so $F(p, q, r)$ equals the above compound of the functions AND, OR, and NOT.

A similar argument applies to any function F whose arguments and values are either 0 or 1—a so-called *Boolean* function. F is a composite of the special Boolean functions AND, OR, and NOT, and one such composite may be read directly from the lines in the truth table for F. It follows that F may also be compounded from mod 2 addition and multiplication, though this result is not so easy to see directly.

Symbolism

It is convenient to abbreviate AND, OR, and NOT by $\wedge$, $\vee$, and $\neg$, respectively. The symbols $\wedge$ and $\vee$ are chosen because of their analogy with symbols $\cap$ and $\cup$ for set intersection and union, and also because they reflect an important relationship between AND and OR—called *duality*—that ordinary language does not.

An example of the duality between $\wedge$ and $\vee$ is the following pair of equations:

$$\neg(p \wedge q) = (\neg p) \vee (\neg q),$$
$$\neg(p \vee q) = (\neg p) \wedge (\neg q).$$

Both are true for all values of p and q. Thus we can take an identity between two Boolean functions, interchange $\wedge$ and $\vee$, and get another identity between Boolean functions. This is true under quite general conditions, so AND and OR have a certain kind of interchangeability, which our notation reflects.

Another important Boolean function that has its own symbol is the implication function "if p then q," which is denoted by $\Rightarrow$. The truth table for this function is:

p	q	$p \Rightarrow q$
0	0	1
0	1	1
1	0	0
1	1	1

It is easily checked that $p \Rightarrow q$ is the same Boolean function as $(\neg p) \vee q$.

A related function is "p if and only if q," or $(p \Rightarrow q) \wedge (q \Rightarrow p)$. This function is denoted by $p \Leftrightarrow q$ and its truth table is:

p	q	$p \Leftrightarrow q$
0	0	1
0	1	0
1	0	0
1	1	1

Notice that $p \Leftrightarrow q$ is the same Boolean function as $p + q + 1$, where $+$ is the mod 2 sum.

9.2 Tautologies, Identities, and Satisfiability

In logic we are particularly interested in *valid* formulas, that is, formulas that are true for all values of the variables. Valid formulas of propositional logic are known as *tautologies*. A simple example of a tautology is $p \vee (\neg p)$, which has the value 1 for all values (0 or 1) of p. In arithmetic we are similarly interested in *identities*—equations that hold for all values of the variables. Tautologies obviously correspond to identities in mod 2 arithmetic. For example, $p \vee (\neg p)$ corresponds to the identity $p \vee (\neg p) = 1$, which we can rewrite in terms of the mod 2 sum and product as

$$p(p+1) + p + (p+1) = 1,$$

or more simply as the equivalent identity $p(p+1) = 0$.

Truth tables enable us to calculate the value of any formula $f(p, q, r, \ldots)$ for any values of the variables p, q, r, ... Hence we can decide whether $f(p, q, r, \ldots)$ is a tautology simply by substituting all possible values of p, q, r, ... If there are n variables, then there are 2^n values of the sequence p, q, r, ..., so this problem is finite and hence solvable. Simple as the solution may be, in principle, it suffers from the practical defect that 2^n is infeasibly large for quite small values of n, say $n = 50$. Thus it may be infeasible to decide whether $f(p, q, r, \ldots)$ is a tautology, even though the formula occupies only a couple of lines.

We do not yet know whether there is a feasible solution to the problem of recognizing tautologies; that is, a solution that can be found in time roughly comparable to the length of the formula f. In fact, we do not even know a feasible solution to the *satisfiability problem*: the problem of recognizing, for any formula f, whether $f(p, q, r, \ldots) = 1$ for *some* values of the variables $p, q, r, \ldots$. The latter problem is especially frustrating because the value of $f(p, q, r, \ldots)$ can be feasibly computed (by truth tables) for any particular values of $p, q, r, \ldots$. But even *verifying* satisfiability seems feasible only if we assume the magical ability to make a "lucky guess" for satisfying values of the variables.

The difficulty of satisfiability seems surprising, particularly when it is viewed as a problem about mod 2 sums and products, which one would expect to be well understood. But, as we observed in section 3.6, mod 2 arithmetic is not as easy as it looks. Deciding whether a polynomial in many variables has a solution in mod 2 arithmetic is an NP problem which is not known to be in P. In fact, as mentioned in section 3.10, finding solutions of polynomial equations mod 2 is as hard as any NP problem—it is NP-*complete*—so it has no polynomial time solution unless *all* NP problems are in P.

This unexpected difficulty has sparked a reassessment of large swaths of mathematics. Many problems, in many fields of mathematics, turn out to have the same NP characteristics as the satisfiability problem.

- The problem consists of infinitely many questions, and there is a method for answering each of them, which takes finite time for each question.
- The time needed to verify a positive answer (if there is one) to a question of length n is "short," in the sense that it is bounded by a polynomial function of n (typically n^2 or n^3).
- But the time to find even one positive answer is generally long—exponentially long relative to the length of the question.

The satisfiability problem has these characteristics because:

- There are infinitely many formulas $f(p, q, r, \ldots)$, and the truth table method allows us to test each one for satisfiability.

- For given values of $p, q, r, \ldots$, finding the value of $f(p, q, r, \ldots)$ takes time roughly equal to the length of the formula.
- But we may have to test all of the 2^n sets of values of the n variables $p, q, r, \ldots$ to find one that satisfies—and n can be roughly as large as the length of the formula.

Recognizing tautologies seems just as hard as recognizing satisfiable formulas, because we have to ensure that *all* sets of values of $p, q, r, \ldots$ satisfy the formula, not just one. In any case, the truth table method does not seem to be an enlightened method for finding tautologies $f(p, q, r, \ldots)$, because it substitutes values of $p, q, r, \ldots$ mechanically without regard to the structure and meaning of $f(p, q, r, \ldots)$. One would hope for a method that not only finds tautologies but *proves* them in mathematical style—starting with some obvious tautologies, such as $p \vee (\neg p)$, and *deducing* further tautologies in a natural manner.

There is such a method, which we give in section 10.8. But, alas, it is not substantially faster than the truth table method in the worst cases. Thus it seems that even the simplest form of logic harbors some deep mysteries.

9.3 Properties, Relations, and Quantifiers

Propositional logic is an indispensable part of logic—and not trivial either—but it is not expressive enough for mathematics. The variables in propositional logic can take only two values, false and true (or 0 and 1), whereas in mathematics we want variables that can take values that are numbers, or points, or sets, and so on.

Moreover, we want to speak about *properties* of x, or *relations* between x and y (or even between three or more variables). This calls for a more expressive form of logic called *predicate logic*. "Predicates" can be properties or relations, and they are denoted by symbols such as

$$P(x), \quad \text{read "x has property P,"}$$
$$R(x, y), \quad \text{read "x and y are in relation R."}$$

Thus "x is prime" is an example of a property, and "$x < y$" is an example of a relation. Notice that the formulas "x is prime" and

"$x < y$" are neither true nor false, because their variables are free to take different values. Formulas acquire truth values either by substituting values for the variables—for example, "4 is prime" is false—or by binding the variables by the *quantifiers*

$$\forall x \quad \text{"for all } x\text{,"}$$
$$\exists x \quad \text{"there is an } x\text{,"}$$

for example, when x, y range over the natural numbers

$$\forall x (x \text{ is prime}) \quad \text{is false,}$$
$$\exists x (x \text{ is prime}) \quad \text{is true,}$$
$$\forall x \exists y (x < y) \quad \text{is true.}$$

(The latter formula is read "for all x there is a y such that $x < y$.")

As these examples suggest, the language of predicate logic can conveniently express typical mathematical statements. In fact, the language is arguably flexible enough to express *all* of mathematics. (This becomes easier if the language is enhanced by including the equality symbol $=$ and symbols for functions.) We will discuss the specific cases of arithmetic and set theory in the sections that follow.

It is quite striking how an awareness of quantifiers brings clarity to otherwise fuzzy concepts, such as limits and continuity. It might be said, in fact, that the foundations of calculus (analysis) become clear only with proper attention to quantifiers.

Take, for example, the vague concepts of "large" and "small" numbers. Our intuition is that $1/n$ is "small" when n is "large," but there is really no such property as "largeness." If a natural number n is "large" then surely $n - 1$ is "large" too, but then we are forced to the absurd conclusion that *every* natural number is "large." What we really mean, when we say that "n large $\Rightarrow 1/n$ small" is that $1/n$ can be made as small as we please by choosing n sufficiently large. This is still not precise, but we are getting closer. It is clearer still to say that we can make $1/n$ less than any prescribed ε by choosing n greater than a suitable N, depending on ε . Using the quantifiers $\forall$ and $\exists$ this statement about ε and N is concisely expressed as follows:

$$\forall(\varepsilon > 0)\exists N(n > N \Rightarrow 0 \leq 1/n < \varepsilon).$$

(To *prove* this statement one could take N as the first integer greater than $1/\varepsilon$.)

Here are some other examples of statements from analysis, written precisely with the help of quantifiers.

1. The sequence $a_1, a_2, a_3, \ldots$ has limit l.

$$\forall(\varepsilon > 0)\exists N(n > N \Rightarrow |a_n - l| < \varepsilon).$$

2. The function f is continuous at $x = a$.

$$\forall(\varepsilon > 0)\exists\delta(|x - a| < \delta \Rightarrow |f(x) - f(a)| < \varepsilon).$$

3. The function f is continuous for x with $a \le x \le b$.

$$\forall x \forall x' \forall(\varepsilon > 0)\exists\delta(a \le x, \ x' \le b \text{ and}$$
$$|x - x'| < \delta \Rightarrow |f(x) - f(x')| < \varepsilon).$$

4. The function f is uniformly continuous for x with $a \le x \le b$.

$$\forall(\varepsilon > .0)\exists(\delta > 0)\forall x \forall x'(a \le x, \ x' \le b \text{ and}$$
$$|x - x'| < \delta \Rightarrow |f(x) - f(x')| < \varepsilon).$$

After the concept of continuity was precisely defined, around 1820, a few more decades passed before it was realized that uniform continuity is a distinctly different concept. For example, $f(x) = 1/x$ is continuous for $0 < x \le 1$, but it is *not* uniformly continuous on this domain. There is no $\delta > 0$ for which $|x - x'| < \delta$ guarantees $\left|\frac{1}{x} - \frac{1}{x'}\right| < \varepsilon$ for a given ε. Whatever δ we choose (say $\delta = 1/1000$), $\left|\frac{1}{x} - \frac{1}{x+\delta}\right|$ grows indefinitely as x approaches 0.

Failure to distinguish between continuity and uniform continuity is at least partly due to the difficulty of grasping the quantifier prefix $\forall\varepsilon\exists\delta\forall x\forall x'$. Even saying it takes some thought. One normally says: "for all ε there is δ such that, for all x and x', ..." It seems psychologically difficult for humans to grasp *alternation* of quantifiers, as in $\forall\exists\forall\exists\cdots$ or $\exists\forall\exists\forall\cdots$. Outside of artificially constructed sentences, mathematics seldom throws up quantifier prefixes worse than $\forall\exists\forall$.

The presence of quantifiers obviously makes it difficult to find all the valid formulas of predicate logic. Each formula now has infinitely many possible interpretations, so we cannot simply check all possible interpretations, as in propositional logic. Nevertheless, the method of proving tautologies mathematically can be extended to a method for proving all the valid formulas of predicate logic. One such method, whose success depends on the Kőnig infinity lemma, is given in section 10.8.

9.4 Induction

In section 2.1 we noticed how Euclid used induction in its "descent" form to prove results such as existence of prime factorization and termination of the Euclidean algorithm. Descent is a natural style of argument in these two cases, since they produce descending sequences of positive integers in a natural way. In other cases, "ascent" is more natural. For example, the numbers in Pascal's triangle are produced by starting with a small number, 1, and growing bigger ones, by adding adjacent numbers in one row to form a new number in the next row. Here it is natural to prove properties by an ascending style of induction, and indeed Pascal (1654) did precisely this in his treatise *The Arithmetic Triangle*.

Pascal's proofs were not absolutely the first to use the ascending style of induction—Levi ben Gershon (1321) is an earlier example—but Pascal's proofs are so numerous and clear as to leave no doubt about their structure. To prove that a property $P(n)$ holds for all natural numbers n above a certain base value b (usually 0 or 1) it suffices to prove:

Base step. That $P(n)$ holds for the base value $n = b$.

Induction step. That if $P(k)$ holds, then $P(k+1)$ holds.

Over the next few centuries, induction became a standard tool in number theory, in both ascending and descending forms. However, it was not one of the foundations of the subject. As late as the mid-nineteenth century, eminent number theorists such as Dirichlet still appealed to geometric intuition to justify basic properties of addition and multiplication such as $a + b = b + a$ and $ab = ba$.

Then Grassmann (1861) made a remarkable breakthrough: the addition and multiplication functions can be *defined*, and their basic properties *proved*, by induction.[1] Thus induction is the very foundation of arithmetic.

Assuming only the existence of the *successor function* $S(n)$, the addition function $+$ is defined inductively by

$$m + 0 = m, \quad m + S(k) = S(m + k).$$

The first equation defines $m + n$ for all m and for $n = 0$. The second equation defines $m + n$ for $n = S(k)$, given that $m + n$ is already defined for $n = k$. It then follows, by induction on n, that $m + n$ is defined for all natural numbers m and all natural numbers n (these being the numbers that can be reached from 0 by applying the successor function).

Given the definition of $+$, the multiplication function $\cdot$ is defined inductively by

$$m \cdot 0 = 0, \quad m \cdot S(k) = m \cdot k + m.$$

Again, the first equation defines the function for all m and for $n = 0$; the second defines $m \cdot S(k)$, given that $m \cdot k$ and the $+$ function are already defined. And again it follows by induction that $m \cdot n$ is defined for all natural numbers m and all natural numbers n.

With these inductive definitions of $+$ and $\cdot$ in hand, we can proceed to give inductive proofs of their fundamental properties. In principle, the proofs are straightforward consequences of the definitions, but the sequence of proofs is quite long, so it takes some experimentation to develop them in the right order. Grassmann (1861) arrived at $ab = ba$ only in his Proposition 72! It would be tedious to give all of these proofs, so I will do only some of the more immediate ones by way of example.

---

[1] Today it is common to speak of *definition by recursion* and *proof by induction*. I see no harm in using the word "induction" for both.

Successor is +1. *For all natural numbers n,* $S(n) = n + 1$.

Proof. The number 1 is defined to be $S(0)$, so

$$n + 1 = n + S(0)$$
$$= S(n + 0) \qquad \text{by definition of } +,$$
$$= S(n) \qquad \text{since } n + 0 = n \text{ by definition of } +. \quad \square$$

Commutativity of adding 1. *For all natural numbers n,* $1 + n = n + 1$.

Proof. Since $S(n) = n + 1$ by the previous proposition, it suffices to prove $S(n) = 1 + n$. We do this by induction on n.

For the base step $n = 0$ we have

$$S(0) = 1 = 1 + 0 \qquad \text{by definition of } +.$$

For the induction step we assume $S(k) = 1 + k$, so $k + 1 = 1 + k$, and consider $S(S(k))$:

$$S(S(k)) = S(k + 1) \qquad \text{by the previous proposition,}$$
$$= S(1 + k) \qquad \text{by the induction hypothesis,}$$
$$= 1 + S(k) \qquad \text{by definition of } +.$$

This completes the induction step, so $S(n) = 1 + n$ for all natural numbers n. $\qquad \square$

Next, to deal with sums of three or more terms, we need the associative law for addition.

Associativity of addition. *For all natural numbers l, m, n,*

$$l + (m + n) = (l + m) + n.$$

Proof. We prove this for all l and m by induction on n.

For the base step we want $l + (m + 0) = (l + m) + 0$. This is true because

$$l + (m + 0) = l + m \qquad \text{because } m + 0 = m \text{ by definition of } +,$$
$$(l + m) + 0 = l + m \qquad \text{for the same reason.}$$

For the induction step, suppose $l + (m + k) = (l + m) + k$ and consider $l + (m + S(k))$:

$$
\begin{aligned}
l + (m + S(k)) &= l + S(m + k) && \text{by definition of } +, \\
&= S(l + (m + k)) && \text{by definition of } +, \\
&= S((l + m) + k) && \text{by induction hypothesis,} \\
&= (l + m) + S(k) && \text{by definition of } +.
\end{aligned}
$$

This completes the induction step, so $l + (m + n) = (l + m) + n$ for all natural numbers l, m, n. □

Now we are ready to prove the commutative law for addition. This is quite complex, since even the base step requires an induction.

Commutativity of addition. *For all natural numbers m and n,*

$$
m + n = n + m.
$$

Proof. We prove this, for all m, by induction on n.

The base step $n = 0$ depends on proving $0 + m = m$, which we do by induction on m.

For $m = 0$, $0 + m = 0 + 0 = 0 = m$ by the definition of $+$.

For the induction step, we assume $0 + k = k$ and consider

$$
\begin{aligned}
0 + S(k) &= 0 + (k + 1) && \text{because successor is } +1, \\
&= (0 + k) + 1 && \text{by associativity of } +, \\
&= k + 1 && \text{by induction hypothesis,} \\
&= S(k) && \text{because successor is } +1.
\end{aligned}
$$

This completes the induction to prove $0 + m = m$. It follows by the definition of $+$ that $m + 0 = m = 0 + m$, so we have done the base step of induction on n to prove $m + n = n + m$.

The induction step is to assume that $m + k = k + m$ and to consider $m + S(k)$. Now

$$m + S(k) = m + (k + 1) \qquad \text{because successor is } +1,$$
$$= (m + k) + 1 \qquad \text{by associativity of } +,$$
$$= (k + m) + 1 \qquad \text{by induction hypothesis,}$$
$$= 1 + (k + m) \qquad \text{by commutativity of adding 1,}$$
$$= (1 + k) + m \qquad \text{by associativity of } +,$$
$$= (k + 1) + m \qquad \text{by commutativity of adding 1,}$$
$$= S(k) + m \qquad \text{because successor is } +1.$$

This completes the induction step, so $m + n = n + m$ for all natural numbers m and n. □

Many readers may wonder, at this point, what we gain by hard formal proofs of seemingly obvious facts such as $m + n = n + m$. I would answer that $m + n = n + m$ may seem obvious because of some habitual mental image of $m + n$, such as rods of lengths m and n placed end to end. But most facts about numbers, such as the infinitude of primes are not "obvious" in the same way, so we need to find the underlying *logical* principles that make the infinitude of primes as certain as $m + n = n + m$. Induction is the principle that underlies both these facts—and infinitely many others—so it is worth understanding *how* it underlies even simple facts such as $m + n = n + m$.

9.5 *Peano Arithmetic

Grassmann's discovery that induction is the basis of arithmetic did not immediately impress the mathematical community—in fact, it seems to have gone unnoticed. Decades passed before the idea was rediscovered by Dedekind (1888), apparently unaware of Grassmann's work. Then Peano (1889), acknowledging Grassmann's contribution, built induction into an *axiom system* for arithmetic, now known as *Peano arithmetic* (PA).

Today, the Peano axioms are usually written in the language of predicate logic, with variables ranging over the natural numbers. The language also has a constant 0 for zero, and function symbols S, $+$, and $\cdot$ whose intended interpretations are successor, sum, and product. We call the language with these constant function symbols, plus equality and logic symbols, the *language of* PA.

The Peano axioms are:

1. $\forall n(\text{NOT } 0 = S(n))$,
   which says that 0 is not a successor.
2. $\forall m \forall n(S(m) = S(n) \Rightarrow m = n)$,
   which says that numbers with the same successor are equal.
3. $\forall m \forall n(m + 0 = m \text{ AND } m + S(n) = S(m + n))$,
   which is the inductive definition of $+$.
4. $\forall m \forall n(m \cdot 0 = 0 \text{ AND } m \cdot S(n) = m \cdot n + m)$,
   which is the inductive definition of $\cdot$.
5. $[\varphi(0) \text{ AND } \forall m(\varphi(m) \Rightarrow \varphi(S(m)))] \Rightarrow \forall n \, \varphi(n)$,
   which says that if φ is a property that holds for 0, and if φ holds for $S(m)$ when it holds for m, then φ holds for all n.

The last is the induction axiom, or more properly the *induction axiom schema*. It actually consists of infinitely many axioms; one for each formula $\varphi(m)$ that can be written in the language of PA.

The arguments in the previous section show that the Peano axioms suffice to prove the associative and commutative properties of $+$, but this is just the beginning. Along the same lines we can prove the associative and commutative laws for $\cdot$, and also the distributive law $a(b + c) = ab + ac$. This validates all the usual calculations with natural numbers, and it is then possible to prove the basic facts about divisibility and prime factorization. Moreover, there are some tricks that make it possible to simulate the use of negative, rational, and algebraic numbers in PA—even some calculus—so essentially all of known number theory is within the scope of PA.

The five Peano axioms capture so much arithmetic in such a simple way that it seems reasonable to take them as the very *definition* of elementary arithmetic. It is much the same as taking the nine field axioms to encapsulate classical algebra in section 4.3. Of course, the Peano axioms have some very hard theorems as consequences, so one cannot say that all their consequences are elementary. They do, however, give an *elementary encapsulation* of arithmetic.

But what of other parts of elementary mathematics, such as combinatorics? A more natural setting for combinatorics would seem to be the theory of finite sets. The beauty of finite set theory is not only that it provides objects like graphs—which are defined as certain finite sets—but also that it provides natural numbers, allowing us to "count."

The surprising and elegant definition of the natural numbers as finite sets first occurs informally in Mirimanoff (1917), but it did not become influential until formalized by von Neumann (1923). As for Peano, for von Neumann the natural numbers arise from 0 by a successor operation. We define 0 to be the smallest possible set, the *empty* set $\emptyset$. Then 1, 2, 3, . . . are defined in turn so that $n+1$ is the set with members 0, 1, 2, . . . , n. Thus

$$0 = \emptyset,$$
$$1 = \{0\},$$
$$2 = \{0, 1\},$$
$$\vdots$$
$$n+1 = \{0, 1, 2, \ldots, n\}.$$

Notice that $n+1$ is the union of the set $n = \{0, 1, 2, \ldots, n-1\}$ with the set $\{n\}$ whose single member is n. So we have

$$n+1 = \{0, 1, 2, \ldots, n-1\} \cup \{n\} = n \cup \{n\}.$$

Since $n+1$ is the successor of n, this amounts to a very concise definition of the successor function,

$$S(n) = n \cup \{n\},$$

in terms of the "native" set operations of union and forming the *singleton set* $\{n\}$ whose member is n.

Even better, the $<$ relation between numbers is simply the *membership* relation, because

$$m < n \Leftrightarrow m \text{ is a member of } n,$$

which we write $m \in n$. As a result, induction follows from a natural assumption about sets, that "infinite descent is impossible for membership." In more positive terms, every set x has a member y that is "$\in$-least," in the sense that no member of x is a member of y. This assumption is called the *axiom of foundation*. We also need some other axioms, asserting the existence of the empty set, and guaranteeing the existence of unions and certain other sets, such as singletons.

The necessary axioms are more complicated than the Peano axioms, so we omit them here. (For more information, including what must be added to capture *infinite* set theory, see section 9.8.) What is important is that the axioms of finite set theory embrace both combinatorics and arithmetic, so it might be felt that finite set theory is superior to PA, because of its greater scope. However, this is not strictly true, because PA is capable of "expressing" all of finite set theory, because it can encode finite sets by numbers.

What makes this possible is that each finite set can be described by a string of symbols using the four characters $\emptyset$, {, }, and the comma. In particular,

0 is described by the string $\emptyset$,

$2 = \{0, 1\}$ is described by the string $\{\emptyset, \{\emptyset\}\}$,

so $\{0, 2\}$ is described by the string $\{\emptyset, \{\emptyset, \{\emptyset\}\}\}$, and so on.

If we now interpret the four characters as the nonzero digits 1, 2, 3, 4 of base 5 numerals, then each string defining a finite set can be interpreted as a number. As we said in section 3.1 (about base 10 or base 2 numerals, but it is the same for base 5), numerals are based on the concepts of addition, multiplication, and exponentiation. We already have the first two functions in PA, and we can define exponentiation by

$$m^0 = 1, \quad m^{S(n)} = m^n \cdot m.$$

Thus PA has the necessary functions to express numerals and hence arbitrary finite sets.

In this sense, arithmetic is the foundation for all discrete mathematics. But what about the *continuous* mathematics we find in geometry and calculus? As we have already seen in chapters 5 and 6, these depend on the real number system $\mathbb{R}$, whose foundations we have not yet discussed. We have postponed this discussion until now for a good reason: the foundations of $\mathbb{R}$ are thoroughly entangled with questions of set theory and logic.

9.6 *The Real Numbers

The theorem that every magnitude that grows
continually, but not beyond all limits, must certainly
approach a limiting value ... investigation convinced
me that this theorem, or any one equivalent to it, can
be regarded in some way as a sufficient basis for
differential calculus. ... It then remained to discover
its true origin in the elements of arithmetic and thus
at the same time to secure a real definition of the
essence of continuity. I succeeded Nov. 24, 1858.

Richard Dedekind (1901), p. 2

The real numbers have been part of elementary mathematics since the time of Euclid's *Elements*, but even then they were the most difficult part. The first sign of trouble was the discovery of irrational quantities such as $\sqrt{2}$. As mentioned in section 5.3, this discovery led the Greeks to distinguish between their "numbers" (which were essentially our rational numbers) and the more general concept of "magnitudes" (essentially our real numbers, but with severely limited algebraic operations). It also led to an elaborate theory of "ratios of magnitudes," in Book V of the *Elements*, which compared magnitudes by comparing their integer multiples.

For example, the length $\sqrt{2}$ might be compared with the unit lengths 1, 2, 3, ... by observing that

$$1 < \sqrt{2} < 2,$$

$$2 < 2\sqrt{2} < 3,$$

$$7 < 5\sqrt{2} < 8,$$

$$16 < 12\sqrt{2} < 17,$$

$$\vdots$$

Or, as we would say today,

$$1 < \sqrt{2} < 2,$$

$$1 < \sqrt{2} < \frac{3}{2},$$

$$\frac{7}{5} < \sqrt{2} < \frac{8}{5},$$

$$\frac{16}{12} < \sqrt{2} < \frac{17}{12},$$

$$\vdots$$

(I got these successive approximations from the continued fraction for $\sqrt{2}$ in section 2.2.)

The existence of increasingly accurate rational approximations to an irrational quantity is in Book V of the *Elements*, where Euclid proves (in effect) that any two distinct magnitudes $a < b$ can be distinguished by finding a rational number between them:

$$a < \frac{m}{n} < b.$$

Thus an irrational magnitude a is *determined by* the rational numbers less than it and the rational numbers greater than it. However, the Greeks stopped short of *defining* an irrational a in terms of the rationals above and below it, because infinitely many such numbers are required to determine a exactly, and the Greeks did not believe that infinite collections were meaningful.

The fear of infinity persisted among many mathematicians until the nineteenth century. One of the first to overcome this fear was Dedekind, who in 1858 finally took the logical next step beyond Euclid's theory of irrational quantities. Dedekind *defined* irrational numbers in terms of infinite sets of rational numbers. Since any real magnitude a is determined by the sets of rationals

$$L_a = \left\{ \frac{m}{n} : \frac{m}{n} \leq a \right\}, \quad U_a = \left\{ \frac{m}{n} : \frac{m}{n} > a \right\},$$

he took the pair of sets L_a, U_a (which he called the *cut* in the rational numbers determined by a) to actually *define* a. Conversely, any separation of the set $\mathbb{Q}$ of rational numbers into a "lower set" L and an "upper set" U, with each member of L less than every member of U, and with U having no least member, was said to define a real number. (Thus rational numbers a are included, when the $\leq$ sign holds.)

The definition of real numbers by Dedekind cuts has several advantages.

1. It is easy to define sum and product of cuts in terms of sums and products of their members, and real numbers then "inherit" the field properties of sum and product from the rational numbers.

2. The definition of product makes it possible to prove $\sqrt{2}\sqrt{3} = \sqrt{6}$, which is quite difficult to do geometrically. (In fact, Dedekind thought it had never been done, though we saw a geometric way to do it in section 5.3.)

3. Numbers are *ordered* by set containment, because $a \leq b \Leftrightarrow L_a \subseteq L_b$.

4. The correspondence between ordering and containment implies that $\mathbb{R}$ is *complete*, as required for calculus. In particular, if

$$[a_1, b_1] \supseteq [a_2, b_2] \supseteq [a_3, b_3] \supseteq \cdots$$

are closed intervals whose lengths $\to 0$ then they have exactly one point in common. (We appealed to this property to prove the zero derivative theorem in section 6.3 and the Bolzano-Weierstrass theorem in section 7.9.)

To see why, we unite the lower cuts L_{a_1}, L_{a_2}, L_{a_3}, ... into a set L, and the upper cuts U_{b_1}, U_{b_2}, U_{b_3}, ... into a set U. Then each member of L is less than every member of U because

$$a_1 \leq a_2 \leq a_3 \leq \cdots \leq b_3 \leq b_2 \leq b_1.$$

Also, since the size of the intervals $\to 0$, each rational number is in one of L or U. Thus L, U is a cut in $\mathbb{Q}$. It defines a number c which clearly belongs to all the intervals.

Another consequence of completeness is that the set $\mathbb{R}$ of real numbers is a good model of the line, because it has *no gaps*. That is,

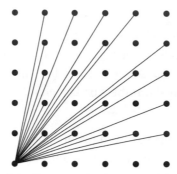

Figure 9.1: Slopes to integer points on the plane.

if $\mathbb{R}$ is separated into two sets $\mathcal{L}, \mathcal{U}$, with each member of $\mathcal{L}$ less than every member of $\mathcal{U}$, then *there is either a greatest member of $\mathcal{L}$ or a least member of $\mathcal{U}$*. To see why, unite the lower cuts of all members of $\mathcal{L}$ into a set L, and the upper cuts of all members of $\mathcal{U}$ into a set U. Then L, U is a cut in $\mathbb{Q}$. It determines a real number c which is either the greatest member of $\mathcal{L}$ or the least member of $\mathcal{U}$.

Thus, for the price of accepting infinite sets as mathematical objects, we can define the system $\mathbb{R}$ of real numbers in such a way as to meet the needs of algebra, geometry, and calculus. This is a major accomplishment, which probably should be considered beyond the bounds of elementary mathematics. Elementary mathematics needs to *use* $\mathbb{R}$, but a full *understanding* of $\mathbb{R}$ depends on some advanced ideas. We will see other ways in which $\mathbb{R}$ calls for advanced ideas in the next section. But first, we attempt to make Dedekind cuts a little more elementary or, at least, more visual.

Visualizing Dedekind Cuts

Dedekind cuts are hard to visualize because the set $\mathbb{Q}$ of rational numbers lies *densely* on the line. Rational points lie in every interval of the line, no matter how small. However, $\mathbb{Q}$ is easier to see if we view the number m/n as the slope of a line through O in the *plane*; namely, the line through O and the integer point (n, m). The integer points in the plane are nicely spread out, so in this way we get a more comprehensible view of $\mathbb{Q}$. Figure 9.1 shows this view of some positive rationals, with m/n associated with the line from O to the point (n, m).

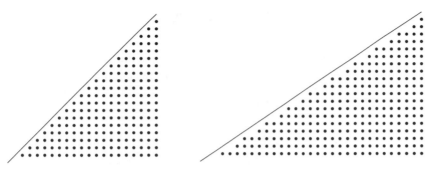

Figure 9.2: The lower Dedekind cuts for 1 and 2/3.

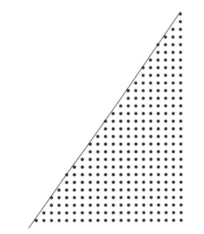

Figure 9.3: The lower Dedekind cut for $\sqrt{2}$.

The ordering of rationals from small to large then corresponds to the ordering of slopes from small to large: the integer points (q, p) *below* the line of slope m/n correspond to rational numbers p/q less than m/n. Figure 9.2 shows the points below the lines of slopes 1 and 2/3, respectively, which we can view as the lower Dedekind cuts for 1 and 2/3.

Figure 9.3 shows the rather more interesting set of integer points below the line of slope $\sqrt{2}$, which we can view as the lower Dedekind cut for $\sqrt{2}$. It differs from the cuts for the rational numbers 1 and 2/3 in having a nonperiodic pattern of "steps" immediately below the line. The step pattern has to be nonperiodic, because if it were periodic then the slope of the line would be rational.

9.7 *Infinity

Now we come to analysis . . . in a sense mathematical
analysis is but a single symphony of the infinite.

David Hilbert (1926)

Just as elementary mathematics needs to use $\mathbb{R}$, without explaining exactly what it is, elementary mathematics needs to use infinite processes without a complete explanation of infinity. This is a symptom of the open-ended nature of mathematics, which prevents us drawing a clear line between the elementary and the advanced parts. In this section, we try to give a glimpse of the advanced ideas that arise when infinity is studied carefully.

The first infinite process we noticed was summing the infinite geometric series, which is involved in questions as simple as expanding $1/3$ as a decimal, or finding the area of a parabolic segment (section 1.5). We are now going to use the geometric series to uncover a truly astonishing property of the set $\mathbb{R}$ of real numbers: $\mathbb{R}$ is not merely infinite; it is *more* infinite than the set $\mathbb{N}$ of natural numbers.

First we should say what it means for a set to be *equinumerous* with $\mathbb{N}$; that is, of the same infinite "size" as $\mathbb{N}$. The members of $\mathbb{N}$ are naturally arranged in an infinite list

$$0, \quad 1, \quad 2, \quad 3, \quad 4, \quad 5, \quad 6, \quad 7, \quad \ldots,$$

with each member occurring at some finite position (for convenience, we say that 0 is in the "zeroth position"). Sets like this used to be called *potentially* infinite, because they arise from a process (start with 0 and keep adding 1) which need not ever be thought complete. Indeed, the classical idea of infinity was just that: a process without end. Now such sets are called *countable* because we can "count" their members, so that each one eventually gets a number:

zeroth member,   first member,   second member,

third member,   . . .

Many other sets of numbers can be viewed in this way, by arranging their elements in a list so that each element occurs at some finite position. In each case there is some process guaranteed to produce any *particular* element of the set in a finite time. Examples are:

1. The set $\mathbb{Z}$ of integers,
   $\mathbb{Z} = \{0, 1, -1, 2, -2, 3, -3, \ldots\}.$
   (After 0, alternating positive and negative of members of the list for $\mathbb{N}$.)

2. The set $\mathbb{Q}^+$ of positive rational numbers,
   $\mathbb{Q}^+ = \{\frac{1}{1}, \frac{1}{2}, \frac{2}{1}, \frac{1}{3}, \frac{3}{1}, \frac{1}{4}, \frac{2}{3}, \frac{3}{2}, \frac{4}{1}, \ldots\}.$
   (Arranging the fractions m/n in order of the sums $m+n$ of their numerator and denominator, and in order of size for each fixed value of $m+n$.)

3. The set $\mathbb{Q}$ of all rational numbers,
   $\mathbb{Q} = \{0, \frac{1}{1}, -\frac{1}{1}, \frac{1}{2}, -\frac{1}{2}, \frac{2}{1}, -\frac{2}{1}, \frac{1}{3}, -\frac{1}{3}, \frac{3}{1}, -\frac{3}{1}, \frac{1}{4}, -\frac{1}{4}, \frac{2}{3}, -\frac{2}{3},$
   $\frac{3}{2}, -\frac{3}{2}, \frac{4}{1}, -\frac{4}{1}, \ldots\}.$
   (After 0, alternating positive and negative of members of the previous list.)

Each such set is called equinumerous with $\mathbb{N}$ because its members can be put in one-to-one correspondence with the members of $\mathbb{N}$. In general, a set X is *equinumerous with $\mathbb{N}$* if its members can be listed

$$x_0, \quad x_1, \quad x_2, \quad x_3, \quad x_4, \quad x_5, \quad x_6, \quad x_7, \quad \ldots,$$

because $x_n \leftrightarrow n$ is then a one-to-one correspondence between X and $\mathbb{N}$.

Now consider the set $\mathbb{R}$, which we can view as a line of infinite length whose points are the members of $\mathbb{R}$. Suppose (looking for a contradiction) that $\mathbb{R}$ is equinumerous with $\mathbb{N}$. In other words, there is an infinite list

$$x_0, \quad x_1, \quad x_2, \quad x_3, \quad x_4, \quad x_5, \quad x_6, \quad x_7, \quad \ldots,$$

whose members are all the real numbers. Viewing the numbers as points on the line, we cover them by intervals of lengths

$$\frac{1}{2}, \quad \frac{1}{4}, \quad \frac{1}{8}, \quad \frac{1}{16}, \quad \frac{1}{32}, \quad \frac{1}{64}, \quad \ldots,$$

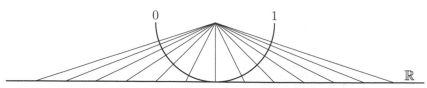

Figure 9.4: Correspondence between $\mathbb{R}$ and the unit interval.

respectively. Then every point of the line is covered, but the total length of the intervals used to cover the line is at most

$$\frac{1}{2} + \frac{1}{4} + \frac{1}{8} + \frac{1}{16} + \frac{1}{32} + \frac{1}{64} + \cdots = 1.$$

Since the line has infinite length, this is a contradiction!

Thus $\mathbb{R}$ is *not* equinumerous with $\mathbb{N}$. It is what we call *uncountable*, and hence not potentially infinite. This discovery, first made by Cantor (1874), was a challenge to all previous thought about infinity. It showed that mathematics cannot avoid infinity by dealing with it only in the potential sense. One of the most important sets in mathematics, $\mathbb{R}$, is a full-blown *actual* infinity.

*Sets of Natural Numbers

The uncountability of $\mathbb{R}$ of course infects all sets equinumerous with $\mathbb{R}$, of which there are many. One of the most important of these is the so-called *power set* of $\mathbb{N}$, $\mathcal{P}(\mathbb{N})$, whose members are all the subsets of $\mathbb{N}$. To show that $\mathbb{R}$ is equinumerous with $\mathcal{P}(\mathbb{N})$ it is convenient to begin with a one-to-one correspondence between $\mathbb{R}$ and the unit interval $(0, 1) = \{x \in \mathbb{R} : 0 < x < 1\}$. Such a correspondence is obvious from figure 9.4 (in which the unit interval has been bent into a semicircle).

Then one can use binary expansions of numbers in $(0,1)$ to set up a correspondence between $(0,1)$ and the infinite sequences of 0s and 1s (we skip the details, which are a little messy because of the exceptional numbers with two different binary expansions). Finally, there is an obvious correspondence between infinite sequences of 0s and 1s and subsets of $\mathbb{N}$: a subset $S \subseteq \mathbb{N}$ corresponds to the infinite sequence with 1 in the nth place if and only if $n \in S$.

s_0	**0**	1	0	1	0	$\cdots$
s_1	0	**1**	1	1	1	$\cdots$
s_2	1	0	**1**	0	1	$\cdots$
s_3	1	1	0	**0**	1	$\cdots$
s_4	0	0	0	0	**0**	$\cdots$
$\vdots$						
s	**1**	**0**	**0**	**1**	**1**	$\cdots$

Figure 9.5: The diagonal construction.

Infinite sequences of 0s and 1s are easier to handle than real numbers in some ways. In particular there is a beautiful direct proof, due to Cantor (1891), that there are uncountably·many of them. (Or, rather, that a countable list of sequences does not include all sequences.) The proof goes as follows.

Suppose (again looking for a contradiction) that $s_0, s_1, s_2, s_3,$ $s_4, \ldots$ is a list of all infinite sequences of 0s and 1s. Figure 9.5 shows part of a diagram tabulating one such hypothetical list of sequences, with their first few digits. Just looking at this table, we can *see* a sequence s that is not on the list.

The nth digit of s is simply the reverse of the nth digit in s_n (shown as a bold digit). Then, for each n, $s \neq s_n$ because they differ in the nth digit. This famous argument is called the *diagonal argument* because it looks at the digits lying along the diagonal of the table.

We have already seen a similar argument, in section 3.8, on the halting problem. There we hit a contradiction by confronting the hypothetical machine T with its own description, $d(T)$, and asking T to halt on $\square$ if and only if it does not halt on $\square$. Here, we hit a contradiction by confronting the hypothetical sequence s with its nth digit, and asking it to be unequal to itself (since $s =$ some s_n if the list of sequences is complete).

The unsolvability of the halting problem is one of several astonishing discoveries made by transporting Cantor's diagonal argument from set theory to the related fields of logic and computation. We discussed some of these discoveries in section 3.9.

9.8 *Set Theory

It should be clear enough from the previous section that infinite sets are an advanced topic. Nevertheless, it is worth describing axioms for set theory, at least informally, in order to show that even advanced mathematics can be encapsulated by a small set of axioms. The axioms are not as simple as those for Peano arithmetic, but we can motivate most of them by examples seen earlier in this chapter. They are called the *Zermelo-Fraenkel* axioms, abbreviated by ZF.

Extensionality. Sets are equal if they have the same members.
   This implies in particular that $\{1, 2\} = \{2, 1\} = \{1, 1, 2\}$ because all of these sets have the same members: 1 and 2.

Empty set. There is a set $\emptyset$ with no members.
   It follows from Extensionality that there is only one empty set. As we saw in section 9.5, $\emptyset$ can serve as the number 0.

Pairing. For any sets x, y there is a set whose members are x and y.
   This is the set we denote by $\{x, y\}$. If $x = y$ it is the singleton set $\{x\}$ by Extensionality. Since $\{x, y\} = \{y, x\}$, by Extensionality again, $\{x, y\}$ is not an *ordered* pair. However, $\{\{x\}, \{x, y\}\}$ can serve as the ordered pair of x and y, because $\{\{x\}, \{x, y\}\} = \{\{y\}, \{y, x\}\}$ only if $x = y$.

Union. For any set x, there is a set whose members are the members of members of x.
   If $x = \{a, b\}$ then the members of members of x make up the set we call $a \cup b$, the "union of a and b." It is also useful to be able to form unions of infinitely many sets. We did this in section 9.6 when we formed the union of the lower Dedekind cuts $L_{a_1}, L_{a_2}, L_{a_3}, \ldots$. This process is validated by applying the union axiom to $x = \{L_{a_1}, L_{a_2}, L_{a_3}, \ldots\}$.

Notice also that Pairing and Union enable us to build finite sets with three or more members. For example, to build the set $\{a, b, c\}$ we use Pairing to build $x = \{a, b\}$ and $y = \{c\}$ then use Union to form $x \cup y = \{a, b, c\}$.

Infinity. There is an infinite set; specifically, there is a set x whose members include $\emptyset$ and, along with any member y, the member $S(y)$. Here $S(y)$ is the successor set $y \cup \{y\}$, so this axiom says there is a set that includes the natural numbers. To get the set $\mathbb{N}$ whose members are *exactly* the natural numbers, we need an axiom that allows us to collect sets with a particular *defining property* into a set. For technical reasons, this axiom is stated in terms of function definitions, in Replacement below.

Power set. For any set x there is a set whose members are the subsets of x.

As we saw in the previous section, this axiom creates a surprisingly large set $\mathcal{P}(x)$ when $x = \mathbb{N}$. In fact, because $\mathcal{P}(\mathbb{N})$ is not countable, there are not enough formulas in the language of ZF to define all its members. This is why we need an axiom to guarantee existence of the power set for $\mathbb{N}$ and other infinite sets.

Replacement (schema). If $\varphi(u, v)$ is a formula defining v as a function $f(u)$, then the range of f for u in a set x is itself a set.

Replacement generalizes the "definable subset" axiom used by Zermelo (1908). Zermelo's axiom was that the elements u of a set x satisfying a formula $\varphi(u)$ form a set. Fraenkel (1922) pointed out that the Replacement schema is needed to obtain sets such as $\{\mathbb{N}, \mathcal{P}(\mathbb{N}), \mathcal{P}(\mathcal{P}(\mathbb{N})), \ldots\}$.

Foundation. Any set has an ε-least member.

As we commented in section 9.5, this serves as an induction axiom.

If we take all of the above axioms *except* Infinity we get a system of finite set theory with induction that has the same strength as the system PA of Peano arithmetic. (Admittedly, the ZF axioms seem more powerful than necessary in the world of finite sets; nevertheless, without Infinity they fail to prove any more than PA.)

When we include the Infinity axiom we get the set $\mathbb{N}$ of natural numbers, and its power set $\mathcal{P}(\mathbb{N})$, which is effectively the set $\mathbb{R}$ of real numbers. From $\mathbb{R}$ we can build the concepts of geometry and analysis, and virtually all of classical mathematics. Thus, ZF encapsulates a vast amount of advanced mathematics. Given that ZF $-$ Infinity is essentially PA, we could say that

$$ZF = PA + \text{Infinity}.$$

Or even more crudely,

advanced mathematics $=$ elementary mathematics $+$ Infinity.

In the next section we discuss how this idea can be refined by introducing more nuanced concepts of "elementary mathematics" and "infinity."

What Has Set Theory Done for Elementary Mathematics?

Set theory has given us a new view of arithmetic and combinatorics—as two ways to look at finite set theory. Whether this will push elementary mathematics in the direction of set theory remains to be seen. However, set theory made a contribution to elementary mathematics a long time ago, when Cantor used it to prove the existence of transcendental numbers.

A real number is called *transcendental* when it is not algebraic; that is, not the solution of a polynomial equation with integer coefficients. The first proof that transcendental numbers exist was given by Liouville (1844), by means of an algebraic theorem on the approximation of algebraic numbers by rationals. Liouville's argument was quite close to elementary, but even so it was outclassed by the argument of Cantor (1874), which involves no algebra at all.

Instead, Cantor used a result learned from Dedekind, that *the set of algebraic numbers is countable.* Using only the fact that each algebraic number is a root of an equation

$$a_n x^n + a_{n-1} x^{n-1} + \cdots + a_1 x + a_0 = 0 \quad \text{for integers } a_0, a_1, \ldots, a_n,$$
$$(*)$$

Dedekind obtained a listing of the algebraic numbers in the following steps:

1. If we define the *height* of the equation (*) by

$$h = n + |a_n| + \cdots + |a_1| + |a_0|,$$

then there are only finitely many equations of height $\leq h$. This is because h bounds both the degree n of (*) and the size of its coefficients.

2. Since $h \geq$ degree, an equation of height h has $\leq h$ roots.

3. Therefore, we can list all algebraic numbers by listing the finitely many equations of height 1, then those of height 2, and so on; and, along with each equation, listing its finitely many roots. By omitting complex roots, we obtain a list of all real algebraic numbers.

The list x_1, x_2, x_3, . . . of real algebraic numbers implies that there is a real number x not on the list, by the uncountability of $\mathbb{R}$. We can explicitly construct such an x by applying the diagonal construction to the decimal expansions of x_1, x_2, x_3, ... We get $x \neq$ each x_n, for example, by letting

$$n\text{th decimal digit of } x = \begin{cases} 1 \text{ if } n\text{th decimal place of } x_n \text{ is not 1,} \\ 2 \text{ if } n\text{th decimal place of } x_n \text{ is 1.} \end{cases}$$

(We avoid digits 0 and 9 in x so that x differs from either expansion of x_n, in case x_n is a number with two decimal expansions, such as $1/2 = 0.500 \cdots = 0.499 \cdots$.)

Thus x is a transcendental number (whose decimal digits could, in principle, be computed). To this day, the above argument is still the most elementary way to prove the existence of transcendental numbers.

9.9 *Reverse Mathematics

When the theorem is proved from the right axioms,
the axioms can be proved from the theorem.

Harvey Friedman (1975)

The idea that advanced mathematics is obtained from ZF – Infinity by adding the Infinity axiom is a very crude way to describe advanced mathematics. The axioms of ZF – Infinity, though they prove no more than PA, contain enormous pent-up energy in the Power Set and Replacement axioms. When this energy is released by adding the Infinity axiom, there is an explosion of new theorems at every level above elementary. We are unable to differentiate between those just above elementary level and those far above it, because they all depend equally on the Infinity axiom.

The idea of *reverse mathematics* is to determine more precisely the axioms on which a given theorem depends. This is done by starting with a "low-powered" elementary system, more like PA, so that adding a simple axiom of infinity does not produce high level theorems. Then one can explore the consequences of *various* axioms of infinity, and sort theorems into levels according to the axioms they depend on. We know exactly what axioms a theorem depends on when we are able to prove the axioms from the theorem. This kind of reversal has been found to occur remarkably often.

Reverse mathematics has been developed by logicians for about 40 years now, and it has been able to rank a large number of classical theorems into five main levels above the level of PA. Since it is a very technical subject, covering many theorems that are too advanced for this book, I will discuss only some of its simplest results. For further information, see the definitive book on the subject, Simpson (2009).

To illustrate the method of reverse mathematics, consider one of its basic axiom systems, ACA_0. (The ACA stands for "arithmetic comprehension axiom.") ACA_0 is essentially PA, but with two types of variables and one extra axiom. It has lower case variables m, n, $\ldots$, x, y, z, $\ldots$ standing for natural numbers as usual, and upper case variables X, Y, Z, $\ldots$ standing for *sets* of natural numbers. Thus the language of ACA_0 can express statements about sets of natural numbers and hence about real numbers (through the correspondence between subsets of $\mathbb{N}$ and real numbers mentioned in section 9.8).

The axioms of ACA_0, with one exception, are just the axioms of PA. Thus the *base theory* for ACA_0 is essentially PA with set variables. To that we add the *arithmetic comprehension axiom* (actually an axiom schema), saying that there is a set X of natural numbers realizing any property $\varphi(n)$ of natural numbers definable in the language of PA:

$$\exists X \forall n (n \in X \Leftrightarrow \varphi(n)). \qquad (*)$$

This axiom schema is effectively an axiom of infinity, because it is the only axiom of ACA_0 asserting the existence of infinite sets.

The effect of the arithmetic comprehension axiom is interesting. It does *not* enable new theorems about natural numbers to be proved in ACA_0: the theorems of ACA_0 about natural numbers are the same as those proved by PA. But arithmetic comprehension gives proofs of many theorems about subsets of $\mathbb{N}$, and hence about real numbers. Among them are three that we have often seen and thought to be just beyond the elementary level:

- The completeness of $\mathbb{R}$.
- A form of the Bolzano–Weierstrass theorem for $\mathbb{R}^n$.
- The Kőnig infinity lemma.

Even more remarkable (and this is where the "reverse" happens), these theorems are each *equivalent* to the arithmetic comprehension axiom (*), and hence to each other. Thus completeness of $\mathbb{R}$ and Bolzano-Weierstrass are, in a reasonable sense, equally advanced, and (*) marks the spot where they lie beyond the elementary level of PA.

Another interesting system, slightly weaker than ACA_0 but with the same base theory, is one called WKL_0. The WKL stands for its added axiom, the "weak Kőnig lemma," which is the special case of the Kőnig infinity lemma where the tree is a subtree of the infinite binary tree. As we saw in section 7.9, this special case occurs in "infinite bisection" arguments, such as the one we used in section 7.9 to prove the Bolzano-Weierstrass theorem for a closed interval of $\mathbb{R}$.

Thus WKL_0, like ACA_0, is essentially PA plus variables for subsets of $\mathbb{N}$ and a set existence axiom—in this case stating the existence of an infinite branch in any infinite subtree of the infinite binary tree. It turns out to be weaker than ACA_0, but nevertheless WKL_0 can prove many important theorems about continuous functions. Among them are:

- Any continuous function on a closed interval takes a maximum value. (This is called the *extreme value theorem*, and we will prove it by an infinite bisection argument in section 10.3.)
- Any continuous function on a closed interval is Riemann integrable.
- The Brouwer fixed point theorem.

And again, we have reversals: each of these theorems implies the weak Kőnig lemma. Thus our suspicion that these theorems about continuous functions are "advanced" is confirmed. They too lie beyond PA, and at the same level as the weak Kőnig lemma.

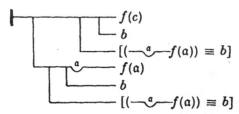

Figure 9.6: Diagram from Frege's *Begriffsschrift*.

9.10 Historical Remarks

As mentioned in sections 3.7 and 3.10, Leibniz dreamed of reducing logic to calculation, but his dream was not realized until the nineteenth century—and then only partially. We have now seen, in section 9.1, how the Boole (1847) idea of an "algebra of logic" fits *propositional* logic very well, revealing it to be essentially the same as mod 2 arithmetic. However, by restricting logic to one that looked like algebra, Boole failed to reach a logic strong enough to express all of mathematics; namely, predicate logic.

Frege (1879) was the first to formulate predicate logic, albeit in a curious diagrammatic system he called the *Begriffsschrift* (concept writing) that was not embraced by mathematicians. It was not embraced by publishers either, due to the difficulty of printing Frege's diagrams. Figure 9.6 shows an example of a proof in Frege's notation. Frege was ahead of his time, because a good understanding of predicate logic was not achieved until the 1920s, culminating in the proof of Gödel (1930) that predicate logic is complete. That is, the standard axiom systems for predicate logic, including Frege's, prove all (and only) the valid formulas.

Gödel's completeness theorem is particularly surprising in the light of his *incompleteness* theorem, proved in Gödel (1931), for systems containing a small amount of mathematics. We sketched how incompleteness arises in section 3.9. The incompleteness theorem shows, among other things, that *mathematics is more than logic*, since the addition of a little mathematics tips the balance from completeness to incompleteness. In fact, logic is "just barely" complete, in the sense that we can mechanically generate all the valid formulas, but not all the invalid ones. It follows that there is no algorithm to decide validity of an arbitrary formula. We say more about this after we give a proof of the completeness theorem in section 10.8. The undecidability of validity was first proved by Church (1935) and Turing (1936).

As we briefly explained in section 3.9, incompleteness arises in axiom systems that are strong enough to encode the operations of Turing machines, typically by encoding steps of computation by operations on numbers. The simplest such system is one introduced by Raphael M. Robinson (1952) and now called *Robinson arithmetic*. The axioms of Robinson arithmetic are the first four Peano axioms listed in section 9.5—thus including the inductive definitions of sum and product but *not* any induction axiom. It takes considerable ingenuity to simulate computation without the induction axiom but it can be done, with the results that

- Robinson arithmetic is *undecidable*. That is, there is no algorithm for deciding, given any formula, whether that formula is a theorem of Robinson arithmetic or not.
- Robinson arithmetic is *incompletable*. That is, whatever axioms are added to Robinson arithmetic (if they are consistent) the resulting theory remains incomplete.

Among the theorems not provable in Robinson arithmetic are the commutative laws for addition and multiplication. This confirms that the induction used to prove these laws in section 9.4 really cannot be avoided. At the same time, it shows that induction is "more advanced" than Robinson arithmetic, in the sense that it enables more theorems to be proved.

Since Peano arithmetic (PA) is the result of adding induction to Robinson arithmetic, its incompleteness follows (assuming that PA is consistent). But the incompleteness of PA is a little puzzling. The theorems known to be unprovable in PA do not yet include any that number theorists *want* to prove, such as the existence of infinitely many primes of the form $2^n - 1$ or $n^2 + 1$. All of the known unprovable theorems of PA are ones devised by logicians.

The situation is more satisfactory for stronger systems, such as the axiom system of ZF set theory described in section 9.8. Many interesting sentences of ZF are now known to be neither provable nor disprovable from the ZF axioms so (since there is insufficient reason to consider them false) they can be taken as new axioms. The most common addition to the ZF axioms is the *axiom of choice* (AC), first formulated by Zermelo (1904). AC says that any set X of nonempty sets x has a *choice function*; that is, a function f such that $f(x) \in x$ for each $x \in X$.

Zermelo introduced AC in order to prove the *well-ordering theorem*, which states that any set Y can be given an ordering under which each subset of Y has a least member. The well-ordering of $\mathbb{N}$ is a theorem equivalent to induction, but well-ordering of uncountable sets, such as $\mathbb{R}$, is not generally provable without AC. In fact, well-ordering of arbitrary sets Y is *equivalent* to AC because it enables us to define a choice function for any set X of nonempty sets. Simply let Y be the union of all the members x of X, and define $f(x)$ to be the *least* member of x in the well-ordering of Y.

To formalize his proof that AC implies well-ordering, Zermelo (1908) gave the first axiom system for set theory. With a later revision by Fraenkel (1922), it became the ZF system we use today. At that time it was not known whether AC was really a new axiom—that is, independent of ZF—but this was proved by a combination of results by Gödel (1938) (consistency of AC) and Cohen (1963). Thus we have been able to say, since 1963, that AC and the well-ordering theorem are more advanced than ZF itself. The same applies to many other sentences of mathematics, as Cohen and others showed. These sentences so far do not include any interesting sentences about the natural numbers, but the following sentences about $\mathbb{R}$ are known to be neither provable nor disprovable in ZF:

- $\mathbb{R}$ can be well-ordered.
- Any infinite subset of $\mathbb{R}$ is either countable or in one-to-one correspondence with $\mathbb{R}$ (the *continuum hypothesis*).
- Any infinite subset of $\mathbb{R}$ contains a countable subset.
- $\mathbb{R}$ is a countable union of countable sets.

What is additionally interesting about AC is that it is actually *equivalent* to many of its consequences, such as:

- The well-ordering of arbitrary sets.
- Comparability of arbitrary sets: for any sets A and B, either A is equinumerous with a subset of B, or B is equinumerous with a subset of A.
- Every vector space has a basis.

So AC is the "right axiom" to prove these theorems.

These results paved the way for reverse mathematics, introduced by Friedman (1975) with the slogan quoted at the beginning of section 9.9. Reverse mathematics refines the situation of AC in set theory—seeking the "right" axioms to prove theorems about $\mathbb{R}$—by starting with a system weaker than ZF (basically PA, with language open to statements about reals) and looking for axioms just strong enough to prove classical theorems of analysis. As Friedman's slogan says, the "right" axiom not only proves the theorem but is equivalent to it. As mentioned in section 9.9, reverse mathematics has succeeded in finding the "right" axioms to prove the completeness of $\mathbb{R}$, the Bolzano-Weierstrass theorem, Riemann integrability of continuous functions, and the Brouwer fixed point theorem relative to a weak system similar to PA. Another theorem for which the "right" axiom has now been found is Gödel's completeness theorem (which we prove in section 10.8). Simpson (2009) gives many other examples.

9.11 Philosophical Remarks

> Symbolic logic may be said to be
> Mathematics become self conscious.
>
> Emil Post (1941), p. 345

The results of mathematical logic, particularly those about unsolvability, incompleteness, and reverse mathematics, are the first to show precisely that some parts of mathematics are "deeper," or more advanced, than others.

It seems indisputable that an unsolvable algorithmic problem is deeper than a solvable one, though we would also like to say that some *solvable* algorithmic problems are deeper than others. There are indeed some results of the latter kind, but what we most want to know about algorithmic problems, of course, is whether $P \neq NP$. If it turns out that $P \neq NP$, then we will be entitled to say, for example, that deciding which formulas of propositional logic are satisfiable is a deeper problem than any problem that can be solved in polynomial time.

We can be confident that an unsolvable algorithmic problem is objectively deep, because solvability of problems is an absolute notion, by the Church-Turing thesis. Provability of theorems is only a relative notion, because of Gödel's incompleteness theorem, but we can still hope to prove that some theorems are deep *relative* to others. As we know from Gödel's incompleteness proof, for any sufficiently strong (and consistent) axiom system A there are theorems T that A cannot prove. It seems reasonable to say that such theorems T are "deeper than," or "more advanced," than A itself. Unfortunately, as mentioned in the previous section, for the system PA that is the best model of elementary mathematics, the "deeper" theorems produced by the Gödel incompleteness argument are not yet of much interest outside logic.

Still, the unprovable theorems of PA and similar systems include a theorem of great interest *inside* logic. This is the statement Con(PA) that expresses the *consistency* of PA—so PA cannot prove its own consistency! The unprovability of Con(PA) is a corollary of Gödel's incompleteness proof, independently noticed by von Neumann and pointed out by him in a letter to Gödel (von Neumann (1930)). In fact, for any sufficiently strong axiom system A—it is enough for A to include PA—the sentence Con(A) expressing consistency of A is not provable in A, if A is consistent. This is why, whenever we claim that a theorem is not provable in some system A, we annoyingly add "if A is consistent." We have to assume that A is consistent because, first, we cannot prove this in A and, second, if A is *not* consistent, then it proves everything—whether true or false.

As we saw in the previous section, the most natural unprovable sentences arise in systems A that can make statements about real numbers. This accords nicely with our experience of elementary mathematics, where the statements bordering on advanced mathematics generally involve real numbers or equivalent concepts, such as infinite sets of natural numbers. Reverse mathematics throws the spotlight on many of these borderline statements by finding the "right" axioms to add to a weak system axiom (essentially PA but with variables for real numbers) in order to prove them. For a large number of classical theorems of analysis we can not only say that they are "deeper than PA," but also assign them to "levels of depth" according to the axioms needed to prove them. Current reverse mathematics distinguishes five different levels of depth, the lower two of which include all the "borderline" advanced theorems about $\mathbb{R}$ considered in this book. This seems reasonable confirmation that these theorems are indeed near the border of elementary mathematics.

10

⤳

Some Advanced Mathematics

PREVIEW

This final chapter contains samples of the eight branches of mathematics discussed in the previous chapters. Each sample takes up a topic seen earlier at an elementary level and carries it further, mainly with the help of principles involving infinity. As we have seen in previous chapters, particularly in the discussion of calculus and logic, crossing the line from elementary to advanced mathematics often involves concepts of infinity.

One borderline advanced concept is the *infinite pigeonhole principle*, stating that if an infinite set is divided into finitely many parts then one of the parts is infinite. We used this principle in section 7.9. Here we use it in section 10.1 to prove the existence of solutions to the Pell equation, in 10.6 to develop some Ramsey theory, and in 10.8 to prove the completeness of predicate logic.

Quite a different use of infinity is in geometry, where the idea of "points at infinity" is needed to formalize the idea that parallel lines "meet at infinity." In section 10.4 we develop this idea in the simplest case: the *real projective line*.

Of the many uses of infinity in calculus or analysis, one enlists properties of continuous functions to prove the fundamental theorem of algebra. We give a proof in section 10.3. Another is the concept of *infinite product*. A famous example is Wallis's product for π, which we derive in section 10.5 and apply in 10.7 to explain why the graph of the binomial coefficients tends to the curve $y = e^{-x^2}$.

Another occurrence of infinity in analysis—the uncountability of $\mathbb{R}$—is barely touched on here. However, it looms over the concepts of unsolvability and incompleteness, which we revisit in sections 10.2 and 10.8.

10.1 Arithmetic: the Pell Equation

In section 2.8 we showed how one solution of the Pell equation $x^2 - my^2 = 1$ can generate infinitely many others. But we left open the problem of finding that one solution. Even for modest values of m, the smallest nontrivial solution of $x^2 - my^2 = 1$ can be hard to find. As mentioned in section 2.9, the smallest nontrivial solution of $x^2 - 61y^2 = 1$ is

$$(x,\ y) = (1766319049,\ 226153980)!$$

Since the smallest nontrivial solution varies quite erratically with m, its existence is unclear in general. However, Lagrange proved in 1768 that *if m is any nonsquare positive integer, the Pell equation $x^2 - my^2 = 1$ has an integer solution $\neq (\pm 1, 0)$*. To smooth the path to solutions, we first explain how the theory of the equation $x^2 - my^2 = 1$ is related to the structure of the algebraic number field $\mathbb{Q}(\sqrt{m})$.

The Pell Equation and the Norm on $\mathbb{Q}(\sqrt{m})$

In section 2.8 we showed that solutions of the Pell equation $x^2 - my^2 = 1$ can be generated with the help of the irrational number $\sqrt{m}$. In particular, if one solution is $x = x_1$, $y = y_1 \neq 0$, then infinitely many solutions $x = x_n$, $y = y_n$ are given by the formula

$$x_n + y_n\sqrt{m} = (x_1 + y_1\sqrt{m})^n, \quad \text{where } n \in \mathbb{Z}.$$

This trick can be better understood with the help of the concept of the *norm on the field $\mathbb{Q}(\sqrt{m})$.*

The field $\mathbb{Q}(\alpha)$ was defined for any algebraic number α in section 4.8, but in the special case $\alpha = \sqrt{m}$ for a nonsquare natural number m it may be defined more simply as

$$\mathbb{Q}(\sqrt{m}) = \{a + b\sqrt{m} : a, b \in \mathbb{Q}\}.$$

It is clear that $\mathbb{Q}(\sqrt{m})$ must include all the numbers $a + b\sqrt{m}$, and that the product of two such numbers is another of the same kind. So, to prove that the numbers $a + b\sqrt{m}$ make up the whole field $\mathbb{Q}(\sqrt{m})$ it suffices to prove that the inverse of $a + b\sqrt{m}$ is another number of the same form. This is so, because

$$\frac{1}{a + b\sqrt{m}} = \frac{a - b\sqrt{m}}{(a + b\sqrt{m})(a - b\sqrt{m})} = \frac{a - b\sqrt{m}}{a^2 - mb^2}$$

$$= \frac{a}{a^2 - mb^2} - \frac{b}{a^2 - mb^2}\sqrt{m},$$

which *is* of the same form, since $\frac{a}{a^2 - mb^2}$ and $\frac{b}{a^2 - mb^2}$ are rational if a and b are.

The *norm* on $\mathbb{Q}(\sqrt{m})$ is defined by

$$\text{norm}(a + b\sqrt{m}) = a^2 - mb^2.$$

Thus the norm is a rational number, and it is an integer if a and b are. What makes the norm particularly useful is the following property, analogous to the multiplicative property of the complex number norm used in section 2.6.

Multiplicative property of the norm. *If* $u = a + b\sqrt{m}$ *and* $u' = a' + b'\sqrt{m}$ *then*

$$\text{norm}(uu') = \text{norm}(u)\text{norm}(u').$$

Proof. Since $u = a + b\sqrt{m}$ and $u' = a' + b'\sqrt{m}$ we have

$$uu' = (a + b\sqrt{m})(a' + b'\sqrt{m}) = (aa' + mbb') + (ab' + ba')\sqrt{m},$$

and therefore

$$\text{norm}(uu') = (aa' + mbb')^2 - m(ab' + ba')^2$$

$$= (aa')^2 + (mbb')^2 - m(ab')^2 - m(ba')^2$$

by cancellation of terms $2maa'bb'$. On the other hand,

$$\text{norm}(u)\text{norm}(u') = (a^2 - mb^2)(a'^2 - mb'^2)$$
$$= (aa')^2 + (mbb')^2 - m(ab')^2 - m(ba')^2,$$

so $\text{norm}(uu') = \text{norm}(u)\text{norm}(u')$ as claimed. $\qquad\square$

The identity that makes this proof work,

$$(a^2 - mb^2)(a'^2 - mb'^2) = (aa' + mbb')^2 - m(ab' + ba')^2,$$

was discovered by the Indian mathematician Brahmagupta around 600 CE. It says, as he realized, that if $(x, y) = (a, b)$ and $(x, y) = (a', b')$ are solutions of the equation $x^2 - my^2 = 1$ then so is $(x, y) = (aa' + mbb', ab' + ba')$—because if a, a', b, b' are integers, then so are $aa' + mbb'$ and $ab' + ba'$.

If we call $a + b\sqrt{m}$ an "integer" of the field $\mathbb{Q}(\sqrt{m})$ when a and b are ordinary integers, then Brahmagupta's discovery can be expressed as follows. If $a + b\sqrt{m}$ and $a' + b'\sqrt{m}$ are "integers" of norm 1, then so is their product

$$(a + b\sqrt{m})(a' + b'\sqrt{m}) = (aa' + mbb') + (ab' + ba')\sqrt{m}.$$

Notice also that if $a + b\sqrt{m}$ is an "integer" of norm 1 then so is its inverse. Indeed, its inverse is $a - b\sqrt{m}$ because

$$(a + b\sqrt{m})(a - b\sqrt{m}) = a^2 - mb^2 = \text{norm}(a + b\sqrt{m}) = 1.$$

Putting these two facts together, we find that if $x_1 + y_1\sqrt{m}$ is an "integer" of norm 1 (so $x = x_1, y = y_1$ is a solution of $x^2 - my^2 = 1$), then so is

$$x_n + y_n\sqrt{m} = (x_1 + y_1\sqrt{m})^n \quad \text{for any integer } n.$$

This extends the infinite collection of solutions of $x^2 - my^2 = 1$ found in section 2.8, by allowing negative values of n. In the next subsection we will see how this mild extension allows us to see why *all* solutions can be obtained in this way.

Obtaining All Solutions from Just One

If $x = x_1$, $y = y_1 \neq 0$ is a positive integer solution of $x^2 - my^2 = 1$, then the number $x_1 + y_1\sqrt{m} > 1$, so all powers $(x_1 + y_1\sqrt{m})^n$ for $n \in \mathbb{Z}$ are positive and distinct. In fact,

$$\cdots < (x_1 + y_1\sqrt{m})^{-1} < 1$$
$$= (x_1 + y_1\sqrt{m})^0 < (x_1 + y_1\sqrt{m})^1 < (x_1 + y_1\sqrt{m})^2 < \cdots.$$

We now suppose that $(x, y) = (x_1, y_1)$ is the smallest such solution; that is, the one for which $x_1 + y_1\sqrt{m}$ is smallest. Then we have:

Positive solutions of the Pell equation. *Every integer solution of* $x^2 - my^2 = 1$ *with* $x, y > 0$ *is of the form* $x = x_n$, $y = y_n$ *for some integer n, where*

$$x_n + y_n\sqrt{m} = (x_1 + y_1\sqrt{m})^n.$$

Proof. Suppose on the contrary that there is a positive solution $(x, y) = (x', y')$ unequal to each (x_n, y_n). Then the positive number $x' + y'\sqrt{m}$ differs from each $x_n + y_n\sqrt{m} = (x_1 + y_1\sqrt{m})^n$ and so we must have

$$(x_1 + y_1\sqrt{m})^n < x' + y'\sqrt{m} < (x_1 + y_1\sqrt{m})^{n+1} \quad \text{for some integer } n.$$

But then, dividing through by $(x_1 + y_1\sqrt{m})^n$, we get

$$1 < (x' + y'\sqrt{m})(x_1 + y_1\sqrt{m})^{-n} < x_1 + y_1\sqrt{m}.$$

Now $(x' + y'\sqrt{m})(x_1 + y_1\sqrt{m})^{-n} = X + Y\sqrt{m}$ for some "integer" $X + Y\sqrt{m}$ of norm 1, since the "integers" of norm 1 are closed under products and inverses.

So $(x, y) = (X, Y)$ is a positive solution of the Pell equation, yet $X + Y\sqrt{m}$ is smaller than $x_1 + y_1\sqrt{m}$, contrary to assumption. This contradiction shows that we were wrong to suppose there are any positive integer solutions of $x^2 - my^2 = 1$ other than $(x, y) = (x_n, y_n)$. $\square$

Thus all positive solutions arise from the smallest one, $x = x_1$, $y = y_1$, and of course the negative solutions are obtained from the positive ones by changing the sign of x.

Existence of a Nontrivial Solution

It remains to show that at least one solution $x = x_1$, $y = y_1 \neq 0$ of $x^2 - my^2 = 1$ exists. An interesting proof of this was given by Dirichlet around 1840, and it was published by Dedekind (1871a) as an appendix to his edition of Dirichlet's lectures on number theory. Dirichlet used what is now called his "pigeonhole principle": if more than k pigeons go into k boxes then at least one box contains at least two pigeons (finite version); if infinitely many pigeons go into k boxes, then at least one box contains infinitely many pigeons (infinite version). We have met the infinite pigeonhole principle already in this book, in the proofs of the König infinity lemma and the Bolzano-Weierstrass theorem in section 7.9, and we will meet it again later in this chapter.

Dirichlet's argument can be divided into the following steps. First, a theorem on the approximation of irrational numbers by rationals:

Dirichlet's approximation theorem. *For any irrational $\sqrt{m}$ and integer $B > 0$ there are integers a, b with $0 < b < B$ and*

$$|a - b\sqrt{m}| < \frac{1}{B}.$$

Proof. For any integer $B > 0$ consider the $B - 1$ numbers $\sqrt{m}$, $2\sqrt{m}, \ldots, (B - 1)\sqrt{m}$. For each multiplier k choose the integer A_k such that

$$0 < A_k - k\sqrt{m} < 1.$$

Since $\sqrt{m}$ is irrational, the $B - 1$ numbers $A_k - k\sqrt{m}$ are strictly between 0 and 1 and they are all different, because an equality between any two of them can be solved for $\sqrt{m}$, implying that $\sqrt{m}$ is rational. Thus we have $B + 1$ different numbers

$$0, \quad A_1 - \sqrt{m}, \quad A_2 - 2\sqrt{m}, \quad \ldots, \quad A_{B-1} - (B - 1)\sqrt{m}, \quad 1$$

in the interval from 0 to 1.

If we then divide this interval into B subintervals of length $1/B$, it follows by the finite pigeonhole principle that at least one subinterval contains two of the numbers. The difference between these two numbers, which is of the form $a - b\sqrt{m}$ for some integers a and b, is therefore irrational and such that

$$|a - b\sqrt{m}| < \frac{1}{B}.$$

Also, $b < B$ because b is the difference of two positive integers less than B. □

The next steps bring in the infinite pigeonhole principle.

1. Since Dirichlet's approximation theorem holds for all $B > 0$, we can make $1/B$ arbitrarily small, forcing new values of a and b to be chosen infinitely often. Therefore, *there are infinitely many integer pairs (a, b) such that $|a - b\sqrt{m}| < 1/B$.* Since $0 < b < B$, we have

$$|a - b\sqrt{m}| < \frac{1}{b}.$$

2. It follows from step 1 that

$$|a + b\sqrt{m}| \leq |a - b\sqrt{m}| + |2b\sqrt{m}| \leq |3b\sqrt{m}|,$$

and, therefore, multiplying the bounds on $|a - b\sqrt{m}|$ and $|a + b\sqrt{m}|$,

$$|a^2 - mb^2| \leq \frac{1}{b} \cdot 3b\sqrt{m} = 3\sqrt{m}.$$

Hence *there are infinitely many $a - b\sqrt{m} \in \mathbb{Z}[\sqrt{m}]$ with norm of absolute value $\leq 3\sqrt{m}$.*

3. By the infinite pigeonhole principle we obtain in turn

   - infinitely many numbers $a - b\sqrt{m}$ with the same norm, N say,
   - among these, infinitely many with a in the same congruence class, mod N,
   - among these, infinitely many with b in the same congruence class, mod N.

4. From step 3 we get two positive numbers, $a_1 - b_1\sqrt{m}$ and $a_2 - b_2\sqrt{m}$, with

- the same norm N,
- $a_1 \equiv a_2 \pmod{N}$,
- $b_1 \equiv b_2 \pmod{N}$.

The final step takes the quotient $a - b\sqrt{m}$ of the two numbers just found. Its norm $a^2 - mb^2$ is clearly 1 by the multiplicative property of norm. It is not so clear that a and b are integers, but this follows from the congruence conditions obtained in step 4.

Nontrivial solution of the Pell equation. *When m is a nonsquare positive integer, the equation $x^2 - my^2 = 1$ has an integer solution $(a, b) \neq (\pm 1, 0)$.*

Proof. Consider the quotient $a - b\sqrt{m}$ of the numbers $a_1 - b_1\sqrt{m}$ and $a_2 - b_2\sqrt{m}$ found in step 4. We have

$$a - b\sqrt{m} = \frac{a_1 - b_1\sqrt{m}}{a_2 - b_2\sqrt{m}} = \frac{(a_1 - b_1\sqrt{m})(a_2 + b_2\sqrt{m})}{a_2^2 - mb_2^2}$$

$$= \frac{a_1 a_2 - mb_1 b_2}{N} + \frac{a_1 b_2 - b_1 a_2}{N}\sqrt{m},$$

where $N = a_2^2 - mb_2^2$ is the common norm of $a_1 - b_1\sqrt{m}$ and $a_2 - b_2\sqrt{m}$. Thus the quotient $a - b\sqrt{m}$ has norm 1 by the multiplicative property of norm.

Since $a_1 - b_1\sqrt{m}$ and $a_2 - b_2\sqrt{m}$ are unequal and positive, their quotient $a - b\sqrt{m} \neq \pm 1$. It remains to show that a and b are integers. This amounts to showing that N divides $a_1 a_2 - mb_1 b_2$ and $a_1 b_2 - b_1 a_2$, or that

$$a_1 a_2 - mb_1 b_2 \equiv a_1 b_2 - b_1 a_2 \equiv 0 \pmod{N}.$$

The first congruence follows from the fact that $a_1^2 - mb_1^2 = N$, which implies

$$0 \equiv a_1^2 - mb_1^2 \equiv a_1 a_1 - mb_1 b_1 \equiv a_1 a_2 - mb_1 b_2 \pmod{N},$$

replacing a_1 and b_1 by their respective congruent values $a_1 \equiv a_2 \pmod{N}$ and $b_1 \equiv b_2 \pmod{N}$ found in step 4.

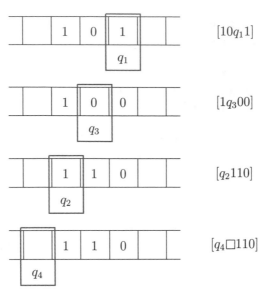

Figure 10.1: Snapshots of a computation and the corresponding words.

The second congruence follows from $a_1 \equiv a_2 \pmod{N}$ and $b_2 \equiv b_1$ $\pmod{N}$ by multiplying, which gives $a_1 b_2 \equiv b_1 a_2 \pmod{N}$, and hence $a_1 b_2 - b_1 a_2 \equiv 0 \pmod{N}$. $\hspace{2em}\square$

10.2 Computation: the Word Problem

Representing a Turing Machine by Word Transformations

The successive stages in a Turing machine computation, illustrated by figure 3.4 in section 3.7, are easily encoded by strings of symbols we call *words*. The word for a given stage is the sequence of symbols on the portion of tape containing all marked squares and the read/write head, with the current state symbol inserted to the left of the symbol for the scanned square. Thus for the snapshots of stages given in figure 3.4, the corresponding words are those in figure 10.1.

We add left and right brackets at the left and right ends of the word to allow the read/write head symbol to "sense" the ends of the marked portion of tape, and to create new blank squares if needed.

The change from one snapshot to the next depends only on the current state q_i and the scanned symbol S_j, so the change from one word to the next depends only on the subword $q_i S_j$ and (when the head moves left) the symbol to the left of q_i. Consequently, the sequence of words encoding a computation can be produced by a sequence of *replacements* of one 2- or 3-letter subword by another word. A Turing machine can therefore be completely described by the *replacement rules* corresponding to its quintuples.

For example, the quintuple $q_1 1 0 L q_3$ corresponds to the replacements

$$1q_1 1 \to q_3 10 \quad \text{and} \quad 0q_1 1 \to q_3 00, \quad \text{and} \quad [q_1 1 \to [q_3 \square 0,$$

according as the symbol to the left of the state symbol is 1 or 0 or [. In the latter case a new blank square is created.

In general, for a machine M with states q_i and symbols S_j the correspondence between quintuples and rules is the following:

quintuple	replacement rules
$q_i S_j S_k R q_l$	$q_i S_j \to S_k q_l$
	$q_i] \to S_k q_l]$ if $S_j = \square$
$q_i S_j S_k L q_l$	$S_m q_i S_j \to q_l S_m S_k$ for each symbol S_m
	$[q_i S_j \to [q_l \square S_k$

It is clear that these rules faithfully produce the words encoding the computation of the machine M from an initial situation encoded by a word w. They produce the unique sequence of words $w_1, w_2, w_3, \ldots$ encoding the results of 1, 2, 3, $\ldots$ steps of computation because at most one replacement rule applies to a word encoding a possible situation for M. And if M halts, the last word contains a subword $q_i S_j$ for which no replacement is supplied, because there is no quintuple of M beginning with $q_i S_j$.

For each such subword $q_i S_j$ that signals a halt, it is convenient to add the replacement rules

$$q_i S_j \to H, \quad HS_m \to H, \quad S_m H \to H, \quad [H \to H, \quad H] \to H$$

for each symbol S_m of M. Then, when halting occurs, the latter rules create a symbol H that can "swallow" all symbols to its left and right, leaving just the 1-letter word H.

We call the system of rules derived from a Turing machine M in this way the *system* Σ_M. It follows by construction of Σ_M that

M eventually halts, starting from the situation encoded by the word w, if and only if the word w is convertible to H by the system Σ_M.

Then it follows, from the unsolvability of the halting problem proved in section 3.8, that we have the following:

Unsolvable word transformation problem. *The problem of deciding, given a system Σ_M and initial word w, whether Σ_M converts w to H is unsolvable.* □

In fact, given the existence of a universal machine U (section 3.9), the word transformation problem is unsolvable for the *fixed* system Σ_U, because the halting problem is unsolvable for the fixed machine U.

*The Word Problem

Thue (1914) introduced a simple problem about words, rather like the one introduced in the subsection above, except that replacement of subwords can take place in either direction. That is, we are given a system T of *equations* between words,

$$u_1 = v_1, \quad u_2 = v_2, \quad \ldots, \quad u_k = v_k,$$

and we have to decide whether given words w, w' can be converted to each other by replacement of subwords by words equal to them. Since u_i can be replaced by v_i, and v_i can be replaced by u_i, the replacement rules of Thue (1914) are "two-way," rather than the "one-way" rules used above to simulate Turing machines. (By a happy coincidence, the name Thue is actually pronounced "two-way.")

Since equations are rather more natural than one-way transformations, it would be nice to prove that the word *equality* problem is also unsolvable, like the word transformation problem. This was done independently by Post and Markov in 1947. It was the first example of a problem from ordinary mathematics to be proved unsolvable. Here we will follow the method of Post (1947), which simply takes the one-way transformations used to simulate Turing machines and makes them two-way.

What can go wrong when we do this? We can certainly get equations $w = w'$, where the words w, w' encode machine situations *not* in the same computation. This is because w, w' can be transformable to the same word v, so $w = w'$ when we allow the transformation $w' \to v$ to be reversed. However, we have the following proposition:

Detection of halting computations. *If w encodes a machine situation, then $w = H$ if and only if there is a halting computation beginning with situation w.*

Proof. If there is a halting computation beginning with situation w, then

$$w \to \text{word containing } H$$

$$\to H \text{ by allowing } H \text{ to "swallow" symbols,}$$

and therefore $w = H$.

Conversely, if $w = H$ then there is a series of equations

$$w = w_1 = w_2 = \cdots = w_n = \text{word containing } H, \qquad (*)$$

where w_n is the last word not containing H. Each equation $w_i = w_{i+1}$ comes from a transformation $w_i \to w_{i+1}$ or the reverse, $w_i \leftarrow w_{i+1}$, so imagine all the equations in (*) rewritten with the appropriate arrows, $\to$ or $\leftarrow$, in place of the $=$ signs.

At least one arrow is $\to$, because if all are $\leftarrow$ we have

$$\text{word containing } H \to w.$$

This is impossible because the replacement rules do not allow H to be destroyed, yet w does not contain it. Thus there is at least one forward arrow. If $w_i \to w_{i+1}$ is the leftmost occurrence of a forward arrow, and there are reverse arrows to its left, we have

$$w_{i-1} \leftarrow w_i \to w_{i+1},$$

which implies $w_{i-1} = w_{i+1}$, because each machine situation w_i has at most one successor. We can then delete $w_{i-1} \leftarrow w_i \to$ from the series, eliminating two opposite arrows. Repeating this argument, we can eliminate all reverse arrows, and hence conclude that

$$w \to \text{word containing } H,$$

which means that w leads to a halting computation. □

Following immediately from this proposition we have:

Unsolvability of the word problem. *If T_U is the Thue system whose equations $u = v$ arise from the word transformations $u \to v$ for a universal Turing machine U, then the problem of deciding, for a given word w, whether $w = H$ in T_U is unsolvable.* □

This problem is also called the *word problem for semigroups* because the classes of equal words, under the operation of word concatenation, form an algebraic structure known as a semigroup. One gets a *group* by adjoining an *inverse letter a^{-1}* for each letter a used to make words, together with the equations

$$aa^{-1} = a^{-1}a = \text{empty word.}$$

For groups we have a similarly stated word problem, which in fact predates Thue's statement of the word problem for semigroups. The word problem for groups was first stated by Dehn (1912). It is actually a more important problem, being related (as Dehn realized) to natural problems in topology, such as the problem of deciding whether a closed curve in a complicated three-dimensional object can be contracted to a point.

Like the word problem for semigroups, the word problem for groups is unsolvable, though this is much harder to prove. One can use equations in a group to simulate Turing machine computations, but the inverse letters make it very hard to control the encoding of computational steps by words. Because of this difficulty, the first proof of unsolvability appeared in Novikov (1955)—a paper of 143 pages. More on the history of the word problem for groups can be found in Stillwell (1982) and a proof of its unsolvability is in Stillwell (1993). The latter corrects an error in the proof given in Stillwell (1982).

10.3 Algebra: the Fundamental Theorem

As emphasized in chapter 4, the fundamental theorem of algebra is not a purely algebraic theorem. Its usual statement—that any polynomial equation $p(x) = 0$ has a solution in the set $\mathbb{C}$ of complex numbers—assumes the existence of the uncountable sets $\mathbb{R}$ and $\mathbb{C}$, and the usual proofs assume the completeness of $\mathbb{R}$ and some consequent property of continuous functions. In this section we will give a proof using only elementary algebra and geometry, plus a standard theorem about continuous functions: the *extreme value theorem*.

*The Extreme Value Theorem

We begin with the simplest version of the extreme value theorem, where the continuous function is on a closed interval $[a, b]$. It will then be clear how to extend the idea to functions on the plane, where we need to work in order to prove the fundamental theorem of algebra.

Extreme value theorem. *If f is a continuous function on $[a, b]$, then f attains both a maximum and a minimum value on $[a, b]$.*

Proof. We first prove the seemingly more modest statement that f is *bounded* on the interval $I_1 = [a, b]$. If it is not, then f is unbounded on at least one half of I_1, either $\left[a, \frac{a+b}{2}\right]$ or $\left[\frac{a+b}{2}, b\right]$. Repeating this argument in (say) the leftmost half I_2 on which f is unbounded, we find a quarter of I_1, I_3, on which f is unbounded, and so on.

Proceeding in this way, we obtain an infinite sequence of nested closed intervals on which f is unbounded:

$$I_1 \supseteq I_2 \supseteq I_3 \supseteq \cdots .$$

The intervals I_k become arbitrarily small, so they have a single common point c, by the completeness of $\mathbb{R}$ (section 9.6). But, when I_k is sufficiently small, it follows from the continuity of f that the values of $f(x)$ for x in I_k lie between $f(c) - \varepsilon$ and $f(c) + \varepsilon$, contradicting the claim that f is unbounded on each I_k.

Thus it is wrong to suppose that f is unbounded on $[a, b]$; its values $f(x)$ have a least upper bound u and a greatest lower bound l, again by the completeness of $\mathbb{R}$.

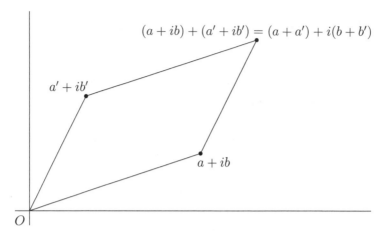

$$(a + ib) + (a' + ib') = (a + a') + i(b + b')$$

$a' + ib'$

$a + ib$

O

Figure 10.2: Adding complex numbers.

Now if f does *not* attain the value u then the function $\frac{1}{u - f(x)}$ is continuous and *unbounded* on $[a, b]$, because the values $f(x)$ come arbitrarily close to their least upper bound u. This contradicts the boundedness of continuous functions just proved, so f does in fact attain a maximum value, u. It similarly attains a minimum value l. □

From the above proof it should be clear how to prove that a continuous function on the plane attains a maximum and minimum value on any closed bounded region, such as a closed disk. One repeatedly divides the region into finitely many closed subregions, so as to obtain nested subregions that become arbitrarily small and hence have a single common point. (We used a similar argument to find a single common point in the plane when proving the Brouwer fixed point theorem in section 7.10.) The rest of the argument is exactly the same.

The Geometry of Complex Numbers

A complex number $z = a + ib$ can be viewed as a point (a, b) of the plane. The *sum* of complex numbers is defined by

$$(a + ib) + (a' + ib') = (a + a') + i(b + b'),$$

which is just *vector addition* in the plane (figure 10.2).

Multiplication of complex numbers also has a geometric interpretation, which is less obvious but more important. It involves both length and angle. Length is involved in the *absolute value* of $z = a + ib$,

$$|z| = \sqrt{a^2 + b^2},$$

which is the distance of z from O. More generally,

$$|z - z'| = \sqrt{(a - a')^2 + (b - b')^2}$$

is the distance *from z to z'*.

The *product* of $z = a + ib$ and $z' = a' + ib'$ is defined by

$$(a + ib)(a' + ib') = (aa' - bb') + i(ab' + ba'),$$

which is what one gets by multiplying according to the usual rules of algebra and assuming $i^2 = -1$. It follows from this definition that *absolute value is multiplicative*; that is,

$$|zz'| = |z||z'|,$$

because

$$|zz'|^2 = (aa' - bb')^2 + (ab' + ba')^2 = (a^2 + b^2)(a'^2 + b'^2) = |z|^2|z'|^2.$$

The middle part $(aa' - bb')^2 + (ab' + ba')^2 = (a^2 + b^2)(a'^2 + b'^2)$ can be checked by multiplying out both sides. (We already observed this, with slightly different notation, in section 2.6. Now we no longer need a, b, a', b' to be integers.)

Since absolute value equals length, it follows that *multiplication of all numbers in the plane by u with $|u| = 1$ preserves all lengths.* Namely, multiplication by u sends any two points z, z' to uz, uz', and the distance between the latter points is

$$|uz - uz'| = |u(z - z')|$$

$$= |u||z - z'| \quad \text{by the multiplicative property}$$

$$= |z - z'| \quad \text{because } |u| = 1,$$

which is the distance between the original points. It is also clear that the zero point O is fixed by any multiplication, so multiplication of all points of the plane by u with $|u| = 1$ is a "rigid motion" of the plane $\mathbb{C}$ of complex numbers that fixes O; that is, a rotation about O.

If we write $u = \cos\theta + i\sin\theta$ then multiplication by u obviously sends 1 to $\cos\theta + i\sin\theta$, which is the point on the unit circle at angle θ. In other words, *multiplication by $u = \cos\theta + i\sin\theta$ rotates the plane $\mathbb{C}$ through angle θ*. More generally, multiplication by an arbitrary complex number, which we can write as

$$v = r(\cos\theta + i\sin\theta) \quad \text{for some real } r,$$

magnifies the plane by r and rotates it through angle θ.

*The Fundamental Theorem of Algebra

To prove the fundamental theorem of algebra we exploit the ability to "rotate" a complex number through an arbitrary angle, and to control its length, by multiplying by a suitable complex number. This ability is the key to the following result, discovered by d'Alembert (1746) and proved by Argand (1806) using the geometric interpretation of complex numbers.

d'Alembert's lemma. *If $p(z)$ is a polynomial in z and $p(z_0) \neq 0$, then there is a complex number Δz such that $|p(z_0 + \Delta z)| < |p(z_0)|$.*

Proof. Suppose that $p(z) = a_n z^n + a_{n-1} z^{n-1} + \cdots + a_1 z + a_0$. Then

$$p(z_0 + \Delta z) = a_n(z_0 + \Delta z)^n + a_{n-1}(z_0 + \Delta z)^{n-1} + \cdots + a_1(z_0 + \Delta z) + a_0$$

$$= a_n(z_0^n + n\Delta z \cdot z_0^{n-1} + \cdots) \quad \text{by the binomial theorem}$$

$$+ a_{n-1}(z_0^{n-1} + (n-1)\Delta z \cdot z_0^{n-2} + \cdots)$$

$$\vdots$$

$$+ a_1(z_0 + \Delta z)$$

$$+ a_0$$

$$= p(z_0) + A \cdot \Delta z + \text{terms in } (\Delta z)^2, (\Delta z)^3, \ldots, (\Delta z)^n,$$

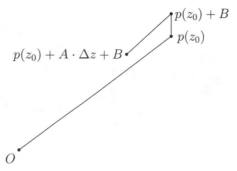

Figure 10.3: d'Alembert's lemma.

where $A = na_n z_0^{n-1} + (n-1)a_{n-1}z_0^{n-2} + \cdots + a_1$ is fixed along with z_0, and we are free to choose the value of Δz. First we choose the size of Δz so small that the sum B of the terms in $(\Delta z)^2, (\Delta z)^3, \ldots, (\Delta z)^n$ is small in comparison with $A \cdot \Delta z$. Next we choose the *direction* of Δz so that $A \cdot \Delta z$ points in the direction from $p(z_0) + B$ towards O. It is then clear (figure 10.3) that $p(z_0) + A \cdot \Delta z + B$ is nearer to O than $p(z_0)$, that is,

$$|p(z_0 + \Delta z)| = |p(z_0) + A \cdot \Delta z + B| < |p(z_0)|. \qquad \square$$

The fundamental theorem now follows easily from d'Alembert's lemma and the two-dimensional extreme value theorem.

Fundamental theorem of algebra. *If $p(z)$ is a polynomial (with real or complex coefficients), then $p(z) = 0$ for some z in $\mathbb{C}$.*

Proof. Suppose on the contrary that $p(z) \neq 0$ for all z in $\mathbb{C}$, so that $|p(z)|$ is real-valued, positive, and continuous on the whole of $\mathbb{C}$. It follows, by the extreme value theorem, that $|p(z)|$ has a minimum value, $m > 0$ on each closed disk of radius R, $\{z : |z| \leq R\}$. It is also clear, since $p(z) \sim a_n z^n$ for large values of $|z|$, that $|p(z)|$ grows beyond all bounds outside disks with sufficiently large radius R.

Therefore, for some value of R, the mimimum value $m > 0$ of $|p(z)|$ for $|z| \leq R$ is in fact the minimum value of $|p(z)|$ over the whole plane $\mathbb{C}$. But this contradicts d'Alembert's lemma, which says we can find Δz with $|p(z + \Delta z)| < |p(z)|$ wherever $|p(z)| > 0$.

Thus the assumption that $p(z) \neq 0$ for all z is false: there is a z in $\mathbb{C}$ with $p(z) = 0$. $\qquad \square$

Figure 10.4: *The Ideal City* by Fra Carnevale.

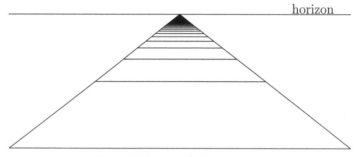

Figure 10.5: An infinite pavement.

10.4 Geometry: the Projective Line

Projective geometry is a branch of geometry that grew from the art of perspective drawing. It gets its name because drawing involves *projecting* a scene onto the plane of the picture. The geometry of perspective was discovered in fourteenth-century Italy, and figure 10.4 shows an example from around 1480, which may be seen in the Walters Museum in Baltimore.

Artists were naturally concerned with the depiction of three-dimensional scenes, but a basic problem of perspective is to draw perspective views of the plane, such as the view shown in figure 10.5, which depicts a pavement of squares between two parallel lines.

Figure 10.5 also shows two key concepts of projective geometry: the *line at infinity*, or horizon, where parallel lines meet, and the common point of two parallel lines, called their *point at infinity*. Thus the "line" in projective geometry (called a *projective line*) is in some way more

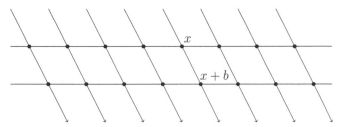

Figure 10.6: Projection realizing the map $x \mapsto x + b$.

complete than the "line" in Euclidean geometry, because of its extra point at infinity. The projective line is also more flexible, because it can undergo projection, under which lengths are distorted. For example, the projections of the squares of pavement in figure 10.5 are obviously not of equal size. The projections of more distant squares are smaller.

Nevertheless, in some sense, the projected squares "look equal." Most people will accept that figure 10.5 depicts a pavement of equal paving stones (though whether they are actually squares is another story). Apparently, some geometric quantity is preserved when we make a projection. We seem to grasp it intuitively, though it is hard to say what it is. The aim of this section is to uncover this preserved quantity, and to do so we make an algebraic study of projections.

Projections of the Line

In this subsection we introduce three kinds of projection, whose combinations include all possible projections. In each case we take the line to be the real number line $\mathbb{R}$ and project one copy of $\mathbb{R}$ onto another. "Projection" can be imagined (almost) literally as shining a light from a point source and viewing the "shadow" of each point x in $\mathbb{R}$ on the second copy of $\mathbb{R}$. We will see that the coordinate of the shadow point is a simple rational function of x.

In the first case the light source is at infinity and the two copies of $\mathbb{R}$ are parallel. Then the light rays are parallel and the shadow of each $x \in \mathbb{R}$ is $x + b$ for some constant b (figure 10.6). This kind of projection preserves all distances because the distance between points $p, q \in \mathbb{R}$ is $|p - q|$ and the distance between their shadows is $|(p + b) - (q + b)| = |p - q|$.

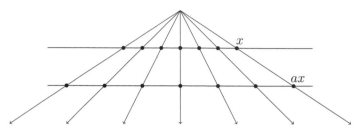

Figure 10.7: Projection realizing the map $x \mapsto ax$.

In the second example, figure 10.7, the two copies of $\mathbb{R}$ are again parallel but the light source is at a finite distance. In this case distances are *not* preserved—they are multiplied by some constant $a \neq 0, 1$. (To get $a < 1$ think of reversing the rays, and to get $a < 0$ put the light source between the lines.)

However, the *ratio* of distances is preserved. If p, q, r are three points on the line then

$$\frac{\text{distance from } p \text{ to } q}{\text{distance from } q \text{ to } r} = \frac{q - p}{r - q}.$$

Similarly, the ratio of distances of their shadows is

$$\frac{\text{distance from } ap \text{ to } aq}{\text{distance from } aq \text{ to } ar} = \frac{aq - ap}{ar - aq} = \frac{q - p}{r - q}.$$

In particular, equal lengths on one line map to equal lengths of the other line.

In the third example the copies of $\mathbb{R}$ are perpendicular and the light source is at a point equidistant from them both. Then if we take the origin on each line to be opposite to the light source, as shown in figure 10.8, x on one line is sent to $1/x$ on the other, as can be checked by comparing suitable similar triangles. (Notice that the "point" and "shadow" change places where the lines cross.) This map does not even preserve the ratio of distances, because some equal lengths on one line map to unequal lengths on the other.

However, a certain "ratio of ratios" is preserved. This is the so-called *cross-ratio*, defined for any four points p, q, r, s by

$$[p, q; r, s] = \frac{(r - p)(s - q)}{(r - q)(s - p)}.$$

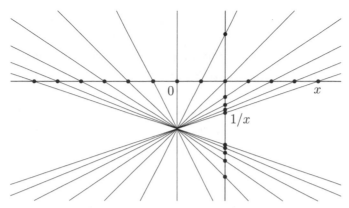

Figure 10.8: Projection realizing the map $x \mapsto 1/x$.

The map $x \mapsto 1/x$ sends p, q, r, s to $1/p$, $1/q$, $1/r$, $1/s$, respectively, and hence their cross-ratio goes to

$$\frac{\left(\frac{1}{r}-\frac{1}{p}\right)\left(\frac{1}{s}-\frac{1}{q}\right)}{\left(\frac{1}{r}-\frac{1}{q}\right)\left(\frac{1}{s}-\frac{1}{p}\right)} = \frac{\frac{p-r}{pr} \cdot \frac{q-s}{qs}}{\frac{q-r}{qr} \cdot \frac{p-s}{ps}}$$

$$= \frac{(p-r)(q-s)}{(q-r)(p-s)}, \qquad \text{multiplying top and bottom by } pqrs,$$

$$= \frac{(r-p)(s-q)}{(r-q)(s-p)} = [p, q; r, s].$$

It is easy to see that the cross-ratio $[p, q; r, s]$ is also preserved by the maps $x \mapsto x + b$ and $x \mapsto ax$ for $a \neq 0$, hence it is preserved by any combination of the three maps above.

The Point at Infinity

The projection that realizes $x \mapsto 1/x$ confronts us with the need for a point at infinity, because the point 0 has nowhere to go in $\mathbb{R}$. The ray from the point of projection through the point 0 is parallel to the vertical copy of $\mathbb{R}$ and hence does not meet it, unless …we take a cue from perspective drawing and say that parallels meet at a point at infinity.

Then we can extend the map $x \mapsto 1/x$ to $\mathbb{R} \cup \{\infty\}$ by declaring that $1/0 = \infty$. By the same token, we ought to have $1/\infty = 0$, because the horizontal line through the point of projection passes through 0 on the vertical copy of $\mathbb{R}$ and is parallel to the horizontal copy, hence they should meet at infinity.

Both of these requirements are satisfied by defining the *real projective line* to be the set $\mathbb{R} \cup \{\infty\}$ and declaring that $1/0 = \infty$ and $1/\infty = 0$. Thus division by zero and infinity is "legalized" on $\mathbb{R} \cup \{\infty\}$, provided we confine arithmetic on $\mathbb{R} \cup \{\infty\}$ to operations that correspond to projections to one copy of $\mathbb{R} \cup \{\infty\}$ onto another. We will say precisely what these operations are in the next subsection, and we will make them part of the definition of the real projective line.

Linear Fractional Transformations

The basic transformations $x \mapsto x + b$ ("add"), $x \mapsto ax$ for $a \neq 0$ ("multiply"), and $x \mapsto 1/x$ ("invert"), realized by the projections above, and all combinations of them, are examples of *linear fractional transformations*

$$x \mapsto \frac{ax + b}{cx + d}.$$

Also, for transformations realized by projections we always have $ad - bc \neq 0$. The reason is that if $ad = bc$ then $a/c = b/d$ and therefore

$$\frac{ax + b}{cx + d} = \text{constant, namely } a/c.$$

This is impossible because each basic projection maps different points to different points.

Conversely, if a, b, c, $d \in \mathbb{R}$ satisfy $ad - bc \neq 0$ then we can realize the transformation $x \mapsto \frac{ax+b}{cx+d}$ by a combination of the basic projections. We see how by deconstructing the expression $\frac{ax+b}{cx+d}$ as follows:

$$\frac{ax + b}{cx + d} = \frac{\frac{a}{c}cx + b}{cx + d} = \frac{\frac{a}{c}(cx + d) + b - \frac{ad}{c}}{cx + d} = \frac{a}{c} + \frac{b - \frac{ad}{c}}{cx + d}.$$

Then if $c \neq 0$ we can build the latter expression by the following series of basic transformations, each of which is an "add," "multiply," or "invert":

$$x \mapsto cx \mapsto cx + d \mapsto \frac{1}{cx+d} \mapsto \frac{b - \frac{ad}{c}}{cx+d} \mapsto \frac{a}{c} + \frac{b - \frac{ad}{c}}{cx+d}.$$

And if $c = 0$ then $ad \neq 0$ (because $ad - bc \neq 0$), so $d \neq 0$ too, and it is then easy to build the required function $\frac{ax+b}{d}$ by basic transformations.

Thus the linear fractional transformations $x \mapsto \frac{ax+b}{cx+d}$ with $ad - bc \neq 0$ are *precisely* the combinations of basic transformations, and hence they can be realized by projections. It remains to show, conversely, that any projection of one copy of $\mathbb{R} \cup \{\infty\}$ onto another is expressible by a linear fractional transformation.

We already know that this is true when the two copies of $\mathbb{R} \cup \{\infty\}$ are parallel. Thus it remains to consider projection of one line onto another that intersects it at a finite point. We can assume (by use of "add" transformations) that the point of intersection is 0 on each line. It is then a straightforward calculation in coordinate geometry to check that x on one line is mapped to $\frac{ax+b}{cx+d}$ on the other, for suitable a, b, c, d.

To sum up: the *real projective line* is the set $\mathbb{R} \cup \{\infty\}$ under the transformations

$$x \mapsto \frac{ax+b}{cx+d}, \quad \text{where } a, b, c, d \in \mathbb{R} \text{ and } ad - bc \neq 0.$$

These transformations are what make the line "projective," because they realize all changes in the line that can be induced by projection. The *geometry* of the projective line is captured by the *invariant* quantity called the *cross-ratio*,

$$[p, q; r, s] = \frac{(r - p)(s - q)}{(r - q)(s - p)},$$

because this quantity remains the same under all projections. (It can be shown quite easily, though I will not do so here, that *any* quantity invariant under projection is a function of the cross-ratio.)

10.5 Calculus: Wallis's Product for π

In chapter 6 we made a tour of (mostly) elementary calculus, culminating in the infinite series for π first found by Indian mathematicians in the fifteenth century:

$$\frac{\pi}{4} = 1 - \frac{1}{3} + \frac{1}{5} - \frac{1}{7} + \cdots .$$

The crux of the proof was some "infinitesimal geometry" of the circle, in which the derivative of arctan y was found by studying the behavior of a small triangle as its size tends to zero.

In this section we make a briefer tour of the same neighborhood, beginning with the derivatives of sine and cosine and leading to an infinite *product* for π discovered by Wallis (1655):

$$\frac{\pi}{2} = \frac{2 \cdot 2}{1 \cdot 3} \cdot \frac{4 \cdot 4}{3 \cdot 5} \cdot \frac{6 \cdot 6}{5 \cdot 7} \cdot \frac{8 \cdot 8}{7 \cdot 9} \cdots .$$

This formula for π is the key to understanding certain *finite* products, such as the binomial coefficients $\binom{2m}{m}$. In section 10.7 we will explain the connection, and use it to show how the graph of the binomial coefficients morphs into the graph of $y = e^{-x^2}$.

Derivatives of Sine and Cosine

The sine and cosine functions are best viewed as a pair, since $\cos \theta$ and $\sin \theta$ are the coordinates of the point P at angle θ on the unit circle (figure 10.9). In the same picture, the angle θ is best viewed as the arc length from the x-axis to P.

Now we observe the respective changes, $\Delta \cos \theta$ in $\cos \theta$ and $\Delta \sin \theta$ in $\sin \theta$, as θ changes by $\Delta \theta$. Figure 10.10 shows the relevant part of the previous diagram.

Increasing θ by the amount $\Delta \theta$ moves us from P to Q, where $\Delta \theta$ is the length of the arc PQ. With this move, the sine increases by the amount

$$\Delta \sin \theta = QB$$

and the cosine increases by the amount

$$\Delta \cos \theta = -BP.$$

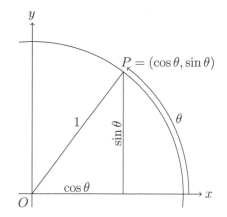

Figure 10.9: The meaning of $\cos\theta$ and $\sin\theta$.

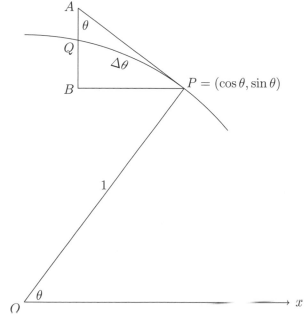

Figure 10.10: Comparing $\Delta\cos\theta$ and $\Delta\sin\theta$ with $\Delta\theta$.

Now the angle at A is θ, the same as the angle at O, because the tangent PA is perpendicular to the radius OP. Also, as $\Delta\theta \to 0$, the ratio between it and the length AP tends to 1. We write this as

$$AP \sim \Delta\theta.$$

We also have, using the same notation,

$$AB \sim QB$$

because the arc PQ tends toward the tangent PA. It follows that

$$\Delta \sin \theta = QB \sim AB = AP \cos \theta \sim \Delta\theta \cos \theta, \tag{1}$$

and

$$\Delta \cos \theta = -BP = -AP \sin \theta \sim -\Delta\theta \sin \theta. \tag{2}$$

In other words,

$$\frac{\Delta \sin \theta}{\Delta\theta} \to \cos \theta \quad \text{and} \quad \frac{\Delta \cos \theta}{\Delta\theta} \to -\sin \theta.$$

It follows, by definition of the derivative, that (1) and (2) give

$$\frac{d}{d\theta} \sin \theta = \cos \theta \quad \text{and} \quad \frac{d}{d\theta} \cos \theta = -\sin \theta.$$

Integrals Involving Sine and Cosine

The formulas just obtained tell us, not just the derivatives *of* sine and cosine, but also the functions whose derivatives *are* sine and cosine. This enables us to easily integrate many functions involving sine and cosine. To ease the process further, we introduce the method of *integration by parts*.

Recall from section 6.3 the *product rule* for differentiation:

$$\frac{d}{dx}(uv) = u\frac{dv}{dx} + v\frac{du}{dx}.$$

It follows, integrating both sides with respect to x from $x = a$ to $x = b$, that

$$u(b)v(b) - u(a)v(a) = \int_a^b u\frac{dv}{dx}dx + \int_a^b v\frac{du}{dx}dx.$$

This equation in itself says no more than the product rule. But it becomes useful when we know one of $\int_a^b u\frac{dv}{dx}dx$, $\int_a^b v\frac{du}{dx}dx$ but not the other, because then we can find the one we don't know in terms of the one we know. For example, if we know $\int_a^b u \frac{dv}{dx} dx$ we can find

$$\int_a^b v \frac{du}{dx} dx = u(b)v(b) - u(a)v(a) - \int_a^b u \frac{dv}{dx} dx.$$

Wallis's Product

We arrive at Wallis's product by looking at the integral $I(n) = \int_0^\pi \sin^n x \, dx$, using integration by parts to reduce the value of n.

$$I(n) = \int_0^\pi \sin^n x \, dx = \int_0^\pi \sin^{n-1} x \cdot \sin x \, dx$$

$$= \int_0^\pi u \frac{dv}{dx} \, dx \quad \text{where } u = \sin^{n-1} x \text{ and } v = -\cos x$$

$$= 0 - \int_0^\pi v \frac{du}{dx} \, dx \quad \text{integrating by parts,}$$

$$\text{since } u(x)v(x) = 0 \text{ for both } x = 0 \text{ and } x = \pi,$$

$$= \int_0^\pi \cos x \cdot (n-1) \sin^{n-2} x \cdot \cos x \, dx \quad \text{by the chain rule}$$

$$= (n-1) \int_0^\pi \cos^2 x \cdot \sin^{n-2} x \, dx$$

$$= (n-1) \int_0^\pi (1 - \sin^2 x) \cdot \sin^{n-2} x \, dx$$

$$= (n-1)I(n-2) - (n-1)I(n).$$

It follows from this equation that $I(n) = \frac{n-1}{n} I(n-2)$, so the problem eventually reduces to finding $I(0)$ and $I(1)$. These are easy.

$$I(0) = \int_0^\pi 1 \, dx = \pi,$$

$$I(1) = \int_0^\pi \sin x \, dx = -\cos \pi + \cos 0 = 2.$$

From the values of $I(0)$ and $I(1)$, using $I(n) = \frac{n-1}{n} I(n-2)$, we find that

$$I(2m) = \frac{2m-1}{2m} \frac{2m-3}{2m-2} \cdots \frac{3}{4} \frac{1}{2} \pi,$$

$$I(2m+1) = \frac{2m}{2m+1} \frac{2m-2}{2m-1} \cdots \frac{4}{5} \frac{2}{3} \cdot 2,$$

which implies that

$$\frac{I(2m)}{I(2m+1)} = \frac{(2m+1)(2m-1)}{2m \cdot 2m} \frac{(2m-1)(2m-3)}{(2m-2)(2m-2)} \cdots \frac{5 \cdot 3}{4 \cdot 4} \frac{3 \cdot 1}{2 \cdot 2} \frac{\pi}{2}.$$
$$(*)$$

Wallis's product begins to show its face in the equation (*). To bring it fully to light we have to prove that $I(2m)/I(2m+1) \to 1$ as $m \to \infty$. Certainly, since $0 \le \sin x \le 1$ for $0 \le x \le \pi$ we have

$$\sin^{2m+1} x \le \sin^{2m} \le \sin^{2m-1} x \quad \text{for } 0 \le x \le \pi,$$

so

$$I(2m+1) \le I(2m) \le I(2m-1),$$

and therefore

$$1 \le \frac{I(2m)}{I(2m+1)} \le \frac{I(2m-1)}{I(2m+1)}, \quad \text{dividing by } I(2m+1).$$

But

$$\frac{I(2m-1)}{I(2m+1)} = \frac{2m+1}{2m} \frac{I(2m-1)}{I(2m-1)}$$

$$= \frac{2m+1}{2m} \quad \text{because } I(n) = \frac{n-1}{n} I(n-2).$$

So

$$1 \le \frac{I(2m)}{I(2m+1)} \le \frac{2m+1}{2m} \to 1 \quad \text{as } m \to \infty,$$

and therefore $\frac{I(2m)}{I(2m+1)} \to 1$ as required, by the method of exhaustion.

This means that we can extend (*) to the *infinite* product

$$1 = \frac{\pi}{2} \frac{1 \cdot 3}{2 \cdot 2} \frac{3 \cdot 5}{2 \cdot 4} \frac{5 \cdot 7}{4 \cdot 6} \frac{7 \cdot 9}{6 \cdot 8} \frac{9}{8 \cdot 8} \cdots,$$

from which it follows that

$$\frac{\pi}{2} = \frac{2 \cdot 2}{1 \cdot 3} \cdot \frac{4 \cdot 4}{3 \cdot 5} \cdot \frac{6 \cdot 6}{5 \cdot 7} \cdot \frac{8 \cdot 8}{7 \cdot 9} \cdots \cdots .$$

10.6 Combinatorics: Ramsey Theory

Whereas the entropy theorems of probability theory
and mathematical physics imply that, in a large
universe, disorder is probable, certain combinatorial
theorems imply that complete disorder is impossible.

Theodore Motzkin (1967), p. 244

Ramsey theory is a branch of combinatorics often marketed under the slogan "complete disorder is impossible," whose context is the longer quote from Theodore Motzkin given above. A small everyday example of the kind of "order" found by Ramsey theory is: among any six people there are three who all know each other, or else three who all do not know each other.

This fact may be exhibited and proved graphically as follows. Represent the six people by dots; join people who know each other by black lines, and people who do not know each other by gray lines. Figure 10.11 shows one such "acquaintance graph." In this case there is a gray triangle, and we are claiming that *there is always either a black triangle or a gray triangle*; in other words *monochromatic triangles are unavoidable.*

Or, in the language of graph theory, *any 2-coloring of the edges of a K_6 contains a monochromatic K_3.* (The general K_n, called the *complete graph* on n vertices, is a collection of n dots with a line connecting any two of them. We met K_5 in section 7.8.) To see why this is true, consider any one of the six vertices in K_6. It is the endpoint of the five edges to the other vertices, at least three of which must have the same color, say black. In that case we have three black edges shown in figure 10.12, going from the given vertex to three others u, v, w.

If any of u, v, w is joined to another by a black edge then this edge completes a black triangle with edges already drawn. If not, then the three edges joining u, v, w form a gray triangle.

Figure 10.11: An acquaintance graph for six people.

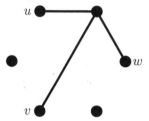

Figure 10.12: Typical vertex of a colored K_6.

This is only a "baby" Ramsey theorem, but more "order" occurs in larger graphs. For example, it is known that any 2-colored K_{18} contains a monochromatic K_4. (This implies that any group of 18 people includes either four mutual acquaintances or else four mutual strangers.) In fact, there is a monochromatic K_m in any sufficiently large 2-colored K_n, though we do not yet know how large n has to be (even for $m = 5$). Surprisingly, it is easiest to see the general picture by going straight to the infinite case.

*An Infinite Ramsey Theorem

The simplest infinite complete graph, K_ω, has a countable infinity of vertices v_1, v_2, v_3, . . ., with an edge joining each pair v_i, v_j of distinct vertices. Notice that K_ω contains many copies of itself, because the complete graph on any infinite subset of v_1, v_2, v_3, . . . is likewise a K_ω. This makes possible the following theorem, first proved by Ramsey (1930):

Infinite Ramsey theorem. *Any 2-colored K_ω contains a monochromatic K_ω.*

Proof. The proof is by applying the infinite pigeonhole principle infinitely often, much as we did in the proof of the Bolzano-Weierstrass theorem in section 7.9. First, we are going to construct an infinite subset $W = \{w_1, w_2, w_3, \ldots\}$ of the vertex set $V = \{v_1, v_2, v_3, \ldots\}$ with the property that each member w_i of W is connected to its successors $w_{i+1}, w_{i+2}, \ldots$ by edges of the same color.

The edges out of v_1 have two possible colors so, by the infinite pigeonhole principle, an infinite set of them have the same color. We let W_1 be the set of endpoints of these[1] same-colored edges, and let w_1 be the first member of W_1 on the list $v_1, v_2, v_3, \ldots$.

Next, the edges out of w_1 to other members of W_1 likewise have two possible colors, so an infinite subset of them have the same color. We let W_2 be the set of endpoints of these same-colored edges, and let w_2 be the first member of W_2.

Since W_2 is infinite we can continue this process indefinitely. There is an infinite set W_3 of endpoints of same-colored edges out of w_2; we let w_3 be the first member of W_3, and so on. Thus we get an infinite set $W = \{w_1, w_2, w_3, \ldots\}$ with the property claimed above: each member w_i is connected to all of $w_{i+1}, w_{i+2}, \ldots$ by edges of the same color.

However, we are not quite done because we could have, say, w_1 connected to $w_2, w_3, \ldots$ by black edges, and w_2 connected to $w_3, w_4, \ldots$ by gray edges. We want a graph with edges all of the same color. To overcome this difficulty we apply the infinite pigeonhole principle one more time.

Some of the w_i may be connected to $w_{i+1}, w_{i+2}, \ldots$ by black edges, other w_i may be connected to $w_{i+1}, w_{i+2}, \ldots$ by gray edges. But *one* of these colors must occur for infinitely many different w_i—call them $x_1, x_2, x_3, \ldots$. Then the edges between any x_j and x_k are all of the same color, and hence they form a monochromatic K_ω. $\qquad\square$

*A Finite Ramsey Theorem

We can now return to the questions arising from the baby Ramsey theorem above. How do we know that any sufficiently large 2-colored K_n contains a monochromatic K_4, or a monochromatic K_5, and so on?

---

[1] To avoid ambiguity, take the black edges if there are infinitely many of them; otherwise take the gray edges.

There is evidently something hard about these questions, since we do not yet know *how* large n must be to ensure there is a monochromatic K_5 in a 2-colored K_n. The beauty of the infinite Ramsey theorem is that it gives a kind of Olympian view from which all finite cases may be derived, without worrying about the details of particular K_m and K_n. We still do not learn how big n must be to get a monochromatic K_5 in K_n, but we do not need to know—the n must exist by virtue of the infinite Ramsey theorem.

A second virtue of the proof is that it uses infinity in a familiar context: the Kőnig infinity lemma. As we saw in section 7.12, Kőnig (1927) introduced his lemma as "a way of reasoning from the finite to the infinite," but it can also be used in the opposite direction, as here.

Finite Ramsey theorem. *For each m there is an n such that all 2-colored K_N with $N > n$ contain a monochromatic K_m.*

Proof. Before looking at any particular value of m we build an infinite tree C of all 2-colorings of finite complete graphs. Figure 10.13 shows the first two levels of C. The top vertex leads to the 2-colorings of K_2, which has a single edge. Consequently, there are two 2-colorings of K_2, which we place in vertices on a level below the top vertex. On the next level there are eight vertices, containing the possible 2-colorings of K_3. These are attached to the 2-colorings of K_2 that they *extend*, when we build K_3 by adding a vertex to the 2-colored K_2 in the level above. Thus there are four extensions of the K_2 with the black edge, and four extensions of the K_2 with the gray edge.

The tree C continues downward in this fashion, with the 2-colorings of K_{n+1} at level n attached to the 2-colorings of K_n (at level $n - 1$) that they extend. Since there are only finitely many 2-colorings of each K_n, each vertex of the tree has finite valency. This allows us to apply the Kőnig infinity lemma. Here is how we do it.

Suppose that for some m it is *not* true that each sufficiently large 2-colored K_n contains a monochromatic K_m. In that case, infinitely many vertices of C are 2-colorings not containing a monochromatic K_m. And *these vertices form a tree D contained in C*, because if a 2-coloring at level $n + 1$ does not contain a monochromatic K_m, then neither does the 2-coloring at level n that it extends.

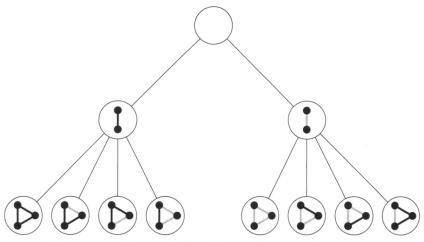

Figure 10.13:   The tree C of 2-colorings of finite complete graphs.

But then D contains an infinite branch B, by the Kőnig infinity lemma. Since each coloring along this branch extends the one above it, together they give a 2-coloring of the K_ω whose vertices are the vertices at all levels.

The infinite Ramsey theorem says that this K_ω contains a monochromatic K_ω. In particular, K_ω contains a monochromatic K_m, whose vertices necessarily occur at some finite level n. This contradicts the construction of the branch B, which avoided monochromatic K_ms all the way down. Therefore, we were wrong to suppose that the finite Ramsey theorem is false. $\qquad\square$

10.7 Probability: de Moivre's Distribution

The Middle Binomial Coefficient and π

We begin with a closer look at Wallis's product for π, derived in section 10.5. Inverting it gives

$$\frac{2}{\pi} = \frac{1\cdot3}{2\cdot2}\,\frac{3\cdot5}{4\cdot4}\,\frac{5\cdot7}{6\cdot6}\,\cdots$$

$$= \lim_{m\to\infty}\left[\frac{1}{2}\,\frac{3\cdot3}{2\cdot4}\,\frac{5\cdot5}{4\cdot6}\,\frac{7\cdot7}{6\cdot8}\,\cdots\,\frac{(2m-1)(2m-1)}{(2m-2)2m}\right].$$

Because of the many repeated factors, the bracketed product on the right-hand side is almost a square. We now manipulate it so as to separate the square part from the rest:

$$\frac{2}{\pi} = \lim_{m \to \infty} \left[\frac{1 \cdot 1}{2 \cdot 2} \frac{3 \cdot 3}{4 \cdot 4} \frac{5 \cdot 5}{6 \cdot 6} \frac{7 \cdot 7}{8 \cdot 8} \cdots \frac{(2m-1)(2m-1)}{2m \cdot 2m} \cdot 2m \right]$$

$$= \lim_{m \to \infty} \left[\left(\frac{1}{2} \frac{3}{4} \frac{5}{6} \frac{7}{8} \cdots \frac{2m-1}{2m} \right)^2 \cdot 2m \right]. \tag{*}$$

Now we compare this expression with the scaled binomial coefficient

$$\binom{2m}{m} \bigg/ 2^{2m} = \frac{(2m)!}{m! m!} \frac{1}{2^{2m}} = \frac{1 \cdot 2 \cdot 3 \cdot 4 \cdots (2m-1) \cdot 2m}{1 \cdot 2 \cdot 3 \cdots m \cdot 1 \cdot 2 \cdot 3 \cdots m} \frac{1}{2^{2m}}$$

$$= \frac{1 \cdot 3 \cdot 5 \cdots (2m-1) \cdot 2 \cdot 4 \cdot 6 \cdots 2m}{1 \cdot 2 \cdot 3 \cdots m \cdot 1 \cdot 2 \cdot 3 \cdots m} \frac{1}{2^{2m}}$$

$$= \frac{1 \cdot 3 \cdot 5 \cdots (2m-1)}{2 \cdot 4 \cdot 6 \cdots 2m} \frac{2 \cdot 4 \cdot 6 \cdots 2m}{2 \cdot 4 \cdot 6 \cdots 2m}$$

$$= \frac{1 \cdot 3 \cdot 5 \cdots (2m-1)}{2 \cdot 4 \cdot 6 \cdots 2m}.$$

This is the same as the term in (*) whose square, times $2m$, has limit $2/\pi$. It follows, taking square roots, that

$$\sqrt{2m} \binom{2m}{m} \bigg/ 2^{2m} \to \sqrt{2/\pi} \quad \text{and hence} \quad \frac{\sqrt{2m}}{\sqrt{2/\pi}} \binom{2m}{m} \bigg/ 2^{2m} \to 1$$

as $m \to \infty$. Since $\sqrt{2/\pi}/\sqrt{2m} = \frac{1}{\sqrt{\pi m}}$, we write this relation as

$$\binom{2m}{m} \bigg/ 2^{2m} \sim \frac{1}{\sqrt{\pi m}}, \tag{**}$$

where the symbol $\sim$ (read "is asymptotic to") means that the ratio of the two sides tends to 1 as $m \to \infty$.

Where e^{-x^2} Comes From

In section 8.5 we saw that the graph of the binomial coefficients $\binom{n}{k}$, as k varies from 1 to n, seems to approach the shape of the bell curve $y = e^{-x^2}$ when n is large. In this subsection we explain why the exponential function appears, by finding an approximation to the size of the general binomial coefficient $\binom{2m}{m+l}$ relative to that of the middle binomial coefficient $\binom{2m}{m}$.

Approximation to $\binom{2m}{m+l}$. *For fixed l, as $m \to \infty$,*

$$\binom{2m}{m+l} \sim \binom{2m}{m} e^{-l^2/m}.$$

Proof. If we substitute the formulas for $\binom{2m}{m}$ and $\binom{2m}{m+l}$ then we find, making some obvious cancellations, that

$$\binom{2m}{m} \bigg/ \binom{2m}{m+l} = \frac{(2m)!}{m!m!} \bigg/ \frac{(2m)!}{(m+l)!(m-l)!}$$

$$= \frac{(m+l)!(m-l)!}{m!m!}$$

$$= \frac{(m+l)(m+l-1)\cdots(m+1)}{m(m-1)\cdots(m-l+1)}$$

$$= \left(1 + \frac{l}{m}\right)\left(1 + \frac{l}{m-1}\right)\cdots\left(1 + \frac{l}{m-l+1}\right).$$

We now replace the product on the right by a sum, by taking the natural log of both sides:

$$\ln\left(\binom{2m}{m} \bigg/ \binom{2m}{m+l}\right) = \ln\left(1 + \frac{l}{m}\right) + \ln\left(1 + \frac{l}{m-1}\right) + \cdots$$

$$+ \ln\left(1 + \frac{l}{m-l+1}\right). \tag{ln}$$

Since l is fixed we can choose m much larger than l, in which case each of the l terms on the right-hand side is very close to $\ln\left(1 + \frac{l}{m}\right)$. We also have, from the series for $\ln(1+x)$ in section 6.7, that

$$\ln(1+x) = x + (\text{an error of size} < x^2) \quad \text{when } x \text{ is small.}$$

We will write this relation $\ln(1+x) \approx x$ for x small, and similarly use the $\approx$ sign in what follows.

First, by choosing m sufficiently large, we get

$$\frac{l}{m} \approx \ln\left(1 + \frac{l}{m}\right) \approx \ln\left(1 + \frac{l}{m-1}\right) \approx \cdots \approx \ln\left(1 + \frac{l}{m-l+1}\right),$$

and substituting these values in (ln) gives

$$\ln\left(\binom{2m}{m} \bigg/ \binom{2m}{m+l}\right) \approx l \cdot \frac{l}{m} = \frac{l^2}{m}.$$

Then, taking exponentials of both sides gives

$$\binom{2m}{m} \bigg/ \binom{2m}{m+l} \approx e^{l^2/m},$$

and therefore

$$\binom{2m}{m+l} \sim \binom{2m}{m} e^{-l^2/m}. \qquad \square$$

The above argument does not really depend on l being fixed, but only on its being *sufficiently small* [2] *compared with* m, so that we can say

$$\frac{l}{m} \approx \ln\left(1 + \frac{l}{m}\right) \approx \ln\left(1 + \frac{l}{m-1}\right) \approx \cdots \approx \ln\left(1 + \frac{l}{m-l+1}\right),$$

and therefore

$$\ln\left(\binom{2m}{m} \bigg/ \binom{2m}{m+l}\right) \approx l \cdot \frac{l}{m} = \frac{l^2}{m}.$$

---

[2] Bear in mind that the graph of $y = e^{-x^2}$ is infinitely wide, so the bell shape one sees is actually the limit of an increasingly narrow central part of the binomial coefficient graph. See figure 8.5, for example, where $m = 100$ and most of the "bell" is between $l = 40$ and $l = 60$.

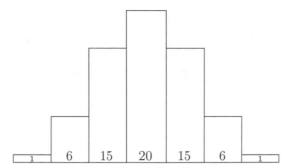

Figure 10.14: Bar graph of the binomial coefficients $\binom{6}{k}$.

In particular, the argument remains valid if we let

$$l = \text{nearest integer to } x\sqrt{m}$$

for a fixed value of x. For in this case

$$\ln\left(1+\frac{l}{m}\right) = \frac{l}{m} + \left(\text{error} < \frac{l^2}{m^2} \approx \frac{x^2}{m}\right),$$

so the $l \approx x\sqrt{m}$ errors amount to $\approx x^3/\sqrt{m}$, which is small in comparison with

$$l \cdot \frac{l}{m} = \frac{l^2}{m} \approx x^2$$

as $m \to \infty$, because x is fixed. We can therefore conclude that

$$\binom{2m}{m+l} \sim \binom{2m}{m} e^{-x^2} \qquad\qquad (***)$$

when $l = \text{nearest integer to } x\sqrt{m}$ and $m \to \infty$.

The Limit of the Graph of Binomial Coefficients

Finally, we explain why the graph of binomial coefficients approaches the shape of $y = e^{-x^2}$, using the above approximations to the coefficients and suitable scaling on the x and y axes. Our starting point is the bar graph of binomial coefficients $\binom{2m}{m+l}$ for $l = -m$ to $l = m$, shown for the case $m = 3$ in figure 10.14.

For this graph to approach a limiting shape we need the height of the bar for the middle coefficient $\binom{2m}{m}$ to approach a constant, which certainly requires scaling in the vertical direction. After choosing a suitable constant, we also need to scale in the horizontal direction, as we explain below, so as to keep the area of the bar graph equal to 1. Notice that, without any vertical scaling, and with the width of each bar equal to 1, the total area is

$$\binom{2m}{0} + \binom{2m}{1} + \binom{2m}{2} + \cdots + \binom{2m}{2m} = (1+1)^{2m} = 2^{2m}.$$

To find our way from the binomial coefficients $\binom{n}{k}$ to the function e^{-x^2} our first signpost is the value of the middle binomial coefficient when $n = 2m$, found above:

$$\binom{2m}{m} \sim \frac{2^{2m}}{\sqrt{\pi m}}.$$

This value points to the value $1/\sqrt{\pi}$ as the "right" height at the middle of the limit curve. So, our first step is to scale the binomial coefficient graph in the vertical direction by dividing the values of all the binomial coefficients $\binom{2m}{m+l}$, as l runs from $-m$ to m, by $2^{2m}/\sqrt{m}$.

To see how to scale in the horizontal direction, we make the constraint that the area of the bar graph of binomial coefficients should equal 1. This is because we wish to interpret the area of the bar for $\binom{2m}{m+l}$ as the probability of throwing $m + l$ heads in $2m$ tosses. As l varies from $-m$ to m, these probabilities add up to 1. Since we have just made

$$\text{height of bar for } \binom{2m}{m+l} = \binom{2m}{m+l} 2^{-2m} \sqrt{m},$$

and since the sum of all the terms $\binom{2m}{m+l}$ is 2^{2m}, we need to set

$$\text{width of bar for } \binom{2m}{m+l} = \frac{1}{\sqrt{m}}.$$

Limit of the binomial coefficient graph. *With the scaling just de-scribed, the limit of the binomial coefficient graph is*

$$y = \frac{1}{\sqrt{\pi}} e^{-x^2}.$$

Proof. When we replace the horizontal coordinate l of the binomial coefficient graph by $l/\sqrt{m}$ we get the coordinate

$$x = l/\sqrt{m}.$$

Thus, for given $x \neq 0$, we choose numbers l and m so that $x = l/\sqrt{m}$ and let $m \to \infty$.

In the previous subsection we found the formula

$$\binom{2m}{m+l} \sim \binom{2m}{m} e^{-x^2}, \tag{***}$$

where $l =$ nearest integer to $x\sqrt{m}$ and $m \to \infty$. So when we replace the vertical coordinate $\binom{2m}{m+l}$ by $\binom{2m}{m+l} 2^{-2m} \sqrt{m}$, and take the limit, we get

$$y = \lim_{m \to \infty} \binom{2m}{m+l} 2^{-2m} \sqrt{m}$$

$$= e^{-x^2} \lim_{m \to \infty} \binom{2m}{m} 2^{-2m} \sqrt{m} \qquad \text{by (***)}$$

$$= \frac{1}{\sqrt{\pi}} e^{-x^2} \qquad \text{by (**)}$$

as required. □

A corollary of this calculation is the astonishing result[3]

$$\int_{-\infty}^{\infty} e^{-x^2} dx = \sqrt{\pi}.$$

---

[3] There is a story about William Thomson, the nineteenth-century British mathematical physicist, who brought up this formula in the course of a lecture in which he praised his French colleague Joseph Liouville. Thomson wrote on the board

$$\int_{-\infty}^{\infty} e^{-x^2} dx = \sqrt{\pi},$$

then told his students, "A mathematician is one to whom that is as obvious as that twice two are four is to you. Liouville was a mathematician."

This follows because the area of each binomial coefficient bar graph is 1, hence so is the area under their limit curve, $y = \frac{1}{\sqrt{\pi}}e^{-x^2}$, which runs from $-\infty$ to ∞. Multiplying by $\sqrt{\pi}$ gives the area under $y = e^{-x^2}$. There are other ways to find the value of this integral, but all of them involve some ingenuity, because e^{-x^2} is not the derivative of any elementary function.

10.8 Logic: the Completeness Theorem

> Pure mathematics is the class of all propositions
> of the form "p implies q."
>
> Bertrand Russell (1903),
> *The Principles of Mathematics*, p. 3

In chapter 3 we saw that the definition of computation leads inevitably, by the diagonal argument, to the existence of unsolvable problems and the incompleteness of mathematics. In the broadest sense, incompleteness means there is no algorithm for generating all the truths (and only truths) of mathematics. In particular, *mathematics has no complete axiom system* from which all theorems may be derived systematically by the rules of logic. However, the incompleteness lies in the mathematics, not the logic. What we lack is a universal system of mathematical axioms, not universal rules of logic.

If we give up the idea of an absolute axiom system, and instead take mathematical proof in the relative sense as showing "what follows from what by pure logic," then the burden of completeness is shifted to logic. And if we take "logic" to mean *predicate logic*, discussed in section 9.2 and known to be adequate for the axiom systems most commonly used, then completeness becomes a reality: if a theorem q is a consequence of axioms p, then there is a proof of "p implies q" in an axiom system for logic. As mentioned in section 9.10, a complete system of predicate logic was proposed by Frege (1879), and its completeness was proved by Gödel (1930). To warm up for a proof of Gödel's completeness theorem, we first discuss the case of propositional logic, where completeness is easier to prove.

Propositional Logic

The completeness of propositional logic should not be a surprise, because we have already given a method for finding its valid formulas in section 9.2: the method of truth tables. However, this method does not apply to predicate logic, so we need to rethink our approach. In this subsection we approach logic in the style of mathematics, choosing some simple and obvious formulas as *axioms*, and deriving the remaining valid formulas by some obvious *rules of inference*.

The axioms and rules of inference are found by working backwards. Given any propositional formula φ, we try to *falsify* φ by breaking it into smaller parts, then trying to falsify these, and so on. Eventually φ is broken down to disjunctions of variables and negated variables, such as $p \vee q \vee (\neg p)$. Such a disjunction is *unfalsifiable* (that is, valid) if and only if it contains both p and $\neg p$ for some variable p, as in the example. In this case we can get a *proof* of φ by starting with axioms such as $p \vee (\neg p)$ and applying *rules of inference* that are simply the falsification rules in reverse.

Here is an example: let φ be $(p \Rightarrow q) \Rightarrow ((\neg q) \Rightarrow (\neg p))$.

To attempt the falsification of φ we first write it in terms of $\neg$ and $\vee$, which is possible for any propositional formula by section 9.2. In this case we replace each subformula of the form $A \Rightarrow B$ by $(\neg A) \vee B$, obtaining

$$\varphi = \neg((\neg p) \vee q) \vee ((\neg \neg q) \vee (\neg p)).$$

Our first falsification rule says that $\neg(A \vee B)$ can be falsified if and only if $\neg A$ *or* $\neg B$ can be falsified, which is clear because $\neg(A \vee B) = (\neg A) \wedge (\neg B)$. The rule is more useful in the following more general form, which carries an arbitrary disjunct C along:

$\neg\vee$ **falsification rule.** To falsify $\neg(A \vee B) \vee C$, falsify $(\neg A) \vee C$ or $(\neg B) \vee C$.

We write this rule in graphic form like this:

$$\neg(A \vee B) \vee C$$

$$(\neg A) \vee C \quad (\neg B) \vee C$$

Applying the rule to φ, where $C = (\neg\neg q) \vee (\neg p)$, gives:

$$\neg((\neg p) \vee q) \vee ((\neg\neg q) \vee (\neg p))$$

$$(\neg\neg p) \vee (\neg\neg q) \vee (\neg p) \qquad (\neg q) \vee (\neg\neg q) \vee (\neg p)$$

Now we apply the more obvious

$\neg\neg$ **falsification rule.** To falsify $(\neg\neg A) \vee C$ falsify $A \vee C$.

The graphic form of this rule is:

$$(\neg\neg A) \vee C$$
$$|$$
$$A \vee C$$

Applying the $\neg\neg$ falsification rule three times to φ then gives:

$$\neg((\neg p) \vee q) \vee ((\neg\neg q) \vee (\neg p))$$

$$(\neg\neg p) \vee (\neg\neg q) \vee (\neg p) \qquad (\neg q) \vee (\neg\neg q) \vee (\neg p)$$
$$| \qquad\qquad\qquad\qquad\qquad\qquad |$$
$$p \vee q \vee (\neg p) \qquad\qquad\qquad (\neg q) \vee q \vee (\neg p)$$

Thus we can falsify φ only if we can falsify $p \vee q \vee (\neg p)$ or $(\neg q) \vee q \vee (\neg p)$, which is impossible. Thus φ is *not* falsifiable, and hence it is valid. Moreover, we can prove φ from the "axioms" $p \vee q \vee (\neg p)$ and $(\neg q) \vee q \vee (\neg p)$ by applying the reverses of the falsification rules as "rules of inference."

$\neg\vee$ **rule of inference.** From $(\neg A) \vee C$ and $(\neg B) \vee C$ infer $\neg(A \vee B) \vee C$.

$\neg\neg$ **rule of inference.** From $A \vee C$ infer $(\neg\neg A) \vee C$.

We can make the system a little tidier by using only the very simplest axioms— $p \vee (\neg p)$, $q \vee (\neg q)$, and so on—if we allow these axioms to be "padded" with extra disjuncts. This amounts to the following rule, which clearly preserves validity:

∨-**padding rule of inference.** From A infer $A \vee B$.

These three rules, together with associative and commutative rules for $\vee$ (used without comment in the example above), then allow any valid formula φ to be proved along the same lines as the one above. Thus we have:

Completeness of propositional logic. *Any valid formula of propositional logic, when written in terms of* $\neg$ *and* $\vee$, *may be proved from the axioms* $p \vee (\neg p)$, $q \vee (\neg q)$, ... *using the three rules of inference above (and the commutative and associative properties of* $\vee$). □

Predicate Logic

As we know from section 9.3, the atomic formulas $p, q, r, \ldots$ of predicate logic have internal structure, involving predicate symbols $P, Q, R, \ldots$, variables $x, y, z, \ldots$, and constants $a, b, c, \ldots$. The atomic formulas can be combined by propositional connectives and they are also subject to the *quantifiers* $\forall x$ ("for all x") and $\exists x$ ("there is an x") for any variable x. As in the previous subsection, we need use only the connectives $\neg$ and $\vee$, and we also need just the universal quantifier $\forall$, because we can define $\exists$ in terms of it:

$$(\exists x)P(x) = \neg(\forall x)\neg P(x).$$

We now follow the same strategy as in the previous subsection: look for a complete set of falsification rules, then reverse them to obtain a complete set of rules of inference. We will use the falsification rules already discovered for $\neg$ and $\vee$, so it remains to find falsification rules for $\forall$. The following rules, shown with their diagrams, turn out to be sufficient.

$\forall$ **falsification rule.** To falsify $(\forall x)A(x) \vee B$, falsify $A(a) \vee B$ for any constant a not free (unquantified) in B.

$$(\forall x)A(x) \vee B$$
$$|$$
$$A(a) \vee B$$

$\neg\forall$ **falsification rule.** To falsify $\neg(\forall x)A(x) \vee B$, falsify $\neg A(a) \vee$ $\neg(\forall x)A(x) \vee B$ for any constant a.

$$\neg(\forall x)A(x) \vee B$$
$$|$$
$$\neg A(a) \vee \neg(\forall x)A(x) \vee B$$

The latter rule does not shorten the formula, so the falsification process may not terminate. This is unfortunate, but not fatal, because we need the process to terminate only for *un*falsifiable formulas—the ones we hope to prove by reversing the process. One thing we can say immediately is that if the falsification process does not terminate for a formula φ then the falsification tree for φ has an infinite branch, by the Kőnig infinity lemma. We now show that, by applying the falsification rules with care, we can ensure that any infinite branch falsifies φ, because it defines an interpretation that makes φ false.

The idea is not to miss any opportunity to apply a rule, so the formulas on the infinite branch have the following properties:

1. If $\neg\neg A$ occurs as a disjunct on a branch, so does A (by applying the $\neg\neg$ falsification rule).
2. If $\neg(A \vee B)$ occurs as a disjunct on the branch, so does $\neg A$ or $\neg B$ (by applying the $\neg\vee$ falsification rule).
3. If $(\forall x)A(x)$ occurs as a disjunct on the branch, so does $A(a_i)$ for some constant a_i (by applying the $\forall$ falsification rule).
4. If $\neg(\forall x)A(x)$ occurs as a disjunct on the branch, so does $\neg A(a_i)$ for every constant a_i (by applying the $\neg\forall$ falsification rule for each a_i in turn).

Since we seize every opportunity to break the formula into shorter ones, on an infinite branch every sentence is eventually broken down to *atoms*; that is, terms of the form $R(a_i, a_j, \dots)$. They may be accompanied by quantified terms, carried along by the $\neg\forall$ rule, but the point is that $R(a_i, a_j, \dots)$ and $\neg R(a_i, a_j, \dots)$ cannot both occur.

If they do, the disjunction containing them cannot be falsified and the branch will terminate.

Therefore we are free to assign the value false to every atomic formula, so *every formula on the branch is falsified, including the formula φ at the top.* Conversely, *if φ is not falsifiable then every branch in the falsification tree terminates.* Then, by reversing the falsification process, we get:

Completeness of predicate logic. *Any valid formula of predicate logic, written in terms of ¬, ∨, and ∀, may be proved from axioms of the form $R(a_i, a_j, \ldots) \vee (\neg R(a_i, a_j, \ldots))$ using the reverses of the falsification rules as rules of inference.* □

10.9 Historical and Philosophical Remarks

*The Word Problem: Semigroups and Groups

As mentioned in section 10.2, the word problem was first posed for *groups* rather than semigroups, by Dehn (1912). Even though the concept of semigroup is more general than that of group, it seems that the group concept lies at a more useful level of generality. Groups are easier to handle, they have a more satisfying theory, and they capture more of the situations that mathematicians have (so far) wished to study.

Dehn himself was interested in the word problem for groups because it exactly captures an algorithmic problem in topology: deciding whether a given closed curve in a space S can be continuously contracted to its (fixed) initial point while remaining in S. In the case where S is a surface, Dehn was able to solve this problem, but for higher-dimensional spaces the contractibility problem is difficult, and in fact known to be unsolvable in certain spaces of four or more dimensions. This follows from the unsolvability of the word problem proved by Novikov (1955).

A related problem, also posed by Dehn (1912), is the *isomorphism problem* for groups: given "presentations" of groups, by listing generators and equations that they satisfy, decide whether the two groups are the same. Adyan (1957) proved that the isomorphism problem is unsolvable, and this led to the unsolvability of the *homeomorphism problem* of topology. This is the problem of deciding, given (finitely described) spaces S and T, whether there is a continuous bijection $S \to T$ with a continuous inverse. The latter problem was proved unsolvable by Markov (1958), in fact for four-dimensional spaces S and T.

All of these results are quite difficult to prove, because they depend on the unsolvability of the word problem. Despite a lot of work—improving considerably on the Novikov (1955) proof—the word problem for groups has yet to be related to computation as simply and elegantly as the word problem for semigroups was by Post (1947) and Markov (1947).

What makes groups and semigroups hard, as already claimed in section 4.11, is *noncommutativity*. As we saw in section 10.2, computation is easily modeled by noncommutative semigroups, and hence it is easy to find unsolvable problems about semigroups. The downside is that noncommutative semigroups are hard to understand. The presence of inverses in groups makes them easier to understand, and to develop their general theory. It is correspondingly *harder* to find unsolvable problems about groups, though still possible when they are not commutative.

When a group is commutative (and not highly infinite), its structure is quite easy to understand. The commonest commutative groups—such as the nonzero rationals—are so well understood that it is unnecessary to mention that they *are* groups.

A more interesting commutative group occurred in section 10.1, where we studied the numbers $a + b\sqrt{m}$ for which $x = a$, $y = b$ is a solution of $x^2 - my^2 = 1$. In effect, we showed that these numbers form a group, because the product of any two of them, and the inverse of any one of them, is another number with the same property. Since multiplication of numbers is commutative, we were able to discover the structure of this group: it consists of all the powers $(x_1 + y_1\sqrt{m})^n$, where $n \in \mathbb{Z}$ and $x = x_1$, $y = y_1$ is the smallest integer solution of $x^2 - my^2 = 1$ with $y \neq 0$.

*The FTA and Analysis

Argand's proof of the fundamental theorem of algebra (FTA), when reinforced by a rigorous treatment of continuous functions, seems to be the simplest known. However, it depends on the extreme value theorem for continuous functions on a *two*-dimensional domain, and we might wonder: how far can we reduce the use of analysis in proofs of the FTA? The answer seems to be: to the intermediate value theorem for odd-degree polynomial functions of one real variable.

Gauss (1816) gave the first such proof, and an exposition of it may be found in Dawson (2015), Chapter 8. Given an equation $p(x) = 0$, where p is a real polynomial of degree $n = 2^m q$, where q is odd, Gauss reduced it to an equation of degree $n/2$, and hence ultimately to an equation of odd degree q. At this stage the proof is completed by an appeal to the intermediate value theorem for odd-degree polynomials. Gauss's reduction process is purely algebraic, but very complicated. Among other things, it uses the so-called *fundamental theorem for symmetric polynomials*, which was discovered in special cases by Newton and proved in generality by Lagrange (1771). The proof of this theorem is a rather lengthy induction, which we will not give here, but it is worth saying what the theorem is about.

A polynomial $p(x_1, x_2, \ldots, x_n)$ in n variables is said to be *symmetric* if it is unchanged by any permutation of the variables $x_1, \ldots, x_n$. For example, $p(x_1, x_2) = x_1^2 + x_2^2$ is a symmetric polynomial in two variables because $p(x_1, x_2) = p(x_2, x_1)$. The *elementary symmetric polynomials* are the coefficients of $x^{n-1}, \ldots, x^0$ in the function $(x - x_1)(x - x_2) \cdots (x - x_n)$, namely,

$$s_1 = -(x_1 + x_2 + \cdots + x_n), \quad \ldots, \quad s_n = (-1)^n x_1 x_2 \cdots x_n,$$

and the fundamental theorem says that any symmetric polynomial in $x_1, \ldots, x_n$ is a polynomial function $P(s_1, \ldots, s_n)$ of the elementary symmetric polynomials. For example, the symmetric polynomial $x_1^2 + x_2^2$ is a polynomial in s_1 and s_2 because

$$x_1^2 + x_2^2 = (x_1 + x_2)^2 - 2x_1 x_2 = s_1^2 - 2s_2.$$

Actually a simpler proof of FTA along these lines had already been sketched by Laplace in 1795. It may be found in Ebbinghaus et al. (1990), pp. 120–122. Laplace's proof was incomplete at the time because it assumed what we called the "algebraist's FTA" in section 4.11: the existence of a field, for each single-variable polynomial p, in which each $p(x) = 0$ has a solution. With the proof of the latter theorem, Laplace's proof became viable, showing that the FTA actually follows from the "algebraist's FTA" with the help of the intermediate value theorem.

*Groups and Geometry

Geometry is one of the most convincing cases where the group concept captures a phenomenon that mathematicians wish to study. At the same time, it is a sign of the depth of the group concept that its relationship with geometry was not uncovered until geometry had been studied for over 2000 years. One does not notice the group concept in geometry until several kinds of geometry have come to light—most importantly, projective geometry. It was projective geometry in particular that led Klein (1872) to notice the role of groups in geometry, and to *define* geometry as the study of *groups and their invariants*.

The real projective line $\mathbb{R} \cup \{\infty\}$, studied in section 10.4, involves perhaps the simplest example of an interesting group and an interesting invariant. The group is the group of linear fractional functions,

$$f(x) = \frac{ax+b}{cx+d} \quad \text{where } a, b, c, d \in \mathbb{R} \text{ and } ad - bc \neq 0,$$

under the operation of function composition. That is, given functions

$$f_1(x) = \frac{a_1 x + b_1}{c_1 x + d_1} \quad \text{and} \quad f_2(x) = \frac{a_2 x + b_2}{c_2 x + d_2},$$

we form the function $f_1(f_2(x))$, which corresponds to performing the projection corresponding to f_2, then the projection corresponding to f_1. This group is not commutative. For example, if

$$f_1(x) = x + 1 \quad \text{and} \quad f_2 = 2x$$

then

$$f_1(f_2(x)) = 2x + 1, \quad \text{whereas} \quad f_2(f_1(x)) = 2(x+1) = 2x + 2,$$

so $f_1 f_2 \neq f_2 f_1$. Nevertheless, we can find the invariant of the linear fractional transformations without knowing much about their group structure. It suffices to know that they are generated by the simple functions $x \mapsto x + b$, $x \mapsto ax$ for $a \neq 0$, and $x \mapsto 1/x$.

As we saw in section 10.4, traditional geometric quantities such as length, or the ratio of lengths, are *not* invariant under all linear fractional transformations. However, some simple computations with the generating transformations show the invariance of the cross-ratio

$$[p, q; r, s] = \frac{(r - p)(s - q)}{(r - q)(s - p)}$$

for any four points of the real projective line. The invariance of the cross-ratio under projection was already known to Pappus, and was rediscovered by Desargues around 1640. However, its *algebraic* invariance had to wait for the identification of the appropriate group by Klein.

With hindsight, we can also see how length, and the ratio of lengths, are also algebraic invariants. The length of the line segment from p to q, $|p - q|$, is the invariant of the group of *translations* $x \mapsto x + b$ of $\mathbb{R}$. We call $\mathbb{R}$ the *Euclidean line* when it is subject to these transformations, because they make any point "the same" as any other point, as Euclid intended.

The ratio of lengths, $\frac{p-r}{p-q}$, for any three points p, q, r, is the invariant of the group of *similarities* $x \mapsto ax + b$ where $a \neq 0$. This group is generated by combining translations with *magnifications* $x \mapsto ax$ for $a \neq 0$. These transformations are called *affine*, and when $\mathbb{R}$ is subject to these transformations it is called the *affine line*.

The deep difference between the Euclidean and affine lines and the projective line is of course the point at infinity, ∞. The point ∞ arises on the projective line because the point 0 otherwise has nowhere to go under the map $x \mapsto 1/x$. It is appropriate to call this point "infinity" because 0 is the limit of $1/n$ as $n \to \infty$, so the image of 0 under $x \mapsto 1/x$ ought to be the "limit" of n as $n \to \infty$. Nevertheless, it is also appropriate to view the projective line as a finite object; namely, the circle.

This is because 0 is also the limit of $-1/n$ as $n \to \infty$, so the image of 0 should be the "limit" of $-n$ as $n \to \infty$. Thus we "approach ∞" as we travel along the line in either direction. We can realize the common "limit" of n and $-n$ by an actual point if we map $\mathbb{R}$ into a circle as shown in figure 10.15.

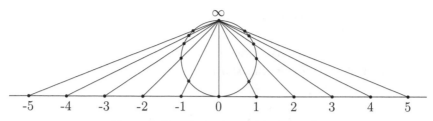

Figure 10.15: The projective line as a circle.

The topmost point on the circle neatly corresponds to the "limit" of both n and $-n$, because it is the actual limit of their images on the circle. So if we let the topmost point correspond to ∞ we can view the circle as a continuous and bijective image of the real projective line.

Affine Geometry

The affine transformations of the line $\mathbb{R}$, mentioned above, extend to the projective line $\mathbb{R} \cup \{\infty\}$ simply by sending the point ∞ to itself. Similarly, there are affine transformations of the plane, and of the projective plane. They are the projections that send finite points to finite points, and points at infinity to points at infinity. In particular, *they send parallels to parallels*. The affine geometry of the plane studies the images of the plane obtained by such projections. The term "affine" was introduced by Euler (1748b), motivated by the idea that images related by affine transformations have an "affinity" with one another.

Like projective geometry, affine geometry has an artistic counter-part. It may be seen in classical Japanese woodblock prints of the eigh-teenth and nineteenth centuries, such as Harunobu's *Evening Chime of the Clock* shown in figure 10.16. This print, which dates from around 1766, clearly shows the preservation of parallel lines, since all parallel lines in the scene actually look parallel. This makes the scene look "flat," though perfectly consistent. In fact, the picture shows how the scene would appear when viewed from infinitely far away, with infinite magnification.

The art of representing parallel lines consistently demands some skill, as it is possible to go badly wrong. Figure 10.17 shows an example, *The Birth of St Edmund*. I have previously used this example—for example, in Stillwell (2010), p. 128—as a failure of projective geometry,

Figure 10.16:   Affine geometry in art. Image courtesy of www.metmuseum.org.

because the parallel lines do not converge towards any horizon. But now I think it is better seen as a failure of affine geometry. The artist really *wants* the parallel lines to look parallel, but he has utterly failed to do so consistently.

Since affine maps pair finite points with finite points, *an affine image cannot include the horizon.* Most Japanese prints comply with this condition, but sometimes even an eminent artist slips up. Figure 10.18 shows the print *Cat at Window* by the nineteenth-century master Hiroshige. The interior is fine, but it is not compatible with the exterior, which includes the horizon.

The Simplest Formulas for π

Wallis's product for π, so important for the approximation of binomial coefficients and the de Moivre distribution, can also be related to the

Figure 10.17: Failed affine geometry in art. The original of this picture is in a
handpainted manuscript in the British Library, *The Lives of Sts. Edmund and Fremund*
by John Lydgate, thought to be completed in 1434. The original painting is somewhat
fuzzy, so I have used a nineteenth-century copy (due to Henry Ward), which is faithful
to the lines in the original but much sharper. The copy is in the Wellcome Library,
London.

formula found in chapter 6,

$$\frac{\pi}{4} = 1 - \frac{1}{3} + \frac{1}{5} - \frac{1}{7} + \frac{1}{9} - \cdots .$$

Wallis found his product by some inspired guesswork, not by calculus
as we would understand it today. But his discovery led his colleague

Figure 10.18: Subtle failure of affine geometry.

Lord Brouncker to another remarkable formula, the continued fraction

$$\frac{\pi}{4} = \cfrac{1}{1 + \cfrac{1^2}{2 + \cfrac{3^2}{2 + \cfrac{5^2}{2 + \cfrac{7^2}{2 + \cdots}}}}}.$$

How Brouncker made this discovery is a mystery, because even today we do not know a simple way to pass from Wallis's product to Brouncker's continued fraction. However, there is quite a simple connection between the continued fraction and the series $1 - \frac{1}{3} + \frac{1}{5} - \frac{1}{7} + \frac{1}{9} - \cdots$. It was discovered by Euler (1748a), p. 311, where it follows immediately from the more general equation

$$\frac{1}{A} - \frac{1}{B} + \frac{1}{C} - \frac{1}{D} + \cdots = \cfrac{1}{A + \cfrac{A^2}{B - A + \cfrac{B^2}{C - B + \cfrac{C^2}{D - C + \cdots}}}}$$

Euler's result may be proved by converting finite series to fractions, and going to the limit. The first step is to check that

$$\frac{1}{A} - \frac{1}{B} = \cfrac{1}{A + \cfrac{A^2}{B - A}}.$$

When $\frac{1}{B}$ on the left side of this equation is replaced by $\frac{1}{B} - \frac{1}{C}$, which similarly equals $\cfrac{1}{B + \frac{B^2}{C-B}}$, the B on the right should be replaced by $B + \frac{B^2}{C-B}$. Hence

$$\frac{1}{A} - \frac{1}{B} + \frac{1}{C} = \cfrac{1}{A + \cfrac{A^2}{B - A + \cfrac{B^2}{C - B}}}.$$

So when we modify the "tail end" of the series (replacing $\frac{1}{B}$ by $\frac{1}{B} - \frac{1}{C}$), only the "tail end" of the continued fraction is affected. Consequently, we can repeat the process, obtaining a continued fraction for a series of n terms. Letting $n \to \infty$ gives the infinite continued fraction.

Thus the simplest known ways of expressing π in terms of integers—by an infinite product, infinite continued fraction, and infinite series—are all related to each other. The resemblance between the three formulas can be enhanced by dividing each side of Wallis's formula by 2 and rearranging as follows:

$$\frac{\pi}{4} = \frac{2}{1 \cdot 3} \cdot \frac{4 \cdot 4}{3 \cdot 5} \cdot \frac{6 \cdot 6}{5 \cdot 7} \cdot \frac{8 \cdot 8}{7 \cdot 9} \cdots$$

$$= \frac{2 \cdot 4}{3 \cdot 3} \frac{4 \cdot 6}{5 \cdot 5} \frac{6 \cdot 8}{7 \cdot 7} \frac{8 \cdot 10}{9 \cdot 9} \cdots$$

$$= \left(1 - \frac{1}{3^2}\right) \left(1 - \frac{1}{5^2}\right) \left(1 - \frac{1}{7^2}\right) \left(1 - \frac{1}{9^2}\right) \cdots .$$

This shows that all three formulas are based on little more than the sequence of odd numbers.

*Ramsey Theory and the Kőnig Infinity Lemma

The proof of the infinite Ramsey theorem given in section 10.6 so closely resembles that of the Bolzano-Weierstrass theorem in section 7.9 that one suspects that the two theorems have the same strength. Indeed, it is a result of reverse mathematics that (a slight generalization of) the infinite Ramsey theorem and Bolzano-Weierstrass are *equivalent*, to each other and to the Kőnig infinity lemma.

The finite Ramsey theorem can be proved in PA, so it is a consequence of axioms we consider elementary. Nevertheless, there is something hard about the finite Ramsey theorem, because, as we mentioned in section 10.6, we do not yet know how large n must be in order to ensure that a 2-colored K_n contains a monochromatic K_5. This example suggests that, however elementary the PA axioms may be, their consequences are not necessarily elementary.

Another sense in which the finite Ramsey theorem seems hard is that a slight variation of it is *not* provable in PA. The variant is called the *Paris-Harrington theorem*, and it was discovered by the logicians Paris and Harrington (1977). To explain the Paris-Harrington theorem, we must first explain a more general version of the finite Ramsey theorem. Until now, we have been talking about K_n, the edges of which are pairs of numbers from an n-element set we may take to be $\{1, 2, 3, \ldots, n\}$. Thus we are assigning *two* colors to *pairs* from the set $\{1, 2, 3, \ldots, n\}$, and asking how large n must be to ensure the existence of a subset M

of $\{1, 2, 3, \ldots, n\}$ with m elements, each pair of which have the same color.

In the general version, we assign k colors to the l-element subsets of the set $\{1, 2, 3, \ldots, n\}$, and ask how big n must be to ensure there is an m-element subset M of $\{1, 2, 3, \ldots, n\}$ which is *monochromatic* in the sense that all l-element subsets of M have the same color. The general finite Ramsey theorem states that, given any k, l, and m, there is an n such that, for any coloring of the l-element subsets of $\{1, 2, 3, \ldots, n\}$ in k colors, there is a monochromatic subset M of $\{1, 2, 3, \ldots, n\}$ with at least m members. This more general version of the finite Ramsey theorem is also provable in PA.

But now, the Paris-Harrington theorem adds the condition that the number of members of M is greater than or equal to its smallest member. With this simple extra condition, the theorem becomes too hard to prove in PA! (It is provable, however, from an infinite Ramsey theorem.)

Completeness, Unsolvability, and Infinity

The proof of the completeness theorem in section 10.8 suggests that the completeness of logic balances on a knife edge: the valid formulas luckily fall on the side where termination of the proof process is certain; the falsifiable formulas fall on the side where termination is uncertain. This may remind you of the halting problem (section 3.8), where there is a terminating process that finds all the valid cases—namely, let the machine run until it halts—but no process that terminates for all the invalid (that is, non-halting) cases. If there were, we could solve the halting problem by running both processes simultaneously. In fact, the problem of deciding which formulas of predicate logic are valid is not just vaguely similar to the halting problem; it is essentially the *same* problem! The problems are the "same" in the sense that each is reducible to the other.[4]

The problem of deciding validity can be reduced to the halting problem by building a Turing machine T to implement the proof process for predicate logic, arranging that T halts for an input formula

---

[4] In technical language, they are of the same *degree of unsolvability* or the same *Turing degree*.

φ if and only if a proof of φ is found. Conversely, Turing (1936) showed how to express the operations of a Turing machine by formulas of predicate logic. For each machine M and input I there is a formula $\varphi_{M,I}$ which is valid if and only if M on input I eventually halts. In this way, the halting problem is reduced to the problem of deciding validity in predicate logic.

It follows, in particular, that the validity problem is *unsolvable*, because the halting problem is. This explains the asymmetry of the completeness proof, with the valid formulas on the "halting" side, and the falsifiable formulas on the "non-halting" side.

The unsolvability of deciding validity is reflected by the infinite ingredient in the completeness proof—the Kőnig infinity lemma—and it indicates that *any* proof of completeness for predicate logic must involve an infinite process. In fact, as a result of reverse mathematics we know that the completeness theorem is essentially *equivalent* to the weak Kőnig infinity lemma. This puts the completeness theorem in the company of other fundamental theorems, such as the extreme value theorem and the Brouwer fixed point theorem—an impressive demonstration of the unity of mathematics!

↫ Bibliography ↫

Abbott, S. (2001). *Understanding Analysis*. Springer-Verlag, New York.

Abel, N. H. (1826). Démonstration de l'impossibilité de la résolution algébrique des équations générales qui passent le quatrième degré. *Journal für die reine und angewandte Mathematik 1*, 65–84. In his *Oeuvres Complètes* 1: 66–87.

Adyan, S. I. (1957). Unsolvability of some algorithmic problems in the theory of groups (Russian). *Trudy Moskovskogo Matematicheskogo Obshchestva 6*, 231–298.

Agrawal, M., N. Kayal, and N. Saxena (2004). PRIMES is in P. *Annals of Mathematics (2) 160*(2), 781–793.

Appel, K. and W. Haken (1976). Every planar map is four colorable. *Bulletin of the American Mathematical Society 82*, 711–712.

Argand, J. R. (1806). *Essai sur une manière de représenter les quantités imaginaires dans les constructions géométriques*. Paris.

Beltrami, E. (1868). Teoria fondamentale degli spazii di curvatura costante. *Annali di Matematica Pura ed Applicata, series 2*, no. 2, 232–255. In his *Opere Matematiche* 1: 406–429, English translation in Stillwell (1996), pp. 41–62.

Berlekamp, E. R., J. H. Conway, and R. K. Guy (1982). *Winning Ways for your Mathematical Plays. Vol. 2*. Academic Press, Inc. [Harcourt Brace Jovanovich, Publishers], London-New York.

Bernoulli, D. (1728). Observationes de seriebus. *Commentarii Academiae Scientiarum Imperialis Petropolitanae 3*, 85–100. In Bernoulli (1982), pp. 49–64.

Bernoulli, D. (1982). *Die Werke von Daniel Bernoulli, Band 2*. Birkhäuser, Basel.

Bernoulli, J. (1713). *Ars conjectandi*. In his *Opera* 3: 107–286.

Biggs, N. L., E. K. Lloyd, and R. J. Wilson (1976). *Graph Theory: 1736–1936*. Oxford University Press, Oxford.

Bolyai, J. (1832b). Scientiam spatii absolute veram exhibens: a veritate aut falsitate Axiomatis XI Euclidei (a priori haud unquam decidanda) independentem. Appendix to Bolyai (1832a), English translation in Bonola (1912).

Bolzano, B. (1817). *Rein analytischer Beweis des Lehrsatzes dass zwischen je zwey Werthen, die ein entgegengesetzes Resultat gewähren, wenigstens eine reelle Wurzel der Gleichung liege*. Ostwald's Klassiker, vol. 153. Engelmann, Leipzig, 1905. English translation in Russ (2004), pp. 251–277.

Bombelli, R. (1572). *L'algebra. Prima edizione integrale. Introduzione di U. Forti. Prefazione di E. Bortolotti*. Reprint by Biblioteca scientifica Feltrinelli. 13, Giangiacomo Feltrinelli Editore. LXIII, Milano (1966).

Bonola, R. (1912). *Noneuclidean Geometry*. Chicago: Open Court. Reprinted by Dover, New York, 1955.

Boole, G. (1847). *Mathematical Analysis of Logic*. Reprinted by Basil Blackwell, London, 1948.

Bourgne, R. and J.-P. Azra (1962). *Ecrits et mémoires mathématiques d'Évariste Galois: Édition critique intégrale de ses manuscrits et publications*. Gauthier-Villars & Cie, Imprimeur-Éditeur-Libraire, Paris. Préface de J. Dieudonné.

Brahmagupta (628). *Brâhma-sphuṭa-siddhânta*. Partial English translation in Colebrooke (1817).

Brouwer, L. E. J. (1910). Über eineindeutige, stetige Transformationen von Flächen in sich. *Mathematische Annalen 69*, 176–180.

Cantor, G. (1874). Über eine Eigenschaft des Inbegriffes aller reellen algebraischen Zahlen. *Journal für die reine und angewandte Mathematik 77*, 258–262. In his *Gesammelte Abhandlungen*, 145–148. English translation by W. Ewald in Ewald (1996), Vol. II, pp. 840–843.

Cardano, G. (1545). *Ars magna*. 1968 translation *The great art or the rules of algebra* by T. Richard Witmer, with a foreword by Oystein Ore. M.I.T. Press, Cambridge, MA-London.

Cauchy, A.-L. (1821). *Cours d'Analyse de l'École Royale Polytechnique*. Paris. Annotated English translation by Robert E. Bradley and C. Edward Sandifer, *Cauchy's Cours d'analyse: An Annotated Translation* , Springer, 2009.

Chrystal, G. (1904). *Algebra: An Elementary Text-book for the Higher Classes of Secondary Schools and for Colleges*. 1959 reprint of 1904 edition. Chelsea Publishing Co., New York.

Church, A. (1935). An unsolvable problem of elementary number theory. *Bulletin of the American Mathematical Society 41*, 332–333.

Clagett, M. (1968). *Nicole Oresme and the Medieval Geometry of Qualities and Motions*. University of Wisconsin Press, Madison, WI.

Cohen, P. (1963). The independence of the continuum hypothesis I, II. *Proceedings of the National Academy of Sciences 50, 51*, 1143–1148, 105–110.

Colebrooke, H. T. (1817). *Algebra, with Arithmetic and Mensuration, from the Sanscrit of Brahmegupta and Bháscara*. John Murray, London. Reprinted by Martin Sandig, Wiesbaden, 1973.

Cook, S. A. (1971). The complexity of theorem-proving procedures. *Proceedings of the 3rd Annual ACM Symposium on the Theory of Computing*, 151–158. Association of Computing Machinery, New York.

Courant, R. and H. Robbins (1941). *What Is Mathematics?* Oxford University Press, New York.

d'Alembert, J. l. R. (1746). Recherches sur le calcul intégral. *Histoire de l'Académie Royale des Sciences et Belles-lettres de Berlin 2*, 182–224.

Davis, M. (Ed.) (1965). *The Undecidable. Basic papers on Undecidable Propositions, Unsolvable Problems and Computable Functions*. Raven Press, Hewlett, NY.

Dawson, J. W. (2015). *Why Prove It Again? Alternative Proofs in Mathematical Practice*. Birkhäuser.

de Moivre, A. (1730). *Miscellanea analytica de seriebus et quadraturis*. J. Tonson and J. Watts, London.

de Moivre, A. (1733). *Approximatio ad summam terminorum binomii $(a + b)^n$*. Printed for private circulation, London.

de Moivre, A. (1738). *The Doctrine of Chances. The second edition*. Woodfall, London.

Dedekind, R. (1871a). Supplement VII. In Dirichlet's *Vorlesungen über Zahlentheorie*, 2nd ed., Vieweg 1871, English translation *Lectures on Number Theory* by John Stillwell, American Mathematical Society, 1999.

Dedekind, R. (1871b). Supplement X. In Dirichlet's *Vorlesungen über Zahlentheorie*, 2nd ed., Vieweg, 1871.

Dedekind, R. (1872). *Stetigkeit und irrationale Zahlen*. Vieweg und Sohn, Braunschweig. English translation in Dedekind (1901).

Dedekind, R. (1877). *Theory of Algebraic Integers*. Cambridge University Press, Cambridge. Translated from the 1877 French original and with an introduction by John Stillwell.

Dedekind, R. (1888). *Was sind und was sollen die Zahlen?* Vieweg, Braunschweig. English translation in Dedekind (1901).

Dedekind, R. (1894). Supplement XI. In Dirichlet's *Vorlesungen über Zahlentheorie*, 4th ed., Vieweg, 1894.

Dedekind, R. (1901). *Essays on the Theory of Numbers*. Open Court, Chicago. Translated by Wooster Woodruff Beman.

Dehn, M. (1900). Über raumgleiche Polyeder. *Göttingen Nachrichten 1900*, 345–354.

Dehn, M. (1912). Über unendliche diskontinuierliche Gruppen. *Mathematische Annalen 71*, 116–144.

Densmore, D. (2010). *The Bones*. Green Lion Press, Santa Fe, NM.

Descartes, R. (1637). *The Geometry of René Descartes. (With a facsimile of the first edition, 1637.)*. Dover Publications, Inc., New York, NY. Translated by David Eugene Smith and Marcia L. Latham, 1954.

Dirichlet, P. G. L. (1863). *Vorlesungen über Zahlentheorie*. F. Vieweg und Sohn, Braunschweig. English translation *Lectures on Number Theory*, with Supplements by R. Dedekind, translated from the German and with an introduction by John Stillwell, American Mathematical Society, Providence, RI, 1999.

Ebbinghaus, H.-D., H. Hermes, F. Hirzebruch, M. Koecher, K. Mainzer, J. Neukirch, A. Prestel, and R. Remmert (1990). *Numbers*, Volume 123 of *Graduate Texts in Mathematics*. Springer-Verlag, New York. With an introduction by K. Lamotke. Translated from the second German edition by H. L. S. Orde. Translation edited and with a preface by J. H. Ewing.

Edwards, H. M. (2007). Kronecker's fundamental theorem of general arithmetic. In *Episodes in the History of Modern Algebra (1800–1950)*, pp. 107–116. American Mathematical Society, Providence, RI.

Euler, L. (1748a). *Introductio in analysin infinitorum, I*. Volume 8 of his *Opera Omnia*, series 1. English translation by John D. Blanton, *Introduction to the Analysis of the Infinite. Book I*, Springer-Verlag, 1988.

Euler, L. (1748b). *Introductio in analysin infinitorum, II*. Volume 9 of his *Opera Omnia*, series 1. English translation by John D. Blanton, *Introduction to the Analysis of the Infinite. Book II*, Springer-Verlag, 1988.

Euler, L. (1751). Recherches sur les racines imaginaires des équations. *Histoire de l'Académie Royale des Sciences et des Belles-Lettres de Berlin 5*, 222–288. In his *Opera Omnia*, series 1, 6: 78–147.

Euler, L. (1752). Elementa doctrinae solidorum. *Novi Commentarii Academiae Scientiarum Petropolitanae 4*, 109–140. In his *Opera Omnia*, series 1, 26: 71–93.

Euler, L. (1770). *Elements of Algebra*. Translated from the German by John Hewlett. Reprint of the 1840 edition, with an introduction by C. Truesdell, Springer-Verlag, New York, 1984.

Ewald, W. (1996). *From Kant to Hilbert: A Source Book in the Foundations of Mathematics*. Vol. I, II. Clarendon Press, Oxford University Press, New York.

Fermat, P. (1657). Letter to Frenicle, February 1657. *Œuvres* 2: 333–334.

Fibonacci (1202). *Fibonacci's Liber abaci*. Sources and Studies in the History of Mathematics and Physical Sciences. Springer-Verlag, New York, 2002. A translation into modern English of Leonardo Pisano's *Book of calculation*, 1202. Translated from the Latin and with an introduction, notes, and bibliography by L. E. Sigler.

Fibonacci (1225). *The Book of Squares*. Academic Press, Inc., Boston, MA, 1987. Translated from the Latin and with a preface, introduction, and commentaries by L. E. Sigler.

Fischer, H. (2011). *A History of the Central Limit Theorem*. Sources and Studies in the History of Mathematics and Physical Sciences. Springer, New York.

Fowler, D. (1999). *The Mathematics of Plato's Academy* (2nd ed.). Clarendon Press, Oxford University Press, New York.

Fraenkel, A. (1922). Zu den Grundlagen der Cantor-Zermeloschen Mengenlehre. *Mathematische Annalen 86*, 230–237.

Frege, G. (1879). *Begriffsschrift*. English translation in van Heijenoort (1967), pp. 5–82.

Friedman, H. (1975). Some systems of second order arithmetic and their use. In *Proceedings of the International Congress of Mathematicians (Vancouver, B. C., 1974), Vol. 1*, pp. 235–242. Canadian Mathematical Congress, Montreal, Quebec.

Galois, E. (1831). Mémoire sur les conditions de résolubilité des équations par radicaux. In Bourgne and Azra (1962), pp. 43–71. English translation and commentary in Neumann (2011), pp. 104–168.

Gardiner, A. (2002). *Understanding Infinity*. Dover Publications, Inc., Mineola, NY. The mathematics of infinite processes. Unabridged republication of the 1982 edition with list of errata.

Gauss, C. F. (1801). *Disquisitiones arithmeticae*. Translated and with a preface by Arthur A. Clarke. Revised by William C. Waterhouse, Cornelius Greither, and A. W. Grootendorst and with a preface by Waterhouse, Springer-Verlag, New York, 1986.

Gauss, C. F. (1809). *Theoria motus corporum coelestium*. Perthes and Besser, Hamburg. In his *Werke* 7: 3–280.

Gauss, C. F. (1816). Demonstratio nova altera theorematis omnem functionem algebraicum rationalem integram unius variabilis in factores reales primi vel secundi gradus resolvi posse. *Commentationes societas regiae scientiarum Gottingensis recentiores 3*, 107–142. In his *Werke* 3: 31–56.

Gauss, C. F. (1832). Letter to W. Bolyai, 6 March 1832. *Briefwechsel zwischen C. F. Gauss und Wolfgang Bolyai*, eds. F. Schmidt and P. Stäckel. Leipzig, 1899. Also in his *Werke* 8: 220–224.

Gödel, K. (1930). Die Vollständigkeit der Axiome des logischen Funktionenkalküls. *Monatshefte für Mathematik und Physik 37*, 349–360.

Gödel, K. (1931). Über formal unentscheidbare Sätze der Principia Mathematica und verwandter Systeme. I. *Monatshefte für Mathematik und Physik 38*, 173–198.

Gödel, K. (1938). The consistency of the axiom of choice and the generalized continuum hypothesis. *Proceedings of the National Academy of Sciences 25*, 220–224.

Gödel, K. (1946). Remarks before the Princeton bicentennial conference on problems in mathematics. In Davis (1965), pp. 84–88.

Gödel, K. (2014). *Collected Works. Vol. V. Correspondence H–Z.* Clarendon Press, Oxford University Press, Oxford. Edited by Solomon Feferman, John W. Dawson, Jr., Warren Goldfarb, Charles Parsons, and Wilfried Sieg. Paperback edition of the 2003 original.

Grassmann, H. (1844). *Die lineale Ausdehnungslehre.* Otto Wigand, Leipzig. English translation in Grassmann (1995), pp. 1–312.

Grassmann, H. (1847). *Geometrische Analyse geknüpft an die von Leibniz gefundene Geometrische Charakteristik.* Weidmann'sche Buchhandlung, Leipzig. English translation in Grassmann (1995), pp. 313–414.

Grassmann, H. (1861). *Lehrbuch der Arithmetik.* Enslin, Berlin.

Grassmann, H. (1862). *Die Ausdehnungslehre.* Enslin, Berlin. English translation of 1896 edition in Grassmann (2000).

Grassmann, H. (1995). *A New Branch of Mathematics.* Open Court Publishing Co., Chicago, IL. The *Ausdehnungslehre* of 1844 and other works. Translated from the German and with a note by Lloyd C. Kannenberg. With a foreword by Albert C. Lewis.

Grassmann, H. (2000). *Extension theory.* American Mathematical Society, Providence, RI; London Mathematical Society, London. Translated from the 1896 German original and with a foreword, editorial notes, and supplementary notes by Lloyd C. Kannenberg.

Hamilton, W. R. (1839). On the argument of Abel, respecting the impossibility of expressing a root of any general equation above the fourth degree, by any finite combination of radicals and rational functions. *Transactions of the Royal Irish Academy 18*, 171–259.

Hardy, G. H. (1908). *A Course of Pure Mathematics.* Cambridge University Press.

Hardy, G. H. (1941). *A Course of Pure Mathematics.* Cambridge University Press. 8th ed.

Hardy, G. H. (1942). Review of *What is Mathematics?* by Courant and Robbins. *Nature 150*, 673–674.

Hausdorff, F. (1914). *Grundzüge der Mengenlehre.* Von Veit, Leipzig.

Heath, T. L. (1925). *The Thirteen Books of Euclid's Elements.* Cambridge University Press, Cambridge. Reprinted by Dover, New York, 1956.

Heawood, P. J. (1890). Map-colour theorem. *The Quarterly Journal of Pure and Applied Mathematics 24*, 332–338.

Hermes, H. (1965). *Enumerability, Decidability, Computability. An introduction to the theory of recursive functions.* Die Grundlehren der mathematischen Wissenschaften in Einzeldarstellungen mit besonderer Berücksichtigung der Anwendungsgebiete, Band 127. Translated by G. T. Herman and O. Plassmann. Academic Press, Inc., New York; Springer-Verlag, Berlin-Heidelberg-New York.

Hessenberg, G. (1905). Beweis des *Desargues*schen Satzes aus dem *Pascal*schen. *Mathematische Annalen 61*, 161–172.

Hilbert, D. (1899). *Grundlagen der Geometrie.* Leipzig: Teubner. English translation: *Foundations of Geometry*, Open Court, Chicago, 1971.

Hilbert, D. (1901). Über Flächen von constanter Gaussscher Krümmung. *Transactions of the American Mathematical Society 2*, 87–89. In his *Gesammelte Abhandlungen* 2: 437–438.

Hilbert, D. (1926). Über das Unendliche. *Mathematische Annalen 95*, 161–190. English translation in van Heijenoort (1967), pp. 367–392.

Huygens, C. (1657). *De ratiociniis in aleae ludo*. Elsevirii, Leiden. In the *Exercitationum Mathematicarum* of F. van Schooten.

Huygens, C. (1659). Fourth part of a treatise on quadrature. *Œuvres Complètes 14*: 337.

Jordan, C. (1887). *Cours de Analyse de l'École Polytechnique*. Gauthier-Villars, Paris.

Kempe, A. B. (1879). On the geographical problem of the four colours. *American Journal of Mathematics 2*, 193–200.

Klein, F. (1872). *Vergleichende Betrachtungen über neuere geometrische Forschungen (Erlanger Programm)*. Akademische Verlagsgesellschaft, Leipzig. In his *Gesammelte Mathematischen Abhandlungen 1*: 460–497.

Klein, F. (1908). *Elementarmathematik vom höheren Standpunkte aus. Teil I: Arithmetik, Algebra, Analysis*. B. G. Teubner, Leipzig. English translation in Klein (1932).

Klein, F. (1909). *Elementarmathematik vom höheren Standpunkte aus. Teil II: Geometrie*. B. G. Teubner, Leipzig. English translation in Klein (1939).

Klein, F. (1932). *Elementary Mathematics from an Advanced Standpoint; Arithmetic, Algebra, Analysis*. Translated from the 3rd German edition by E. R. Hedrick and C. A. Noble. Macmillan & Co., London.

Klein, F. (1939). *Elementary Mathematics from an Advanced Standpoint; Geometry*. Translated from the 3rd German edition by E. R. Hedrick and C. A. Noble. Macmillan & Co., London.

Kőnig, D. (1927). Über eine Schlussweise aus dem Endlichen ins Unendliche. *Acta Litterarum ac Scientiarum 3*, 121–130.

Kőnig, D. (1936). *Theorie der endlichen und unendlichen Graphen*. Akademische Verlagsgesellschaft, Leipzig. English translation by Richard McCoart, *Theory of Finite and Infinite Graphs*, Birkhäuser, Boston 1990.

Kronecker, L. (1886). Letter to Gösta Mittag-Leffler, 4 April 1886. Cited in Edwards (2007).

Kronecker, L. (1887). Ein Fundamentalsatz der allgemeinen Arithmetik. *Journal für die reine und angewandte Mathematik 100*, 490–510.

Lagrange, J. L. (1768). Solution d'un problème d'arithmétique. *Miscellanea Taurinensia 4*, 19ff. In his *Œuvres 1*: 671–731.

Lagrange, J. L. (1771). Réflexions sur la résolution algébrique des équations. *Nouveaux Mémoires de l'Académie Royale des Sciences et Belles-lettres de Berlin*. In his *Œuvres 3*: 205–421.

Lambert, J. H. (1766). Die Theorie der Parallellinien. *Magazin für reine und angewandte Mathematik (1786)*, 137–164, 325–358.

Laplace, P. S. (1812). *Théorie Analytique des Probabilités*. Paris.

Lenstra, H. W. (2002). Solving the Pell equation. *Notices of the American Mathematical Society 49*, 182–192.

Levi ben Gershon (1321). *Maaser Hoshev*. German translation by Gerson Lange: *Sefer Maasei Choscheb*, Frankfurt, 1909.

Liouville, J. (1844). Nouvelle démonstration d'un théorème sur les irrationalles algébriques. *Comptes Rendus Hebdomadaires des Séances de l'Académie des Sciences, Paris 18*, 910–911.

Lobachevsky, N. I. (1829). *On the foundations of geometry*. Kazansky Vestnik. (Russian).

Markov, A. (1947). On the impossibility of certain algorithms in the theory of associative systems (Russian). *Doklady Akademii Nauk SSSR 55*, 583–586.

Markov, A. (1958). The insolubility of the problem of homeomorphy (Russian). *Doklady Akademii Nauk SSSR 121*, 218–220.

Matiyasevich, Y. V. (1970). The Diophantineness of enumerable sets (Russian). *Doklady Akademii Nauk SSSR 191*, 279–282.

Maugham, W. S. (2000). *Ashenden, or, The British Agent*. Vintage Classics.

Mercator, N. (1668). *Logarithmotechnia*. William Godbid and Moses Pitt, London.

Minkowski, H. (1908). Raum und Zeit. *Jahresbericht der Deutschen Mathematiker-Vereinigung 17*, 75–88.

Mirimanoff, D. (1917). Les antinomies de Russell et Burali-Forti et le problème fondamental de la théorie des ensembles. *L'Enseignement Mathèmatique 19*, 37–52.

Motzkin, T. S. (1967). Cooperative classes of finite sets in one and more dimensions. *Journal of Combinatorial Theory 3*, 244–251.

Neumann, P. M. (2011). *The Mathematical Writings of Évariste Galois*. Heritage of European Mathematics. European Mathematical Society (EMS), Zürich.

Newton, I. (1671). De methodis serierum et fluxionum. *Mathematical Papers*, 3, 32–353.

Novikov, P. S. (1955). On the algorithmic unsolvability of the word problem in group theory (Russian). *Proceedings of the Steklov Institute of Mathematics 44*. English translation in *American Mathematical Society Translations*, series 2, 9, 1–122.

Oresme, N. (1350). *Tractatus de configurationibus qualitatum et motuum*. English translation in Clagett (1968).

Ostermann, A. and G. Wanner (2012). *Geometry by its History*. Undergraduate Texts in Mathematics. Readings in Mathematics. Springer, Heidelberg.

Pacioli, L. (1494). *Summa de arithmetica, geometria. Proportioni et proportionalita*. Venice: Paganino de Paganini. Partial English translation by John B. Geijsbeek published by the author, Denver, 1914.

Paris, J. and L. Harrington (1977). A mathematical incompleteness in Peano arithmetic. In *Handbook of Mathematical Logic*, ed. J. Barwise, North-Holland, Amsterdam.

Pascal, B. (1654). Traité du triangle arithmétique, avec quelques autres petits traités sur la même manière. English translation in *Great Books of the Western World*, Encyclopedia Britannica, London, 1952, 447–473.

Peano, G. (1888). *Calcolo Geometrico secondo l'Ausdehnungslehre di H. Grassmann, preceduto dalle operazioni della logica deduttiva*. Bocca, Turin. English translation in Peano (2000).

Peano, G. (1889). *Arithmetices principia: nova methodo*. Bocca, Rome. English translation in van Heijenoort (1967), pp. 83–97.

Peano, G. (1895). *Formulaire de mathématiques*. Bocca, Turin.

Peano, G. (2000). *Geometric Calculus*. Birkhäuser Boston, Inc., Boston, MA. According to the *Ausdehnungslehre* of H. Grassmann. Translated from the Italian by Lloyd C. Kannenberg.

Petsche, H.-J. (2009). *Hermann Graßmann—Biography*. Birkhäuser Verlag, Basel. Translated from the German original by Mark Minnes.

Plofker, K. (2009). *Mathematics in India*. Princeton University Press, Princeton, NJ.

Poincaré, H. (1881). Sur les applications de la géométrie non-euclidienne à la théorie des formes quadratiques. *Association française pour l'avancement des sciences 10*, 132–138. English translation in Stillwell (1996), pp. 139–145.

Poincaré, H. (1895). Analysis situs. *Journal de l'École Polytechnique.*, series 2, no. 1, 1–121. In his *Œuvres* 6: 193–288. English translation in Poincaré (2010), pp. 5–74.

Poincaré, H. (2010). *Papers on Topology*, Volume 37 of *History of Mathematics*. American Mathematical Society, Providence, RI; London Mathematical Society, London. *Analysis situs* and its five supplements, Translated and with an introduction by John Stillwell.

Pólya, G. (1920). Über den zentralen Grenzwertsatz der Wahrscheinlichkeitsrechnung und das Momentenproblem. *Mathematische Zeitschrift 8*, 171–181.

Post, E. L. (1936). Finite combinatory processes. Formulation 1. *Journal of Symbolic Logic 1*, 103–105.

Post, E. L. (1941). Absolutely unsolvable problems and relatively undecidable propositions. Account of an anticipation. In Davis (1965), pp. 340–433.

Post, E. L. (1947). Recursive unsolvability of a problem of Thue. *Journal of Symbolic Logic 12*, 1–11.

Ramsey, F. P. (1930). On a problem of formal logic. *Proceedings of the London Mathematical Society 30*, 264–286.

Reisch, G. (1503). *Margarita philosophica*. Freiburg.

Robinson, R. M. (1952). An essentially undecidable axiom system. *Proceedings of the International Congress of Mathematicians, 1950*, 729–730. American Mathematical Society, Providence, RI.

Rogozhin, Y. (1996). Small universal Turing machines. *Theoretical Computer Science 168*(2), 215–240. Universal machines and computations (Paris, 1995).

Ruffini, P. (1799). *Teoria generale delle equazioni in cui si dimostra impossibile la soluzione algebraica delle equazioni generale di grade superiore al quarto*. Bologna.

Russ, S. (2004). *The Mathematical Works of Bernard Bolzano*. Oxford University Press, Oxford.

Russell, B. (1903). *The Principles of Mathematics. Vol. I*. Cambridge University Press, Cambridge.

Ryan, P. J. (1986). *Euclidean and non-Euclidean geometry*. Cambridge University Press, Cambridge.

Saccheri, G. (1733). *Euclid Vindicated from Every Blemish*. Classic Texts in the Sciences. Birkhäuser/Springer, Cham, 2014. Dual Latin-English text, edited and annotated by Vincenzo De Risi. Translated from the Italian by G. B. Halsted and L. Allegri.

Simpson, S. G. (2009). *Subsystems of Second Order Arithmetic* (2nd ed.). Perspectives in Logic. Cambridge University Press, Cambridge; Association for Symbolic Logic, Poughkeepsie, NY.

Sperner, E. (1928). Neuer Beweis für die Invarianz der Dimensionzahl und des Gebietes. *Abhandlungen aus dem Mathematischen Seminar der Universität Hamburg 6*, 265–272.

Steinitz, E. (1913). Bedingt konvergente Reihen und konvexe Systeme. (Teil I.). *Journal für die reine und angewandte Mathematik 143*, 128–175.

Stevin, S. (1585a). *De Thiende*. Christoffel Plantijn, Leiden. English translation by Robert Norton, *Disme: The Art of Tenths, or Decimall Arithmetike Teaching*, London, 1608.

Stevin, S. (1585b). *L'Arithmetique*. Christoffel Plantijn, Leiden.

Stifel, M. (1544). *Arithmetica integra*. Johann Petreium, Nuremberg.

Stillwell, J. (1982). The word problem and the isomorphism problem for groups. *Bulletin of the American Mathematical Society (New series) 6*, 33–56.

Stillwell, J. (1993). *Classical Topology and Combinatorial Group Theory* (2nd ed.). Springer-Verlag, New York, NY.

Stillwell, J. (1996). *Sources of Hyperbolic Geometry*. American Mathematical Society, Providence, RI.

Stillwell, J. (2010). *Mathematics and its History* (3rd ed.). Springer, New York, NY.

Stirling, J. (1730). *Methodus Differentialis*. London. English translation Tweddle (2003).

Tartaglia, N. (1556). *General Trattato di Numeri et Misure*. Troiano, Venice.

Thue, A. (1897). Mindre Meddelelser. II. *Archiv for Mathematik og Naturvidenskab 19*(4), 27.

Thue, A. (1914). Probleme über Veränderungen von Zeichenreihen nach gegebenen Regeln. J. Dybvad, Kristiania, 34 pages.

Turing, A. (1936). On computable numbers, with an application to the Entscheidungsproblem. *Proceedings of the London Mathematical Society, series 2,* no. *42*, 230–265.

Tweddle, I. (2003). *James Stirling's Methodus differentialis*. Sources and Studies in the History of Mathematics and Physical Sciences. Springer-Verlag London, Ltd., London. An annotated translation of Stirling's text.

van Heijenoort, J. (1967). *From Frege to Gödel. A Source Book in Mathematical Logic, 1879–1931*. Harvard University Press, Cambridge, MA.

von Neumann, J. (1923). Zur Einführung der transfiniten Zahlen. *Acta litterarum ac scientiarum Regiae Universitatis Hungaricae Francisco-Josephinae, Sectio Scientiarum Mathematicarum 1*, 199–208. English translation in van Heijenoort (1967), pp. 347–354.

von Neumann, J. (1930). Letter to Gödel, 20 November 1930, in Gödel (2014), p. 337.

von Staudt, K. G. C. (1847). *Geometrie der Lage*. Bauer und Raspe, Nürnberg.

Wallis, J. (1655). Arithmetica infinitorum. *Opera 1*: 355–478. English translation *The Arithmetic of Infinitesimals* by Jacqueline Stedall, Springer, New York, 2004.

Wantzel, P. L. (1837). Recherches sur les moyens de reconnaitre si un problème de géométrie peut se resoudre avec la règle et le compas. *Journal de Mathématiques Pures et Appliquées 1*, 366–372.

Weil, A. (1984). *Number Theory. An Approach through History, from Hammurapi to Legendre*. Birkhäuser Boston Inc., Boston, MA.

Whitehead, A. N. and B. Russell (1910). *Principia Mathematica. Vol. I.* Cambridge University Press, Cambridge.

Zermelo, E. (1904). Beweis dass jede Menge wohlgeordnet werden kann. *Mathematische Annalen 59*, 514–516. English translation in van Heijenoort (1967), pp. 139–141.

Zermelo, E. (1908). Untersuchungen über die Grundlagen der Mengenlehre I. *Mathematische Annalen 65*, 261–281. English translation in van Heijenoort (1967), pp. 199–215.

Zhu Shijie (1303). *Sijuan yujian*. (Precious mirror of four elements).

Index

abacus, 78, 98
Abel, Niels Henrik, 111
— and Galois, 141
— unsolvability of quintic, 111
absolute value, 351
— is multiplicative, 351
abstraction, 32
AC, 332
— consistency of, 332
— and definition of continuity, 241
— equivalent to existence of basis, 333
— equivalent to well-ordering, 332
— independence of, 332
— more advanced than ZF, 332
— reverse mathematics of, 333
ACA_0, 299, 328
acceleration, 202
actual infinity, 33, 322
addition
— by abacus, 78
— in base 10, 78
— of complex numbers, 350
— inductive definition, 24, 70, 308
— a linear time problem, 87
— of numerals, 77
— of points, 192
— table, 79
— by Turing machine, 88
— of vectors, 124
adjoining a number to $\mathbb{Q}$, 133
affine
— geometry, 386 ; in art, 386
— line, 385
— transformation, 385
algebra, 7, 106
— abstract, 110
— in calculus, 140
— of complex numbers, 110
— fundamental theorem, 140
— in geometry, 140
— of logic, 330
— origin of word, 107

algebraic
— geometry, 178
— integer, 65
— number, 35
— number theory, 63
— numbers ; are countable, 326
— realization ; of geometric constructions, 164
— topology, 276
algebraic number field, 133
— constructed from polynomials, 134
— finite-dimensional over $\mathbb{Q}$, 137
— rationalizes calculations, 143
— as a vector space, 136
algorithm, 4
— Collatz, 5
— as computing machine, 6
— Euclidean, 5
— origin of word, 107
— as Turing machine, 94
al-Khwārizmi, 107
altitude, 174
analysis, 234
— arises from calculus, 239
— arithmetization of, 234
— deeper than PA, 335
— and reverse mathematics, 333
— a symphony of the infinite, 320
angle, 148
— as arc length on circle, 226
— bisection, 177
— in half-plane model, 182
— inner product definition, 12, 172
— measure, 150
— right, 150 ; inner product definition, 172
— in a semicircle, 152, 174
— sum ; of triangle, 151
— trisection, 176
Appel, Ken, 273
arc length, 226
Archimedes, 13
— area of parabolic segment, 15, 195
— cattle problem, 65